METHODS IN MOLECULAR BIOLOGY

For further volumes:
http://www.springer.com/series/7651

For over 35 years, biological scientists have come to rely on the research protocols and methodologies in the critically acclaimed *Methods in Molecular Biology* series. The series was the first to introduce the step-by-step protocols approach that has become the standard in all biomedical protocol publishing. Each protocol is provided in readily-reproducible step-by-step fashion, opening with an introductory overview, a list of the materials and reagents needed to complete the experiment, and followed by a detailed procedure that is supported with a helpful notes section offering tips and tricks of the trade as well as troubleshooting advice. These hallmark features were introduced by series editor Dr. John Walker and constitute the key ingredient in each and every volume of the *Methods in Molecular Biology* series. Tested and trusted, comprehensive and reliable, all protocols from the series are indexed in PubMed.

SMC Proteins

Methods and Protocols

Edited by

Yasuto Murayama

Department of Chromosome Science, National Institute of Genetics, Mishima, Japan

Yasutaka Kakui

Waseda Institute for Advanced Study, Waseda University, Tokyo, Japan

Humana Press

Editors
Yasuto Murayama
Department of Chromosome Science
National Institute of Genetics
Mishima, Japan

Yasutaka Kakui
Waseda Institute for Advanced Study
Waseda University
Tokyo, Japan

ISSN 1064-3745 ISSN 1940-6029 (electronic)
Methods in Molecular Biology
ISBN 978-1-0716-5004-2 ISBN 978-1-0716-5005-9 (eBook)
https://doi.org/10.1007/978-1-0716-5005-9

This Humana imprint is published by the registered company Springer Science+Business Media, LLC, part of Springer Nature.
The registered company address is: 1 New York Plaza, New York, NY 10004, U.S.A.

Preface

Genomic DNA is a meter-long polymer, yet it is packed into a small cell or nucleus that is only a few micrometers in diameter. Genome folding is not merely the packaging of long genomic DNA but also fundamental to numerous chromosomal functions across the cell cycle, development and differentiation. Proteins belonging to the Structural Maintenance of Chromosomes (SMC) family are one of the major architects of chromosomes and are widely conserved among the three kingdoms of life: Bacteria, Archaea, and Eucarya. Functioning as multisubunit ATPase assemblies, SMC proteins participate in many biological processes, including chromosome organization and segregation, spatiotemporal gene regulation, DNA repair, and viral prevention. This volume of Methods in Molecular Biology provides a variety of cutting-edge methodologies used to reveal the functions and regulations of SMC proteins at molecular and chromosomal levels in prokaryotic, eukaryotic, and archaeal systems. *Part I* describes single-molecule and biophysical approaches for studying the molecular mechanisms of SMC proteins in the regulation of DNA and chromatin dynamics at the level of proteins and cells. *Part II* summarizes methods for mapping the topologies of DNA and SMC proteins, exploring the mechanistic links between DNA folding and the unique closed-ring structures of SMC proteins. *Part III* presents biochemical assays for investigating the functional interplays between SMC proteins and related chromosomal processes, including chromatin remodeling, DNA replication, and DNA repair. *Part IV* describes the use of next-generation sequencing (NGS)-based techniques to explore three-dimensional chromosomal architectures mediated by SMC proteins. The methodologies described in this volume will facilitate further investigation into the SMC proteins and related factors in the regulation of chromosome structures and functions.

Mishima, Japan — *Yasuto Murayama*
Tokyo, Japan — *Yasutaka Kakui*

Contents

Contributors

Hemanta Adhikary • *Department of Cellular and Molecular Medicine, Ottawa Institute of System Biology, University of Ottawa, Ottawa, ON, Canada*

Luis Aragon • *DNA Motors Group, MRC Laboratory of Medical Sciences, London, UK*

Richard J. Bartlett • *DNA Motors Group, MRC Laboratory of Medical Sciences, London, UK*

Céline Bouchoux • *Chromosome Segregation Laboratory, The Francis Crick Institute, London, UK*

James Collier • *MRC Laboratory of Molecular Biology, Cambridge, UK*

Neus Colomina • *Departament de Ciències Mèdiques Bàsiques, Institut de Recerca Biomèdica de Lleida, Universitat de Lleida, Lleida, Spain*

Stephen J. Cross • *Wolfson Bioimaging Facility, University of Bristol, Bristol, UK*

Damien D'Amours • *Department of Cellular and Molecular Medicine, Ottawa Institute of System Biology, University of Ottawa, Ottawa, ON, Canada*

Gemma L. M. Fisher • *DNA Motors Group, MRC Laboratory of Medical Sciences, London, UK*

Stephan Gruber • *Department of Fundamental Microbiology (DMF), Faculty of Biology and Medicine (FBM), University of Lausanne (UNIL), Lausanne, Switzerland*

Torahiko L. Higashi • *Research Division, Chugai Pharmaceutical Co. Ltd., Yokohama, Japan*

Takashi Hishida • *Department of Life Science, Graduate School of Science, Gakushuin University, Toshima-ku, Tokyo, Japan*

Sara Illodo • *Department of Physics and Astronomy, Seoul National University, Seoul, South Korea; Institute for Data Innovation in Science, Seoul National University, Seoul, South Korea*

Yasutaka Kakui • *Waseda Institute for Advanced Study, Waseda University, Tokyo, Japan*

Hyeong Ku Kim • *Department of Physics and Astronomy, Seoul National University, Seoul, South Korea*

Kyoung-Dong Kim • *Department of Systems Biotechnology, Chung-ang University, Anseong, South Korea*

Yoshimi Kinoshita • *Department of Biophysics, Graduate School of Science, Kyoto University, Kyoto, Japan*

Anno I. Koetje • *DNA Motors Group, MRC Laboratory of Medical Sciences, London, UK*

Hon Wing Liu • *Department of Fundamental Microbiology (DMF), Faculty of Biology and Medicine (FBM), University of Lausanne (UNIL), Lausanne, Switzerland*

Neus P. Lorite • *Departament de Ciències Mèdiques Bàsiques, Institut de Recerca Biomèdica de Lleida, Universitat de Lleida, Lleida, Spain; Division of Oncogenomics, The Netherlands Cancer Institute, Amsterdam, The Netherlands*

Juan Méndez • *DNA Replication Group, Molecular Oncology Programme, Spanish National Cancer Research Centre (CNIO), Madrid, Spain*

Masashi Minamino • *Chromosome Segregation Laboratory, The Francis Crick Institute, London, UK; Research Division, Chugai Pharmaceutical Co. Ltd., Yokohama, Japan*

Maxim Molodtsov • *The Francis Crick Institute, London, UK; Department of Physics and Astronomy, University College London, London, UK*

Sofía Muñoz • *Instituto de Biología Funcional y Genómica (IBFG), CSIC-Universidad de Salamanca, Salamanca, Spain*

Yasuto Murayama • *Department of Chromosome Science, National Institute of Genetics, Mishima, Japan; Department of Genetics, Graduate University for Advanced Studies (SOKENDAI), Mishima, Japan*

Tomoko Nishiyama • *Department of Biophysics, Graduate School of Science, Kyoto University, Kyoto, Japan*

Shunsuke Noda • *Department of Chromosome Science, National Institute of Genetics, Mishima, Shizuoka, Japan*

Ken-ichi Noma • *Institute of Molecular Biology, University of Oregon, Eugene, OR, USA; Institute for Genetic Medicine, Hokkaido University, Sapporo, Japan*

Martina Richeldi • *The Francis Crick Institute, London, UK*

Joaquim Roca • *DNA Topology Lab. Molecular Biology Institute of Barcelona (IBMB-CSIC), Barcelona, Spain*

Sara Rodríguez-Acebes • *DNA Replication Group, Molecular Oncology Programme, Spanish National Cancer Research Centre (CNIO), Madrid, Spain*

Je-Kyung Ryu • *Department of Physics and Astronomy, Seoul National University, Seoul, South Korea; Institute for Data Innovation in Science, Seoul National University, Seoul, South Korea; Institute of Applied Physics of Seoul National University, Seoul, South Korea*

Sanjana Saravanan • *DNA Motors Group, MRC Laboratory of Medical Sciences, London, UK*

Snehlata • *Departament de Ciències Mèdiques Bàsiques, Institut de Recerca Biomèdica de Lleida, Universitat de Lleida, Lleida, Spain*

Madhusudhan Srinivasan • *Department of Biochemistry, University of Oxford, Oxford, UK*

Naomichi Takemata • *Department of Biotechnology, College of Life Sciences, Ritsumeikan University, Kusatsu, Shiga, Japan*

Jordi Torres-Rosell • *Departament de Ciències Mèdiques Bàsiques, Institut de Recerca Biomèdica de Lleida, Universitat de Lleida, Lleida, Spain*

Frank Uhlmann • *Chromosome Segregation Laboratory, The Francis Crick Institute, London, UK*

Menelaos Voulgaris • *DNA Motors Group, MRC Laboratory of Medical Sciences, London, UK*

Kodai Yamaura • *Research Organization of Science and Technology, Ritsumeikan University, Kusatsu, Japan*

Part I

Single-Moleucle and Biophysical Assays of SMC Proteins

Chapter 1

Single-Molecule Analyses of Human Cohesin Complex

Yoshimi Kinoshita and Tomoko Nishiyama

Abstract

Cohesin is a ring-shaped protein complex that mediates sister chromatid cohesion during cell division. Recently, cohesin was found to build higher-order chromatin structures during interphase, such as topologically associating domains (TADs), loop domains, and chromatin domains. How cohesin formed these structures remained unclear until the development of single-molecule assays that assess the behavior of individual cohesin and DNA molecules. The DNA loop extrusion assay revealed that cohesin can compact DNA by extruding DNA loops. The translocation assay revealed that cohesin can move along DNA. These single-molecule assays are powerful techniques and help our understanding of how higher-order chromosome structures are formed. Here, we describe the DNA loop extrusion assay and the translocation assay by using recombinant human cohesin complex and λDNA to reconstitute the activity of cohesin on DNA at a single-molecule resolution.

Key words Cohesin, Single-molecule imaging, DNA loop extrusion, Translocation, NIPBL-Mau2

1 Introduction

In eukaryotes, genomic DNA is folded into highly organized structures and undergoes changes in the chromatin architecture [1, 2]. During cell division, genetic information is inherited by two daughter cells through proper sister chromatid cohesion and segregation. Cohesin, one of several SMC (Structural Maintenance of Chromosomes) protein complexes, was originally identified as the factor responsible for sister chromatid cohesion from S phase to metaphase [3, 4]. The ring-shaped complex consists of four subunits: Smc1, Smc3, Scc1, and STAG1/2. Both Smc1 and Smc3 contain ABC (ATP-Binding Cassette) ATPase motifs in their head domains. Cohesin undergoes dynamic conformational changes through the opening and closing of the head regions to carry out its function after being loaded onto DNA with NIPBL [5, 6]. In addition to being responsible for sister chromatid cohesion during cell division, cohesin has been found to be essential for organizing *in-cis* chromatin structures in interphase [7].

Yasuto Murayama and Yasutaka Kakui (eds.), *SMC Proteins: Methods and Protocols*, Methods in Molecular Biology, vol. 2991,
https://doi.org/10.1007/978-1-0716-5005-9_1,

Chromatin analysis methods, such as ChIP-seq, Hi-C, and Micro-C, have revealed that individual chromosomes are structured with topologically associated domains (TADs). TADs are distant regions on a single chromosome that interact *in-cis* and therefore, DNA loops are formed. Cohesin localizes at the boundaries of TADs, suggesting that cohesin can facilitate DNA loops [7–9]. Additionally, DNA loop formations are essential for bringing enhancers and promoters into close proximity for the regulation of gene expression [2, 8–10]. How cohesin facilitates DNA loops had remained unclear until the recent development of certain single-molecule assays. The DNA loop extrusion assay that uses a flow-stretching technique with TIRF (Total Internal Reflection Fluorescence) microscopy allows for the visualization of the activity by SMC complexes, including cohesin [11–13]. In the assay, linear λDNA are immobilized at both ends on a microfluidic surface, and the extrusion of DNA loops by cohesin can be visualized in the presence of the cohesin loader NIPBL-Mau2 and ATP [5, 12, 13]. In addition to DNA loop extrusion assay, the translocation assay allows for the visualization of the dynamic movement of cohesin on DNA [14–16]. The translocation activity of cohesin is thought to be required for cohesin to localize at specific sites on the genome [2, 16]. Overall, these single-molecule assays have helped our understanding of cohesin in organizing chromatin structures.

In this chapter, we describe in detail the methods for the two single-molecule assays in measuring DNA loop extrusion and translocation activities of recombinant human cohesin complex. First, we explain how to prepare biotinylated λDNA and a microfluidic flow cell. Then, we describe the methods to visualize cohesin activities using imaging techniques. Using these single-molecule assays, we also describe the functional role of the N-terminus of NIPBL in regulating cohesin activity on DNA.

2 Materials

2.1 Biotinylated λDNA

1. Thermal cycler.
2. λDNA.
3. Biotinylated primer A: GGGCGGCGACCT (Bio). The complementary sequence to the 5′ overhang of the antisense strand of λDNA and is also biotin-labeled at the 3′ end.
4. Biotinylated primer B: AGGTCGCCGCCC (Bio). The complementary sequence to the 5′ overhang of the sense strand of λDNA and is also biotin-labeled at the 3′ end.
5. T4 polynucleotide kinase (PNK).

6. PNK reaction buffer: 70 mM Tris-HCl (pH 7.6), 10 mM $MgCl_2$, and 5 mM dithiothreitol (DTT).
7. T4 DNA ligase.

2.2 Preparation of Microfluidic Flow Cell

1. Glass coverslips: 22 mm × 22 mm with a thickness of 0.16–0.19 mm (*see* **Note 1**).
2. Cover glass staining rack.
3. Ultrasonic bath.
4. 1 N potassium hydroxide (KOH).
5. Methanol.
6. 2% 3-aminopropyltriethoxysilane (APTES): Dissolved in 100% methanol.
7. 1 M sodium bicarbonate ($NaHCO_3$, pH 8.15).
8. mPEG-succinimidyl valerate (mPEG-SVA, MW 5000).
9. Biotin-mPEG-SVA (MW 5000).
10. Biotin-mPEG solution: 6.6 mg/mL biotin-mPEG-SVA and 100 mg/mL mPEG-SVA in 100 mM sodium bicarbonate.
11. Sealer.
12. Microscope glass slide: 76 mm × 26 mm, modified with holes drilled in.
13. Superglue.
14. Double-sided adhesive silicone sheet with a thickness of 0.12 mm.
15. Polyethylene tubings as inlet tubings: Length of 15 cm, inner diameter of 0.35 mm, and outer diameter of 1.05 mm.
16. Polyethylene tubings as outlet tubings: Length of 30 cm, inner diameter of 0.80 mm, and outer diameter of 1.20 mm.
17. Ultrapure bovine serum albumin (BSA).
18. Streptavidin solution: 40 μg/mL streptavidin, 20 mM Tris-HCl (pH 7.5), and 1 mg/mL ultrapure BSA.

2.3 Microscope Setup

1. TIRF microscope.
2. Anti-vibration table.
3. A semiconductor laser with a maximum wavelength of 488 nm (50 mW power).
4. A semiconductor laser with a maximum wavelength of 561 nm (50 mW power).
5. A dichroic mirror which reflects over 94% photons in the range of 471–491 nm and 559–568 nm (corresponding to the lasers' excitation wavelengths) and transmits over 93% photons in the range of 503–543 nm and 582–1200 nm (corresponding to emission wavelengths).

6. 100x objective lens for TIRF imaging with 1.49 NA.
7. Immersion oil optimized for TIRF imaging.
8. Dual-view image splitting optics which provide a pair of images from a single camera. The images are separated by a 560 nm single-edge dichroic beam splitter and further filtered using band-pass filters at 512/25 nm and 630/92 nm.
9. Electron multiplying charge coupled device (EMCCD) camera.
10. A cooling water circulation system (*see* **Note 2**).
11. Syringe pump with withdraw settings.

2.4 DNA Loop Extrusion Assay

1. Syringes (5 mL) and needles (0.9 mm × 38 mm).
2. 10 mg/mL Casein: Dissolve in 100 mM HEPES pH 8.0 and filter through 0.2 μm pore size (or centrifuged at the speed of 100 k rpm for 20 min) to remove aggregates. Store at −20 °C.
3. 200 mM ATP: Dissolve in ultrapure H_2O after adjusting the pH to 7.0 with NaOH and store at −20 °C.
4. 10 μM SYTOX orange: Dissolve in DMSO and store at −20 °C.
5. 2M Glucose: Dissolve in ultrapure H_2O, filter through 0.2 μm pore size, and store at −20 °C.
6. 2 mg/mL Catalase: Dissolve in ultrapure H_2O, filter through 0.2 μm pore size, and store at −20 °C.
7. 10 mg/mL Glucose oxidase: Dissolve in ultrapure H_2O, filter through 0.2 μm pore size, and store at −20 °C.
8. 500 μM A/C heterodimerizer AP21967: Dissolve in ethanol and store at −20 °C (*see* **Note 3**).
9. DNA buffer: 20 mM Tris-HCl (pH 7.5), 150 mM NaCl, 0.25 mg/mL BSA, and 10 nM SYTOX orange.
10. Imaging-casein buffer: 50 mM Tris-HCl (pH 7.5), 50 mM NaCl, 5% glycerol, 0.25 mg/mL BSA, and 0.5 mg/mL casein.
11. Imaging buffer 1: 50 mM Tris-HCl (pH 7.5), 50 mM NaCl, 5% glycerol, 0.25 mg/mL BSA, and 10 nM SYTOX orange.
12. Imaging buffer 2: 50 mM Tris-HCl (pH 7.5), 50 mM NaCl, 5% glycerol, 0.25 mg/mL BSA, 30 nM SYTOX orange, 3 mM $MgCl_2$, 5 mM ATP, 20 mM glucose, 20 μg/mL catalase, 100 μg/mL glucose oxidase [GODCAT oxygen scavenger system], and 1 mM DTT.
13. Cohesin tetramer: Recombinant human cohesin (Smc1-FKBP-3HA, Smc3-FRB-3xFLAG, Scc1-Halo, 10xHis-STAG1) is expressed in insect cells and purified using affinity chromatography. Scc1 contains a C-terminal Halo-tag which is covalently labeled with HaloTag ligand Alexa Fluor 488 during the purification of the protein complex (Fig. 1a) [5].

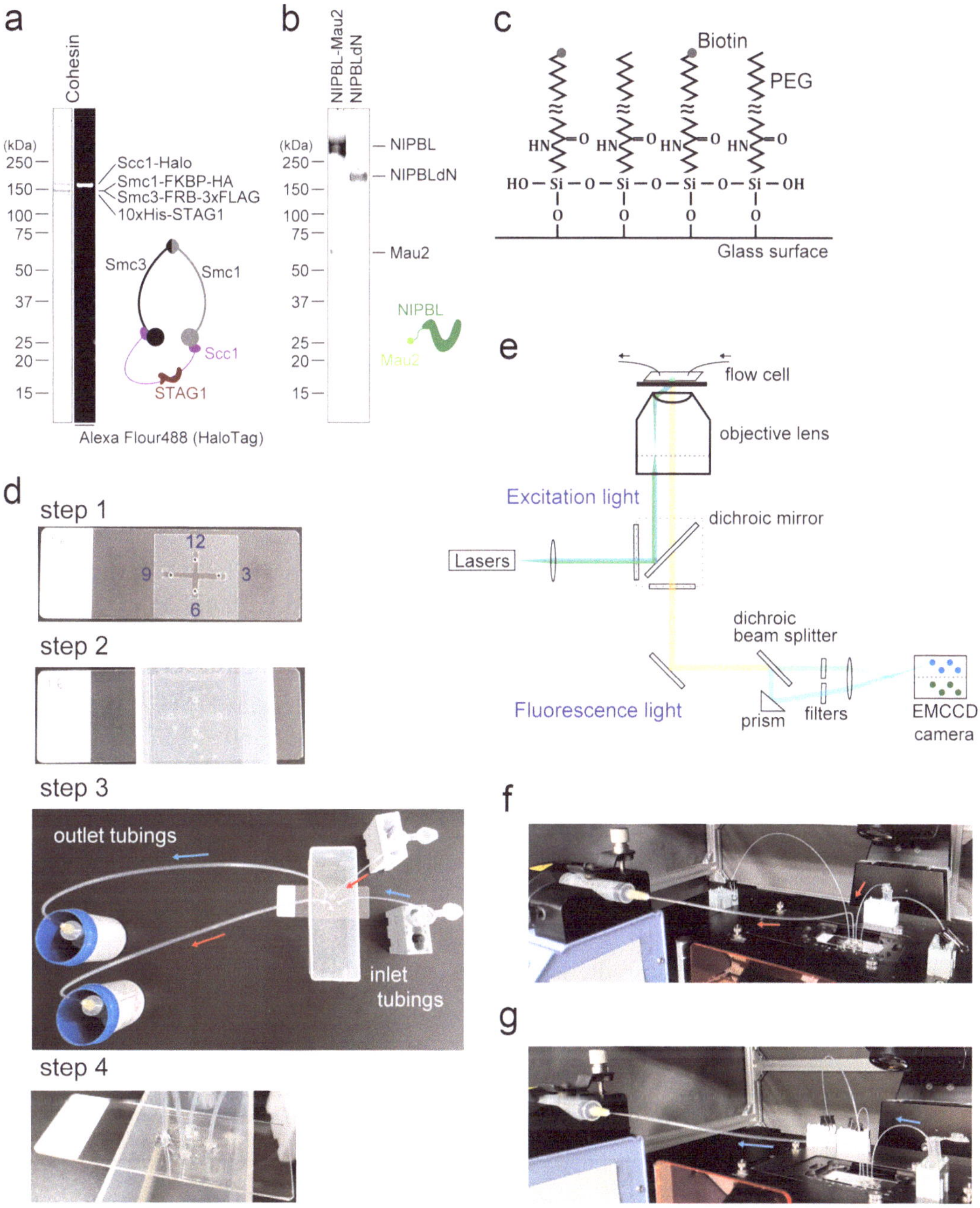

Fig. 1 Preparation of the flow cell, recombinant human cohesin, and NIPBL-Mau2. (**a**) SYPRO™ Ruby staining of recombinant human cohesin tetramer after SDS-PAGE. Alexa Fluor 488 HaloTag ligand covalently bonded to the Halo-tag at the C-terminus of Scc1 was visualized in the SDS-PAGE gel using the chemiluminescence imaging system when excited by a blue light. (**b**) SYPRO™ Ruby staining of NIPBL-Mau2 and NIPBLdN after SDS-PAGE. (**c**) Chemical modification of coverslip surface. APTES and (Biotin-) mPEG-SVA are sequentially coated onto the coverslip. (**d**) Photograph of the flow cell preparation. Step 1: The hollowed-out tape is adhered to the microscope slide. Step 2: The coated coverslip is adhered to the tape. Step 3: Ends of the outlet and inlet tubings are fitted into the holes, syringe, or tubes. Step 4: Each tubing fitted in the holes of the microscope glass are superglued. (**e**) Schematic diagram of TIRF microscope setup. Two types of excitation lasers illuminate the flow cell glass, and fluorescent images are separated by dual-view optics and captured by the same camera. (**f**) Photograph of the flow cell positioned above the microscope stage with fluid withdrawn "vertically" using a syringe. (**g**) Photograph of the flow cell positioned above the microscope stage with fluid withdrawn "horizontally" using a syringe

14. Cohesin loader NIPBL-Mau2: Recombinant full-length human NIPBL and Mau2 (FLAG-Halo-NIPBL-10xHis, Mau2) is expressed in insect cells and purified using affinity chromatography (Fig. 1b) [5].
15. Cohesin loader NIPBLdN: N-terminal IDR (M1-P1162) deletion of recombinant human NIPBL is expressed in insect cells and purified using affinity chromatography (Fig. 1b) [5] (*see* **Note 4**).

2.5 Translocation Assay

1. DNA anneal buffer: 20 mM Tris-HCl (pH 7.5), 150 mM NaCl, 0.25 mg/mL BSA, and 5 mM $MgCl_2$.
2. Imaging buffer 3: 50 mM Tris-HCl (pH 7.5), 50 mM NaCl, 5% glycerol, 0.25 mg/mL BSA, 2 mg/mL casein, 800 units/mL T4 DNA ligase, 1 mM ATP, 2 mM $MgCl_2$, and 10 nM SYTOX orange.
3. Translocation buffer: 50 mM Tris-HCl (pH 7.5), 50 mM NaCl, 5% glycerol, 0.25 mg/mL BSA, 10 nM SYTOX orange, 2.5 mM $MgCl_2$, 1 mM ATP, 20 mM glucose, 20 μg/mL catalase, 100 μg/mL glucose oxidase, and 1 mM DTT.
4. Translocation buffer 500: 50 mM Tris-HCl (pH 7.5), 500 mM NaCl, 5% glycerol, 0.25 mg/mL BSA, 50 nM SYTOX orange, 2.5 mM $MgCl_2$, 1 mM ATP, 20 mM glucose, 20 μg/mL catalase, 100 μg/mL glucose oxidase, and 1 mM DTT.

2.6 Software

1. NIS-Elements (Nikon): Software for taking and analyzing images.
2. FIJI (NIH): Software for analyzing images.
3. KymoResliceWide plugin for FIJI.
4. MATLAB (MathWorks).
5. Excel (Microsoft).
6. Prism (GraphPad.)

3 Methods

3.1 Biotinylated λDNA

λDNA is a linear double-stranded DNA consisting of 48,502 base pairs with 12 nucleotide-long 5′ overhangs at each end. Biotin can be used to label the λDNA at both ends by annealing biotinylated 12-base primers that are complementary to the overhang sequences of the λDNA. The biotinylated λDNA can then be immobilized on streptavidin-coated coverslips.

1. Phosphorylate the 5′ ends of λDNA, biotinylated primer A, and biotinylated primer B individually by incubating each with 5 units of T4 PNK and 1 mM ATP in PNK reaction buffer at

37 °C for 3 h. Typical reactions use 0.5 μg/μL λDNA (0.15 pmol λDNA in 10 μL), 5 μM primer A, and 12 μM primer B.

2. Add 0.02 μM phosphorylated primer A (2 pmol) to 0.15 pmol phosphorylated λDNA into a solution with a final concentration of 50 mM NaCl and 3 mM $MgCl_2$. Anneal the phosphorylated primer A to the λDNA by gradually lowering the temperature from 55 °C to 4 °C for more than 1 h using a thermal cycler.
3. Ligate the primer A to the annealed product at 16 °C for 2 h by adding 10 mM ATP and 300 units of T4 DNA ligase. At this point, only one end of λDNA will be biotinylated.
4. To label the other end of the λDNA with biotin, add 0.1 μM phosphorylated primer B (10 pmol) to the ligated product and anneal the phosphorylated primer B to the λDNA by lowering the temperature gradually as described in **step 2**.
5. Ligate the primer B to the product at 16 °C for 2 h by adding 10 mM ATP and 300 units of T4 DNA ligase. Inactivate the ligase by heating at 65 °C for 10 min and then add 5 mM EDTA to stabilize the DNA. The final concentration of the biotinylated λDNA will be 1.5 nM.

3.2 Preparation of Microfluidic Flow Cell

1. Place the staining rack filled with glass coverslips in a beaker containing 1 N KOH and sonicate the beaker full of coverslips in the ultrasonic bath for 30 min. Organic contaminants on the surfaces of the coverslips will be removed by sonication and the surfaces will be hydrophilized. Ensure that the coverslips are entirely immersed in the solution.
2. Wash the coverslips with ultrapure water.
3. Place the staining rack in a new beaker containing 100% methanol and sonicate the beaker full of coverslips for 15 min.
4. Place the staining rack in a beaker containing 2% APTES in methanol and incubate in the dark at room temperature overnight.
5. Wash the coverslips with ultrapure water and dry the coverslips.
6. Prepare a biotin-mPEG solution immediately before the next step.
7. Place the coverslips on parafilm and divide the coverslips into two groups. Drop 15 μL of biotin-mPEG solution onto each coverslip in one group. Cover the solution on the coverslip with the other coverslip. Ensure that the solution is spread across the inner surfaces between the two coverslips and place the coverslips in a dark, humid condition at 4 °C overnight (or room temperature for 3 h).

8. Carefully separate the two attached coverslips. Wash the entire surfaces of each coverslip with ultrapure water and air-dry them completely, regardless of which surface is functionalized. Note that the biotin-mPEG was adhered to only one side of each coverslip (Fig. 1c).
9. Store the coverslips until use by placing the coverslips individually in cases containing silica gel, then seal the cases, and store them at −80 °C.
10. On a microscope glass slide, drill two 1 mm diameter holes that are 9 mm apart at the 6 and 12 o'clock positions and 14 mm apart at the 3 and 9 o'clock positions. Adjust the hole size with a round diamond file to ensure the polyethylene tubings of outer diameter of 1.05 mm fit through the holes at the 3 and 12 o'clock positions while tubings of outer diameter of 1.20 mm fit through the holes at the 6 and 9 o'clock positions.
11. Cut out a 3 cm × 2.5 cm piece of double-sided adhesive tape and cut out a cross shape with a vertical length of 1.5 cm with width of 2 mm and a horizontal length of 2 cm lateral with width of 2 mm from the center of tape. Attach the 3 cm × 2.5 cm piece of tape with the hollowed out cross onto the microscope slide with holes so that all four holes are inside the hollowed space of the tape (Fig. 1d, **step 1**). Be careful not to cover any holes with the tape (*see* **Note 5**).
12. Attach the biotin-coated coverslip to the adhesive tape on the microscope glass. Make sure that the biotin-coated side of the coverslip is facing the microscope glass (Fig. 1d, **step 2**).
13. Flip the microscope glass so that the holes are facing up and the attached coverslip is facing down. Apply 5 μL of streptavidin solution through one of the holes to coat the inner surface of the coverslip.
14. Fit one end of each polyethylene tubings into their corresponding holes on the microscope slide. The inlet tubings are inserted into the holes at the 3 and 12 o'clock positions, and the outlet tubings are inserted into the holes at the 6 and 9 o'clock positions. Be careful not to push the tubings too far inside as it will dislodge or break the bottom coverslip (Fig. 1d, **step 3**). Use superglue to fix the tubings in place and to seal any cracks (Fig. 1d, **step 4**). Fit the syringe needle attached to syringes into the other end of the outlet tubings (outer diameters of 1.20 mm). Wait for about 30 min for the superglue to harden.

3.3 DNA Loop Extrusion Assay

1. On the microscope, adjust the optical axis and aperture of the laser beam and set the total reflection angle (Fig. 1e) (*see* **Note 6**).

2. Place a tube containing 1 mL DNA buffer and insert the inlet tubings in the tube. Manually withdraw the plungers of both syringes fitted in the outlet tubings to confirm that DNA buffer solution flows through the inlet tubings (outer diameters of 1.05 mm) without air entering the flow cell or the tubings.
3. Place the flow cell on the stage of the microscope with the coverslip facing down. Seal the ends of the horizontal tubings (3 and 9 o'clock positions) with superglue so that air cannot enter through them. Set a syringe onto the syringe pump and fit the syringe needle through the end of the vertical outlet tubing (Fig. 1f). Place the vertical inlet tubing in a tube with 1 mL DNA buffer. Set the syringe pump to pull back the plunger to apply DNA buffer at a constant flow rate of 30 μL/min for 2 min through the vertical inlet tubing and into the flow cell.
4. Replace the DNA buffer tube with a tube containing 5 pM biotinylated λDNA in DNA buffer. Apply biotinylated λDNA at a constant flow rate of 30 μL/min for 50 s and then wait 5–10 min to allow λDNA to tether onto the inner surface of the coverslip (*see* **Note 7**).
5. Replace the biotinylated λDNA-containing tube with a tube containing Imaging buffer 1. Wash out the unbound λDNA with the buffer at a constant flow rate of 30 μL/min for 1 min. During the flow, check to ensure that multiple double-tethered, U-shaped DNA can be observed in the flow cell.
6. Remove the syringe and the Imaging buffer 1-containing tube from the vertical tubings and seal the ends of the vertical tubings (6 and 12 o'clock positions) with superglue so that air cannot enter from the tubings. Cut the glued tips of the horizontal tubing. Insert the syringe on the syringe pump into the end of the outlet tubing (9 o'clock position) while placing the end of the inlet tubing (3 o'clock position) in a tube containing Imaging buffer 1 (Fig. 1g).
7. Apply Imaging buffer 1 at a constant flow rate of 30 μL/min for 2 min through the horizontal tubings and confirm that multiple U-shaped DNAs are observed in the flow cell.
8. Replace the Imaging buffer 1-containing tube with a tube containing Imaging-casein buffer. Apply the buffer at a constant flow rate of 30 μL/min for 50 s and wait for 5 min. This step will prevent nonspecific binding of cohesin and NIPBL during loop extrusion.
9. Replace the casein-containing tube with a tube containing Imaging buffer 1. Apply Imaging buffer 1 at a constant flow rate of 30 μL/min for at least 3 min. During this time, select areas with multiple U-shaped DNA to record (capture areas). It

is preferable to capture images at the central region of the flow cell.

10. Replace the Imaging buffer 1-containing tube with a tube containing Imaging buffer 2. Apply the buffer at a constant flow rate of 30 μL/min for 2 min.
11. Prepare a tube containing 100 μL Imaging buffer 2 solution that includes 1 nM cohesin, 2 nM NIPBL-Mau2 or NIPBLdN, and 1.5 nM AP21967 (optional). Incubate at 16 °C for 10 min (*see* **Note 8**).
12. Replace the Imaging buffer 2-containing tube with the tube containing cohesin–NIPBL. Apply the cohesin–NIPBL mixture at a constant flow rate of 32 μL/min at room temperature (23 °C). Start recording with the camera to take images after the first 10 s. After recording for 2 min, add the remaining prepared Imaging buffer 2 (836 μL) directly to the tube containing cohesin–NIPBL (*see* **Note 9**). The solution is continuously applied at a constant flow rate of 32 μL/min, and the successive images are captured every 5 s for 25 min at multiple capture areas with an exposure time of 100 ms for each channel (Fig. 2a).
13. Confirm that the recorded file is saved. Microscope slides can be reused. After the experiment, remove the tubing from the flow cell and immerse the microscope slide in acetone. Leave it in for several hours (or overnight) to allow the adhesive tape and superglue to peel off. After peeling off the superglue using tweezers and a paper towel, immerse the microscope slides in 2-propanol for a few minutes to remove as much organic matter and contaminants as possible. Ensure that any superglue or other contaminants do not remain in the holes as well. Finally, wipe the microscope slide with 70% ethanol to completely remove any remaining organic contaminants.
14. Use the NIS-Elements software to enclose the desired area with a rectangle region of interests (ROI) for analysis. Successive images in the ROI are generated in the recorded file. When analyzing, single-tethered DNA and DNA loops that are unclear are excluded.
15. DNA loop extrusion is assessed by whether the tip region of U-shaped DNA (bottom of the U-shape) overlaps for at least 3 frames (15 s) when cohesin is bound. The ratio of extruded DNA loops is calculated by dividing the number of cohesin-bound DNA with extruded loops with the total number of cohesin-bound DNA (Fig. 2b) (*see* **Note 10**). The duration of loop extrusion is calculated from the start of the formation of the DNA loop to the timepoint when the formed loop

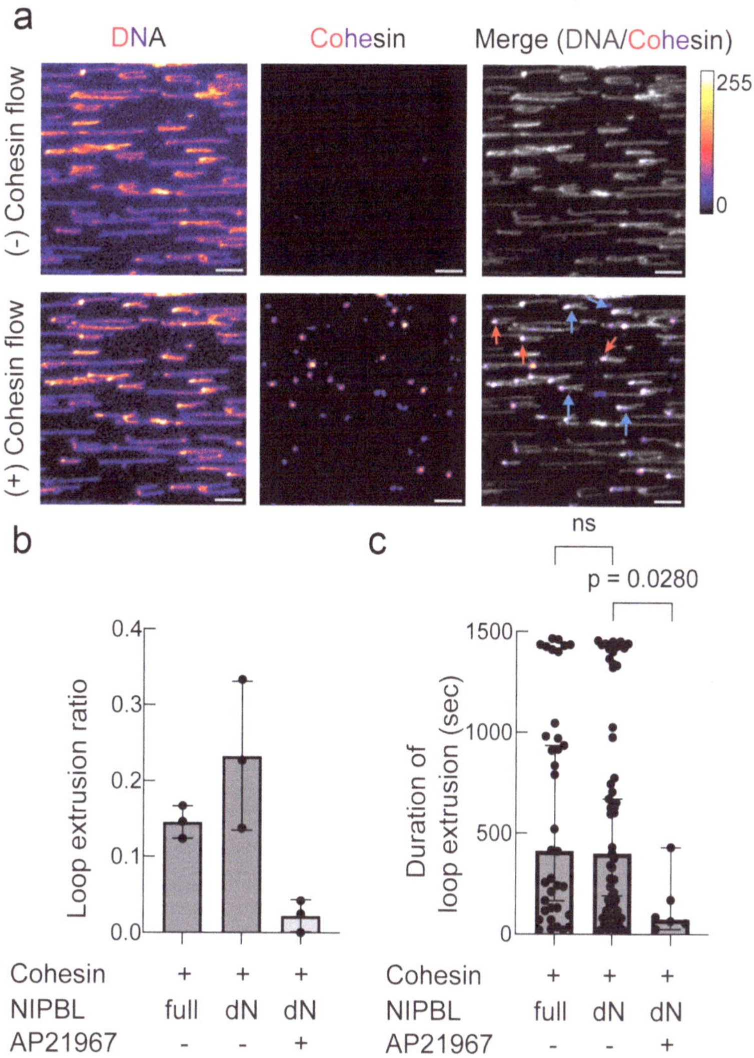

Fig. 2 DNA loop extrusion of recombinant human cohesin complex. (**a**) Fluorescent images of λDNA without or with the addition of cohesin. Bar = 5 μm. Blue arrows show extruded DNA loops, while red arrows show the association of cohesin without extruded DNA loop. (**b**) The distribution of DNA loop extrusion ratio of all U-shaped λDNA in the presence of either NIPBL-Mau2 or NIPBLdN (*n* = 3 flow cells with over 48 DNA observed per flow cell). (**c**) The distribution of maintaining duration of DNA loop extrusion in the presence of either NIPBL-Mau2 or NIPBLdN (median with 95% CI, p-values from Mann-Whitney test)

disappears (Fig. 2c). The loop extrusion rate can be approximated from the fraction of looped DNA over time by measuring the intensities of the entire DNA and non-looped (nonoverlapping) DNA region using FIJI. Through this process, the behavior of fluorescently labeled cohesin is analyzed, as explained in the next section.

3.4 Translocation Assay

1. In this assay, only the horizontally positioned holes (3 and 9 o'clock positions) of the microscope slide is used. The inlet tubing is placed in the 3 o'clock position while the outlet tubing is placed in the 9 o'clock position. The other holes (6 and 12 o'clock positions) are sealed to prevent leakage, as detailed in **Note 5**. Insert the end of the inlet tubing in the tube containing DNA buffer and insert the syringe into the end of the outlet tubing. Withdraw the plunger of the syringe manually to confirm that the DNA buffer flows smoothly without air entering the flow cell or tubings. Place the flow cell on the microscope stage and set a syringe on the syringe pump. Apply DNA buffer at a constant flow rate of 30 μL/min for 2 min.
2. Replace the DNA-buffer-containing tube with a tube containing DNA buffer with 7 pM λDNA, which is biotin-labeled at only one end with biotinylated primer A as prepared in Subheading 3.1, **step 3**. Apply λDNA at a constant flow rate of 30 μL/min for 40 s and then incubate for 5 min.
3. Replace the λDNA-containing tube with a tube containing DNA buffer. Apply DNA buffer at a constant flow rate of 30 μL/min for 2 min to remove any unbound λDNA. During the flow, check that multiple λDNA are single tethered in the desired capture area.
4. Replace the DNA buffer-containing tube with a tube containing 20 nM phosphorylated biotinylated primer B in DNA anneal buffer. Apply primer B at a constant flow rate of 80 μL/min for 12 min to allow it to anneal to the overhang of λDNA and tether onto the coated coverslip.
5. Replace the primer B-containing tube with a tube containing Imaging buffer 3. Apply Imaging buffer 3 at a constant flow rate of 30 μL/min for 40 s. Then incubate for 10 min to prevent nonspecific binding of cohesin and NIPBL and to ligate primer B to λDNA.
6. Replace the Imaging buffer 3-containing tube with a tube containing Imaging buffer 1. Apply the buffer at a constant flow rate of 30 μL/min for 2 min. After 2 min, select the capture areas to record. It is preferable to capture images at the central region of the flow cell.
7. Replace the Imaging buffer 1-containing tube with a tube containing Translocation buffer. Apply the buffer which includes ATP and oxygen scavengers at a constant flow rate of 30 μL/min for 2 min. After stopping the flow, confirm that multiple λDNA are double tethered in the desired captured areas by live capture imaging of the λDNA, using the laser with wavelength 561 nm.

8. Mix 500 pM cohesin and 500 pM NIPBL-Mau2 or 3 nM NIPBLdN in Translocation buffer at 16 °C for 10 min. Replace the Translocation buffer-containing tube with the tube containing cohesin–NIPBL mixture and apply the mixture at a constant flow rate of 30 μL/min for 45 s (*see* **Note 11**).
9. Replace the cohesin–NIPBL-containing tube with a tube containing Translocation buffer. Incubate for 2 min to allow for cohesin–NIPBL to bind to DNA and then apply the buffer at a constant flow rate of 30 μL/min for 2 min.
10. Capture images at room temperature (23 °C) for 1 min at a fast rate (approximately every 104 ms) in multiple areas with the exposure time of 100 ms (*see* **Note 12**).
11. Replace the Translocation buffer-containing tube with a tube containing Translocation buffer 500. Incubate for 2 min and apply the buffer at a constant flow rate of 30 μL/min for 2 min. After the flow ends, capture images as explained in **step 10**.
12. Use the NIS-Elements software to enclose the desired area with a rectangular region of interest (ROI) for tracking analysis and generate successive images within the ROI if necessary (Fig. 3a). The method for generating a kymograph of fluorescently labeled cohesin (Fig. 3b) is explained in **Note 13**. The average end-to-end distance of tethered λDNA (Fig. 3c; 11.9 ± 0.9 μm, error bar = SD) is analyzed by creating a stacked image of the first 10 frames and manually drawing a line along the DNA using the FIJI software (*see* **Note 13**).
13. To detect and track the fluorescent signal of Alexa Fluor 488-labeled cohesin, image smoothing is applied to reduce noise and to introduce slight blurring. This step allows for accurate identification and tracking of weak fluorescence signals.
14. Local contrast adjustment is applied to enhance the difference between bright and dark areas, allowing for accurate detection of weak fluorescent intensity. This step is optional and can be applied if necessary.
15. Perform object tracking using the tracking module in the NIS-Elements software. Set an intensity threshold to define the conditions for signal detection. Apply this threshold to all successive frames to track binary objects throughout the time-lapse series. After detection, generate the identified binary objects and track their movements by determining centroids based on the shapes and areas of the objects (Fig. 3c). Finally, export the tracking data to an Excel file. The intensities of the tracked points are analyzed from a square region centered on the centroid of each timepoint using FIJI.

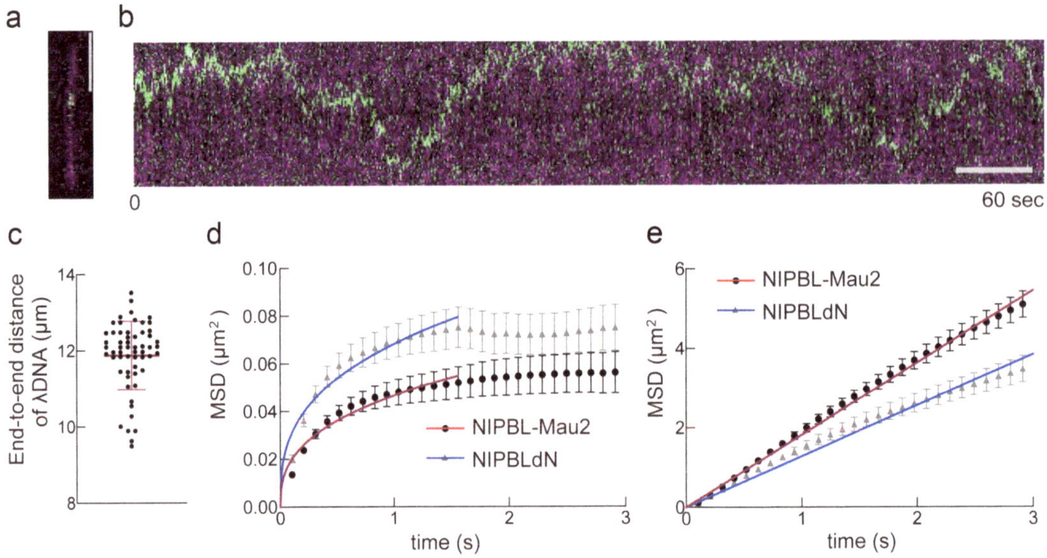

Fig. 3 Translocation assay of recombinant human cohesin complex. (**a**) Fluorescent images of cohesin (green) along λDNA (magenta). Bar = 5 μm. (**b**) The kymograph of image in (**a**) is shown for 1 min. Bar = 5 s. (**c**) The distribution of end-to-end distance of tethered λDNA (mean of 11.9 ± 0.9 μm, error bar = SD, *n* = 58). (**d**) MSD of cohesin complex at 50 mM NaCl. Diffusion coefficient, D, is 0.0127 ± 0.0004 $\mu m^2/s$ (95% CI, $R^2 = 0.94$, $n = 45$) and scaling exponent, a, is 0.37 ± 0.06 for NIPBL-Mau2 (circles and red approximated line). D is 0.0149 ± 0.0006 (95% CI, $R^2 = 0.87$, $n = 24$) and a is 0.27 ± 0.06 for NIPBLdN (triangles and blue approximated line) (from 2 or 3 flow cells with over 11 cohesin complexes observed per flow cell). (**e**) MSD of cohesin complex at 500 mM NaCl. Diffusion coefficient, D, is 0.4539 ± 0.0058 $\mu m^2/s$ (95% CI, $R^2 = 0.995$, $n = 97$) for NIPBL-Mau2 (circles and red approximated line). D is 0.3202 ± 0.0090 (95% CI, $R^2 = 0.974$, $n = 58$) for NIPBLdN (triangles and blue approximated line). Scaling exponents are equal to 1

16. Calculate the mean squared displacement (MSD) of each object by computing the squared displacement between positions at different timepoints using a MATLAB program. The formula for MSD is $MSD(\Delta t) = \langle [r(t + \Delta t) - r(t)]^2 \rangle$, where $r(t)$ is the two-dimensional position vector at time t and Δt is the time difference between the points.
17. The diffusion coefficient (D) is calculated using the MSD plot using the approximation formula $MSD(t) = 4D \cdot t^{\alpha}$, where scaling exponent $\alpha = 1$ is for Brownian *diffusion* and $0 < \alpha < 1$ is for sub-diffusion. The analysis is performed using Prism software (Figs. 3d, e; *see* **Notes 14** and **15**).

4 Notes

1. If the thickness of glass coverslip deviates from 0.17 mm, the incident angle of the TIRF light changes at the glass surface. This leads to insufficient correction of spherical aberration and the captured images may be blurred.

2. An EMCCD camera is cooled down using a cooling water circulation system to suppress thermal noise and to amplify electronic signal of the light. The increase of electron-multiplying (EM) gain enables the detection of objects even at a low intensity where an ordinary CCD camera cannot.
3. A/C heterodimer of DmrA (FR506 binding protein; FKBP-12) and DmrC (FKBP-rapamycin binding domain; FRB) T2098L are dimerized by C16-(S)-7-methylindolerapamycin (AP21967, C16-AiRap). In this experiment, FKBP and FRB T2098L are fused to the C-terminal of Smc1 and Smc3, respectively.
4. Mau2 interacts with the N-terminal region of NIPBL (M1-K38) [17]. Therefore, Mau2 is not included in NIPBLdN.
5. For both the translocation assay and loop extrusion assay, a flow cell with two holes rather than four holes can be used. In the case of a flow cell with two holes, the holes are made 14 mm apart at the 3 and 9 o'clock positions. A 2 cm × 2.5 cm piece of double-sided adhesive tape is used and a 1.2 mm × 1.6 cm rectangle in the center is cut out. Attach the tape and place a half-cut coverslip (11 cm × 22 cm).
6. Because the tethered U-shaped and I-shaped λDNA are slightly suspended from the surface of the coated coverslip, the total reflection angle for observation should be slightly reduced.
7. The volume inside an inlet tubing is 14.5 μL and the half volume of the horizontal flow channel is 1 μL. Therefore, λDNA and other solutions arrive at the central region of the flow cell after 30 s when the flow rate is 32 μL/min. To keep the amount of λDNA binding at a constant level, the λDNA should be applied for 50 s.
8. When the proportion of fluorescently labeled wild-type cohesin binding to λDNA is low (or high), increasing (or decreasing) the cohesin concentration may lead to more successful observations of DNA loops. When high proportion of DNA condense immediately after cohesin binding, reducing the NIPBL concentration may increase the proportion of DNA loops. When most λDNA retain its U-shape despite cohesin binding, increasing the NIPBL concentration may be effective, as it likely indicates that the cohesin tetramer is bound to λDNA in the absence of NIPBL.
9. After directly adding Imaging buffer 2 to the tube containing the remaining cohesin–NIPBL mixture, the concentrations of cohesin and NIPBL are diluted 16-fold. Therefore, the probability of additional cohesin binding to λDNA will be low when the diluted solution is applied.

10. DNA loop extrusion has been observed in both NIPBL-Mau2 and NIPBLdN [5, 12, 13, 18], and Fig. 2b, c show that lacking both the N-terminal IDR (M1-P1162) of NIPBL and the Mau2 protein has no effect on DNA loop extrusion. Additionally, ref. [5] show that when the N-terminal IDR of NIPBL and the Mau2 protein are present, the addition of AP21967 that binds the heads of Smc1-FKBP and Smc3-FRB together does not inhibit DNA loop extrusion [5]. However, DNA loop extrusion is greatly reduced in the presence of AP21967 when N-terminal IDR of NIPBL and the Mau2 protein are lacking. Therefore, to extrude DNA loops and to keep loops stable, the opening of Smc1 and Smc3 heads is essential when the reaction lacks the N-terminal IDR of NIPBL and the Mau2 protein (Figs. 2b, c) [5].
11. In contrast to the DNA loop extrusion assay, the translocation assay is designed to track the behavior of cohesin without buffer flow (Figs. 3a–c). MSD of DNA-bound NIPBLdN is higher than that of NIPBL-Mau2 (Figs. 3d, e), indicating that the N-terminal IDR of NIPBL and Mau2 are required for stable binding of cohesin to DNA.
12. To acquire images at a fast rate, a trigger device is connected to the EMCCD camera and the "no delay" mode is set in the NIS-Element software. This allows for faster responses by the camera.
13. To generate a kymograph of fluorescently labeled cohesin, a straight line with a width of 4 pixels (130 μm/pixel) is manually drawn along the DNA in the stacked image using the FIJI software. The KymoResliceWide plugin (https://github.com/elifesciences-publications/KymoResliceWide) is then applied to generate the kymograph. When both ends of λDNA are tethered, a small tension (less than 0.2 pN in the presence of SYTOX orange) is applied, as explained by the worm-like chain (WLC) model [19]. Under this low tension, DNA remains extended and does not undergo large conformational fluctuations such as condensation, though thermal fluctuations persist. Therefore, a wide line is used to generate the kymograph to account for thermal fluctuation.
14. In low salt conditions (50 mM NaCl), the behavior of both cohesin–NIPBL-Mau2 and cohesin–NIPBLdN complexes are characterized by sub-diffusion, as indicated by the scaling exponents ranging from 0.27 to 0.37. The diffusion coefficient is comparable between the full-length and truncated N-IDR of NIPBL. The cohesin–NIPBL complex strongly associates with DNA and restricts DNA fluctuations in the translocation assay compared to those in the DNA loop extrusion assay. This is due to the flow being halted during image acquisition in the

translocation assay. This restriction is likely to result in an increase in non-isotropic interactions between the cohesin–NIPBL complex and the DNA backbone [16].

15. In high salt conditions (500 mM NaCl), both cohesin–NIPBL-Mau2 and cohesin–NIPBLdN complexes diffuse rapidly along DNA in line with Brownian diffusion. The diffusion coefficient increases by 1.4 times in the presence of the N-terminal IDR of NIPBL and the Mau2 protein. The increased ionic strength reduces electrostatic interactions between the cohesin–NIPBL complex and DNA, allowing complexes to move more freely and exhibit the typical Brownian motion. Removal of the N-terminal IDR of NIPBL and the Mau2 protein appears to weaken electrostatic interactions with DNA, resulting in the increased diffusion.

Acknowledgments

We thank M. M. Yoshida for critical reading of the manuscript. We also thank the Technical Center at Nagoya University for preparing microscope slides with holes. This work was supported by JSPS Grant-in-Aid for Scientific Research/KAKENHI (#24 K18082 to K.Y. and #20H05937 to T.N.).

References

1. Lieberman-Aiden E, van Berkum NL, Williams L et al (2009) Comprehensive mapping of long-range interactions reveals folding principles of the human genome. Science 326:289–293
2. Rao SSP, Huntley MH, Durand NC et al (2014) A 3D map of the human genome at kilobase resolution reveals principles of chromatin looping. Cell 159:1665–1680
3. Michaelis C, Ciosk R, Nasmyth K (1997) Cohesins: chromosomal proteins that prevent premature separation of sister chromatids. Cell 91(1):35–45
4. Guacci V, Koshland DE, Strunnikov A (1997) A direct link between sister chromatid cohesion and chromosome condensation revealed through the analysis of MCD1 in S. cerevisiae. Cell 91(1):47–57
5. Sakata R, Niwa K, Ugarte La Torre D et al (2021) Opening of cohesin's SMC ring is essential for timely DNA replication and DNA loop formation. Cell Rep 35:108999
6. Bauer BW, Davidson IF, Canena D et al (2021) Cohesin mediates DNA loop extrusion by a "swing and clamp" mechanism. Cell 184:1–17
7. Rao SSP, Huang S, Glenn St Hilaire B et al (2017) Cohesin loss eliminates all loop domains. Cell 171:305–320
8. Kagey MH, Newman JJ, Bilodeau S et al (2010) Mediator and cohesin connect gene expression and chromatin architecture. Nature 467:430–435
9. Hsieh TS, Cattoglio C, Slobodyanyuk E et al (2020) Resolving the 3D landscape of transcription-linked mammalian chromatin folding. Mol Cell 78:539–553
10. Thiecke MJ, Wutz G, Muhar M et al (2020) Cohesin-dependent and -independent mechanisms mediate chromosomal contacts between promoters and enhancers. Cell Rep 32:107929
11. Ganji M, Shaltiel IA, Bisht S et al (2018) Real-time imaging of DNA loop extrusion by condensin. Science 360:102–105
12. Davidson IF, Bauer B, Goetz D et al (2019) DNA loop extrusion by human cohesin. Science 366:1338–1345
13. Kim Y, Shi Z, Zhang H et al (2019) Human cohesin compacts DNA by loop extrusion. Science 366:1345–1349

14. Davidson IF, Goetz D, Zaczek MP et al (2016) Rapid movement and transcriptional re-localization of human cohesin on DNA. EMBO J 35(24):2671–2685
15. Kanke M, Tahara E, Huis In't Veld PJ et al (2016) Cohesin acetylation and Wapl-Pds5 oppositely regulate translocation of cohesin along DNA. EMBO J 35(24):2686–2698
16. Stiglar J, Çamdere GÖ, Koshland DE et al (2016) Single-molecule imaging reveals a collapsed conformational state for DNA-bound Cohesin. Cell Rep 15(5):988–998
17. Braunholz D, Hullings M, Gil-Rodríguez MC et al (2012) Isolated NIBPL missense mutations that cause Cornelia de Lange syndrome alter MAU2 interaction. Eur J Hum Genet 20: 271–276
18. Davidson IF, Barth R, Zaczek M et al (2023) CTCF is a DNA-tension-dependent barrier to cohesin-mediated loop extrusion. Nature 616(7958):822–827
19. Marko JF, Siggia ED (1995) Statistical mechanics of supercoiled DNA. Phys Rev E 52(3):2912–2938

Chapter 2

Interaction Between an Individual Cohesin Complex and DNA Revealed by Optical Tweezers and Single-Molecule Microscopy

Martina Richeldi and Maxim Molodtsov

Abstract

Establishment of cohesion between sister chromatids in mitosis is one of the fundamental principles used by cells to ensure accuracy of the chromosome segregation. Cohesion allows spindle microtubules to generate tension across sister chromatids that acts as a signal that correct spindle-chromosome connections have been made. Cohesion is established by cohesin—a ring-shaped protein complex that is thought to link sister chromatids by physically entrapping their DNA. However, molecular mechanisms underlying cohesin interaction with DNA and its ability to form DNA–DNA interactions are not well understood. Here, we describe a protocol to reconstitute this process at the single-molecule level and investigate how single cohesin protein complexes establish DNA–DNA interactions and resist mechanical forces exerted on DNA.

Key words Cohesin, Cell division, Chromosomes, Force, Optical trap, Optical tweezers, Single-molecule, Cohesin–DNA interaction

1 Introduction

The ability to measure mechanical forces generated and counteracted by single molecules revolutionized our understanding of molecular mechanisms that drive rearrangements of intracellular structures. Thanks to the combination of single-molecule microscopy and force spectroscopy, it is now possible to image movement of single molecules and protein complexes and observe their response to the externally applied force in reconstitution experiments in vitro. This approach has been instrumental in revealing molecular mechanisms underlying the action of cytoskeleton motors such as kinesin, dynein, and myosin [1–6].

However, unlike the cytoskeleton, mechanical forces that lead to the rearrangements of DNA, genomes, and chromosomes are much less well understood. Key molecular machines that reorganize chromatin are part of the Structural Maintenance of

Yasuto Murayama and Yasutaka Kakui (eds.), *SMC Proteins: Methods and Protocols*, Methods in Molecular Biology, vol. 2991, https://doi.org/10.1007/978-1-0716-5005-9_2,

Chromosomes (SMC) family of protein complexes. SMC proteins use energy of ATP hydrolysis to establish DNA–DNA interactions, move along the DNA, and rearrange it in 3D space [7, 8]. Understanding how these machines generate and resist mechanical forces is essential for our understanding of how chromatin and chromosomes are built and rearranged as cells grow and divide. However, the progress has been limited so far; assays that were successfully applied to investigate mechanisms of molecular motors cannot be directly translated for studying SMC protein complexes. Some of the challenges include physical properties of the DNA that is several orders of magnitude more flexible in comparison to the cytoskeleton filaments and SMCs themselves that are complex, multi-subunit complexes with a high degree of structural flexibility.

Here, we focus on the SMC protein complex cohesin, responsible for physically tethering sister chromatids and allowing tension to be generated between the chromatids in response to the forces generated by the mitotic spindle. Given the ringlike-shaped architecture of cohesin, it was hypothesized early on that the complex may hold sister chromatids by topologically embracing sister DNAs within the lumen of the ring (Fig. 1a, b) supported by the experiments in which cleaving the Scc1 subunit of cohesin releases DNA and dissolves DNA–DNA interaction [9–11].

To test this model directly at the single-molecule level and investigate its implications for the mechanical stability of cohesin–DNA interaction, we describe a step-by-step protocol to reconstitute the DNA–DNA interaction established and held together by a single cohesin protein complex (Fig. 1c). Through a combination of total internal reflection fluorescence (TIRF) microscopy for single-molecule visualization and optical trapping for force application, this protocol was recently used to verify that a single cohesin molecule can establish DNA–DNA interaction and to measure its mechanical stability [12]. This protocol generates force–distance curves that show rupture events between cohesin and DNA, with the visual verification that those rupture events correspond to the

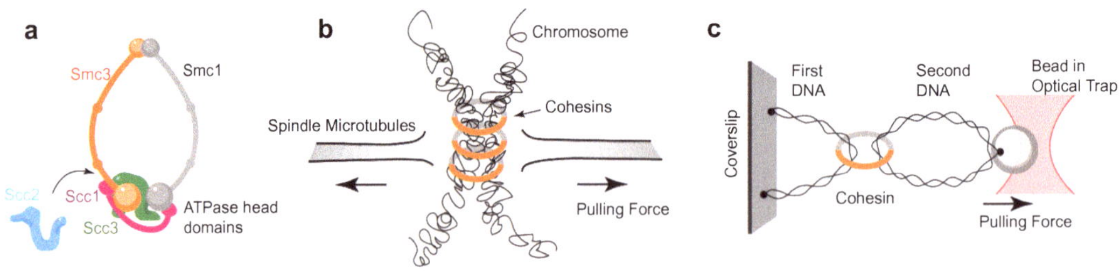

Fig. 1 SMC complex cohesin and the schematics of the experiment. (**a**) Schematics of cohesin protein complex. Individual subunits are labelled using budding yeast nomenclature. (**b**) An illustration of how tension from spindle microtubules is applied to sister chromatids and counteracted by cohesin. (**c**) Schematics of the reconstitution experiment

disengagement of the cohesin complex. Single cohesin rupture forces can be further analyzed and compared between mutants to reveal the mechanics underlying the establishment of DNA–DNA interactions by single cohesin.

2 Materials

2.1 Purified Proteins and Cohesin Labelling

1. The fission yeast Mis4^{Scc2}-Ssl3^{Scc4} cohesin loader complex, which is purified as previously described [12] (*see* **Note 1**).
2. The fission yeast cohesin tetramer (Psm1^{Smc1}, Psm3^{Smc3}, Rad21^{Scc1}, Psc3^{Scc3}) fused with SNAP- and CLIP-tags on the Smc1 and Smc3 subunits (cohesin-SC), which is purified as previously described [12].
3. 8 μM biotin conjugated with benzylcytosine (CLIP-tag).
4. 4 μM TMR or LD655 conjugated with benzylguanine (SNAP-tag).
5. Superose 6 10/300 GL gel filtration column (Cytiva).
6. L buffer: 20 mM Tris-HCl pH 7.5, 200 mM NaCl, 10% (w/v) glycerol, 1 mM tris(2-carboxyethyl)phosphine (TCEP).
7. 10% (v/v) Tween 20.
8. 1 M dithiothreitol (DTT).
9. Coomassie protein staining solution.
10. Ultrafiltration tube with 30 kDa molecular weight cutoff.
11. 7.5% SDS-PAGE apparatus.
12. Spectrophotometer.
13. FPLC system.

2.2 The First DNA Molecule

1. 48.5 kbp-long λ-phage DNA.
2. Oligonucleotide 1: aggtcgccgccc[BiotinTg]
3. Oligonucleotide 2: gggcggcgacct[BiotinTg].
4. Annealing buffer: 10 mM Tris-HCl pH 7.5, 50 mM NaCl, 1 mM EDTA.
5. Micro Bio-Spin P30 column (Bio-Rad).

2.3 The Second DNA Molecule

1. 7.25 kb-long M13mp18 single-stranded DNA.
2. 7.25 kbp-long M13mp18 double-stranded DNA.
3. Label IT MF488 and biotin nucleic acids labelling kits (Mirus).
4. G-50 microspin column.
5. TE buffer: 10 mM Tris-HCl pH 7.5, 1 mM EDTA.

2.4 Optically Trapped Beads

1. Polystyrene streptavidin-coated beads 0.5 μm in diameter (Bands Labs).
2. T buffer: 35 mM Tris-HCl pH 7.5, 130 mM NaCl, 1 mM $MgCl_2$, 15% (w/v) glycerol, 1 mM TCEP, 1% (w/v) glucose, 0.2 mg/mL glucose oxidase, 35 μg/mL catalase (*see* **Note 2**).
3. T buffer containing 0.1 mg/mL β-casein.
4. Sonicator bath.

2.5 Microfluidics

1. Glass slides (26 × 76 × 1.0 mm) [12, 13].
2. Chemically cleaned and silanized glass coverslips (24 × 60 mm, calibrated thickness 170 ± 5 μm) [12, 13].
3. Parafilm.
4. 5% dichloromethylsilane in heptane (v/v).
5. Chloroform.
6. Metal tubing connectors (0.6 mm outer diameter, 12.5 mm long, Type 304 stainless steel, New England Small Tube).
7. Inlet and outlet tubing (Tygon polyethylene tubing with inner diameter 0.5 mm and outer diameter 1.5 mm).
8. Epoxy glue.
9. Glass syringe.
10. Syringe pump.
11. Drill.
12. Tabletop plasma cleaner.
13. Heating plate.

2.6 Single-Molecule Microscopy and Optical Tweezers Experiments

For the buffers with glucose oxidase and catalase, *see* **Note 2**.

1. R buffer: 35 mM Tris-HCl pH 7.5, 50 mM NaCl, 1 mM $MgCl_2$, 15% (w/v) glycerol, 1 mM TCEP, 1% (w/v) glucose, 0.2 mg/mL glucose oxidase, 35 μg/mL catalase.
2. S buffer: 35 mM Tris-HCl pH 7.5, 500 mM NaCl, 1 mM $MgCl_2$, 15% (w/v) glycerol, 1 mM TCEP, 1% (w/v) glucose, 0.2 mg/mL glucose oxidase, 35 μg/mL catalase.
3. T buffer: 35 mM Tris-HCl pH 7.5, 130 mM NaCl, 1 mM $MgCl_2$, 15% (w/v) glycerol, 1 mM TCEP, 1% (w/v) glucose, 0.2 mg/mL glucose oxidase, 35 μg/mL catalase.
4. U buffer: 35 mM Tris-HCl pH 7.5, 75 mM NaCl, 1 mM $MgCl_2$, 15% (w/v) glycerol, 1 mM TCEP, 1% (w/v) glucose, 0.2 mg/mL glucose oxidase, 35 μg/mL catalase.
5. PBS: Dulbecco's phosphate buffer saline, pH 7.4.
6. 1 mg/mL Avidin-DN (VectorLabs) in PBS.
7. 1% Pluronic F-127 in PBS (w/v).

8. 10 mg/mL β-casein in PBS.
9. 5 mM biotin in PBS.
10. 200 nM SYTOX Orange (100× Stock).
11. 50 mM ATP Stock in 100 mM Tris pH 7.5 (100× Stock).
12. T7 DNA polymerase (0.1 unit/μL).
13. 10 mM deoxynucleotide triphosphates mixture (10× Stock).

2.7 Microscope

To apply external force and visualize single cohesin molecules, we used a microscope that combined optical trapping and TIRF microscopy. Necessary components of the setup include:

1. Inverted microscope.
2. Integrated optical trap with quadrant photodetectors for force calibration and measurement (*see* **Note 3**).
3. TIRF module with high-quality TEM00 CW lasers (50 mW 488 nm, 150 mW 561 nm, and 150 mW 647 nm) coupled through a stable 3 mm thick dichroic mirror placed in the standard microscope dichroic cube position (*see* **Note 4**).
4. 100× NA 1.45 Oil TIRF Objective (*see* **Note 5**).
5. EMCCD camera (*see* **Note 6**).
6. Optosplit to image cohesin and DNA simultaneously in two colors on the same chip of the EMCCD camera.

3 Methods

3.1 Cohesin Labelling

For the experiments in which the force was applied directly to cohesin, we used cohesin construct with SNAP- and CLIP-tags for the attachment of the fluorophore for visualization (at Smc1-SNAP) and the biotin tag (at Smc3-CLIP) for force application, respectively (*see* **Notes** 7 and **8**). For the experiments in which the force is applied to the second DNA, cohesin must not be biotinylated (otherwise, it would bind to the bead simultaneously with the second DNA) but have only a fluorophore. To ensure that results from force measurements indeed relate to single molecules, it is important to ensure over 90% efficiency of labelling for both tags. Biotinylation efficiency can be tested by incubating the biotinylated protein with excess avidin and running the sample on an SDS-PAGE gel. Complete shift of the cohesin band due to binding to avidin indicates close to 100% biotinylation.

1. Purify cohesin-SC as previously described [12].
2. Mix 0.2–0.5 uM cohesin-SC with 4 μM of TMR or LD655 conjugated with benzylguanine in L buffer containing 1 mM DTT and 0.003% (v/v) Tween 20.

3. Incubate the mixture at 25 °C for 1 hour for the attachment of the fluorophore at Smc1-SNAP.
4. For biotin labelling at Smc3-CLIP, add 8 μM biotin-benzylcytosine, and incubate the mixture at 4 °C for 16 h. Skip **step 4** if you prepare cohesin with only a fluorophore for an experiment involving second DNA.
5. Apply the mixture to a Superose 6 10/300 GL gel filtration column equilibrated with L buffer to remove unbound dye and biotin.
6. Record an absorbance spectrum between 220 and 800 nm in 1 nm increments using a spectrophotometer to check the efficiency of fluorescent labelling.
7. Pool the peak fractions and concentrate using 100 kDa Amicon Ultra filters if the final concentration is less than 0.2 μM.
8. To confirm biotin labelling, mix the labelled cohesin with an excess amount of avidin at room temperature for 20 minutes, then run the sample on a 7.5% SDS-PAGE gel, and stain with Coomassie.
9. Freeze the aliquots in liquid nitrogen and store at −80 °C.

3.2 The First DNA Molecule

We use the term "first DNA" to refer to the DNA that is introduced the first, bound to the surface of the cover glass and serves as initial template for cohesin loading (Fig. 1c). For this DNA, use a 48.5 kbp-long λ-phage DNA with biotins on both ends.

1. Mix 250 μg/mL 48.5 kbp-long λ-phage DNA with 10 μM of oligonucleotides 1 and 2 in annealing buffer.
2. Incubate the mixture at 72 °C for 30 min.
3. Switch off the heat block and allow the tube to cool to room temperature.
4. Mount the mixture onto a Micro Bio-Spin P30 equilibrated with annealing buffer.
5. Spin the tube at 700 g for 1 min.
6. Determine the DNA concentration by measuring abscondence of 260 nm using a spectrophotometer.
7. Aliquot DNA and store at −80 °C.

3.3 The Second DNA Molecule

We call the "second DNA" circular DNA that is introduced after cohesin was loaded on the first DNA. Second DNA can bind cohesin, which is already bound to the first DNA thus establishing DNA–DNA interaction. For this DNA, we used either a single-stranded 7.25 kbp-long M13mp18 or the analogous double-stranded 7.25 kb M13mp18 I RF DNA. To visualize and apply force to the second DNA using optical trap (as in Fig. 1c), we covalently attached to it biotins and few MF488 fluorophores using Label IT Mirus commercial kit. To reduce the number of biotins to approximately one per DNA, we decreased the amount of the Label IT reagent to 0.01 μL/μg and the incubation time to 5 min.

1. Mix M13mp18 DNA (either single or double stranded) with the MF488 and/or Biotin Label IT reagents at a ratio of 0.01 μL of reagents per 1 μg of DNA.
2. Incubate at 37 °C for 5 min.
3. Mount the mixture onto a G-50 microspin column equilibrated with TE buffer.
4. Spin the tube at 700 g for 1 min.
5. Determine the DNA concentration by measuring abscondence of 260 nm using a spectrophotometer.
8. Aliquot DNA and store at −20 °C.

3.4 Beads for the Force Application

Beads must be attached to cohesin or DNA to apply the force using an optical trap. Larger beads allow application of higher forces, but smaller beads allow much better control of the geometry of the experiment and improved accuracy in measurements. As a compromise, we used polystyrene streptavidin-coated beads 0.5 μm in diameter. It is important to use beads with consistent size and as narrow size distribution as possible because calibration and measurement are typically done on different beads. All buffers containing beads should be supplemented with 0.1 mg/mL *β*-casein, which helps to prevent beads aggregation and sticking to the coverglass. The following steps are used to prepare beads before adding them to the flow cell.

1. Dilute the stock solution of polystyrene streptavidin-coated beads 20-fold in 1 mL of T buffer containing 0.1 mg/mL of β-casein.
2. Spin at 10,000 g at 4 °C for 15 min.
3. Remove supernatant.
4. Resuspend the beads in 1 mL buffer T with 0.1 mg/mL β-casein and repeat **steps 2** and **3**.
5. Resuspend the beads to the final reaction volume of 100 μL.
6. Sonicate the beads on ice in a sonicator bath for 5–10 min to break aggregates.

3.5 Microfluidics Assembly

The experimental flow cell is assembled by sandwiching the glass slide with glued-in metal tubing connectors, a parafilm spacer, and a chemically cleaned and silanized coverslip (Fig. 2). The volume of the resulting flow cell is 10–20 μL. The flow is controlled by creating negative pressure at the outlet using a syringe pump. We recommend using glass syringes to minimize possible jerks of flow

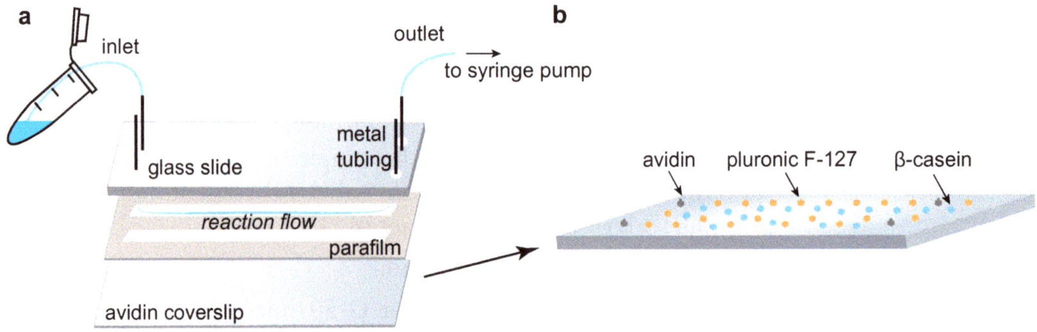

Fig. 2 Graphical illustration of the microfluidics flow cell. (**a**) Components of the two-channel flow cell used for two independent single-molecule experiments. (**b**) Close-up of the hydrophobic coverslip showing the attachment of avidin molecules, followed by passivation using the detergent Pluronic F-127 and β-casein

and unevenness in pressure, which could affect binding of critical components of the experiments, such as DNA or beads (*see* **Note 9**). The inlet tubing is less than 5 cm long to minimize the dead volume, which was ~5 μL. One end of the inlet tube is placed directly into the tube with the solution to be pumped into the flow cell. Coverslips must be chemically cleaned, coupled to avidin and thoroughly passivated to avoid any nonspecific interactions between the reagents in the flow cell and the surface of the coverslip. Many protocols are available for cleaning and passivation process; we used the one from [13] (*see* **Note 10**).

1. Drill holes in a glass slide to accommodate metal connectors for the tubing and glue in connectors using 5-minute 2-component epoxy. Let the epoxy solidify for at least 1 h.
2. Wash the glass slides with epoxy-glued metal connectors: Sonicate in ethanol for 5 min, rinse with water, sonicate in 1 M NaOH for 5 min, rinse with water, and dry on a heating plate to ensure no water is left inside the metal connectors.
3. Prepare the coverslips using the following procedure: Sonicate in 100% acetone for 15 min, sonicate in 100% ethanol for 15 min, and rinse with MilliQ water 10 times.
4. Dry the coverslips with nitrogen gas and plasma clean for 5 min on maximum power setting.
5. Dip the coverslips in 5% dichloromethylsilane dissolved in heptane for 1 h at room temperature to silanize the coverslips.
6. Wash the coverslips by a 5-min sonication in chloroform followed by a 5-min sonication in pure water. Repeat the wash procedure twice.
7. Wash the coverslips by a 5-min sonication in chloroform and then dry up the coverslip with nitrogen gas. **Steps 3–7** are typically done on a separate day to prepare a batch of coverslips that can be stored dry at room temperature for several weeks.

8. Cut parafilm into a rectangular spacer that surrounds the two holes in the glass slide.
9. Sandwich a parafilm spacer between the glass slide and the silanized coverslip to assemble them (Fig. 2a).
10. Heat the assembly by putting it on heating stage set to 90 °C to slightly melt the parafilm and apply light pressure to keep the channels in place.
11. Let the assembly cool down and connect Tygon tubing to the metal connectors.
12. Connect the outlet tubing to a glass syringe pump.
13. After the experiment, the coverslip and tubing are removed and disposed of, but the glass slide with metal connectors can be reused multiple times after repeating the cleaning **step 2**.

3.6 Single-Molecule Microscopy and Optical Tweezers Experiments

Due to the sensitivity of optical traps, they are typically placed inside specially designed rooms. We perform all preparatory steps in a standard wet-lab laboratory while keeping all reagents on ice. When all solutions and the flow cell are ready, move to the optical trapping room while keeping reagents on ice.

In the beginning of the day of the measurement, turn on all optical trapping equipment (at least 2 h before the experiment). Turn lasers and shutters on, place a dummy slide into the microscope, and bring objective into the working position. Turn on the optical trap. Switching on the microscope in advance ensures that all drifts (mostly due to local heating) are stabilized prior to the experiment.

3.6.1 Flow Cell Surface Passivation

Before the beads are added to the flow cell, washing steps can be performed using flow rates up to 100 μL/min; when other flow rates are required, they are specified. All steps are carried out on the microscope at room temperature while all reagents are kept on ice. Alternatively, steps up to loading cohesin on the first DNA can be performed in a standard wet lab, prior to transferring the flow cell and all reagents to the microscope room.

1. Wash the freshly assembled flow cell equipped with a new coverslip with 600 μL of PBS.
2. Flow in 150 μL of 1:50 dilution of Avidin-DN in PBS and incubate for at least 30 min.
3. Wash with 600 μL of PBS.
4. Flow in 200 μL of 1% solution of Pluronic F-127 in PBS and incubate for 10 min.

5. Wash with 200 μL of PBS containing 10 mg/mL β-casein. Incubate for at least 2 hours, but preferably overnight. β-casein helps to passivate further the coverslip after the Pluronic F-127 step.
6. Wash with 500 μL of PBS.

3.6.2 The First DNA Immobilization

1. Flow in 150 μL of a 10 pM solution containing the "first" biotinylated DNA (λ-phage DNA, prepared in Subheading 3.2) in PBS at 20 μL/min. At this rate, most DNA molecules bind via their both ends along the direction of the flow and are stretched to an average length of 9 μm between the tethering points.
2. Wash out excess of DNA with 150 μL of PBS at 50 μL/min.
3. Flow in 100 μL of 5 mM biotin and incubate for 10 min. This blocks all avidin on the surface and prevents it from binding biotinylated cohesin or the second biotinylated DNA.
4. Flow in 150 μL of R buffer. This is the step where PBS is replaced with the reaction buffer.

3.6.3 Loading Cohesin on the First DNA

1. Wash the flow cell with additional 150 μL of R buffer.
2. Mix 0.5 nM of the purified cohesin tetramer (*see* **Note 11**), 1 nM of the Mis4^{Scc2}-Ssl3^{Scc4} cohesin loader complex, 0.5 mM ATP, and 0.1 mg/mL β-casein in R buffer and preincubate the solution for 5 min at room temperature.
3. Add 50 μL of the mixture to the flow cell at a rate of 15 μL/min and incubate for 15 min (*see* **Note 12**).
4. At a 30 μL/min flow rate, wash with 150 μL total of R buffer, followed by 150 μL of S buffer to remove any DNA-bound cohesin that is not salt resistant (*see* **Note 13**).
5. If the next step is imaging, wash with 150 μL of T buffer containing a fresh mixture of 1 mM TCEP, 1% (w/v) glucose, 0.2 mg/mL glucose oxidase, and 35 μg/mL catalase. If DNA needs to be visualized, add 2 nM SYTOX Orange.

3.6.4 Loading the Second DNA

Skip this step in the experiments without the second DNA. When the second DNA is used, it should be biotinylated for attachment to the beads and cohesin used in the previous step should not.

1. Flow in 0.5 nM of the second DNA mixed with 2 nM Mis4^{Scc2}-Ssl3^{Scc4} and 0.5 mM ATP in U buffer at 10 μL/min and incubate for 10 min (*see* **Note 14**).
2. When single-stranded DNA is used, it may be converted into the double stranded DNA after the loading by adding 120 μL of a mixture containing 0.1 unit/μL of T7 DNA polymerase

and 1 mM deoxynucleotide triphosphates in U buffer. Incubate for 15 min.

3. Wash with 150 μL of T buffer.

3.6.5 Attaching the Beads

We recommend performing experiments without staining the DNA, which can mechanically alter the DNA and affect force–distance curve measurements. Beads tethered to DNA undergo distinct and recognizable movement and therefore, the experiment can be performed without DNA visualization. If the staining is required, exchange the imaging buffer supplemented with extra 2 nM SYTOX Orange.

1. Flow in the washed and diluted beads at 5 μL/min for a duration of ~10 min.
2. Incubate the beads without flow for an extra 5 min and wash out excess beads with T buffer at 5 μL/min (*see* **Notes 15** and **16**).
3. After the first wash, there will still be some sparse unbound beads floating in the flow channel. Capture one of the freely floating beads and use it to calibrate the optical trap stiffness (*see* **Note 17**).

3.6.6 The Force–Distance Recording and Cohesin Rupture Force Measurement

1. Make sure the optical trap is off. Using brightfield illumination, identify a bead that seems tethered (either directly to cohesin or to the second DNA; *see* **Note 18**). While this can be done easier with 2 nM SYTOX Orange to visualize DNA, we recommend doing most experiments without the DNA stain. Correctly attached bead can be identified by its tethered motion, where fluctuations in the axis perpendicular to the DNA are typically higher than fluctuations along the DNA.
2. Visualize cohesin in the fluorescence channel (561 nm for TMR or 647 for LD655), and ensure it colocalizes with the bead observed in the brightfield channel.
3. Image the fluorophore on cohesin until it bleaches in one step to ensure the bead is bound to a single cohesin complex (*see* **Notes 19** and **20**). Adjust the imaging laser intensity such that bleaching takes 100–300 seconds.
4. Move the optical trap over the bead tethered to cohesin/DNA and open the shutter. The movement of the bead should be immediately constrained by the trap. Ensure that the coordinates of the bead are shown as zero, suggesting no external force is applied to the bead. For further verification, topologically loaded cohesin bound to the bead can be slid along the DNA without any resistance.
5. Position the bead to be 50–150 nm above the coverslip surface (*see* **Note 21**).

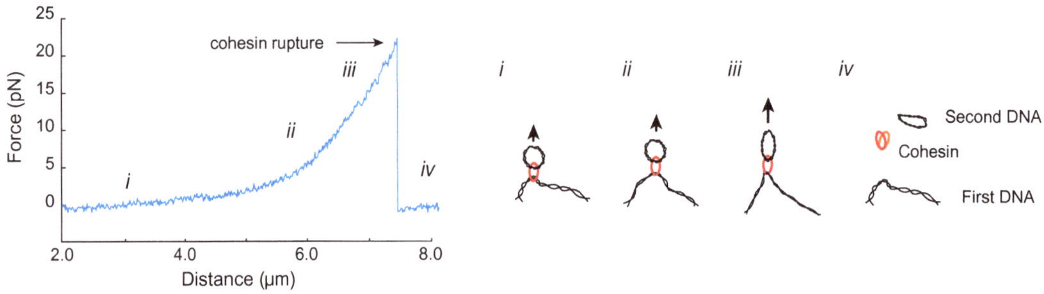

Fig. 3 An example of the force–distance curve showing cohesin rupture. Schematics on the right illustrates DNA–cohesin–DNA interactions at the increasing forces as the system goes through the states shown by roman numerals

6. Record the force–distance curve by moving the optical trap relative to the coverslip perpendicular to the direction along which the DNA is tethered until the bead detaches and coordinates of the bead jump to zero indicating no more forces is applied (*see* **Notes 22** and **23**, Fig. 3).

4 Notes

1. Cohesin requires the Mis4^{Scc2}-Ssl3^{Scc4} cohesin loader complex for loading on DNA. This complex is expressed and purified separately and added to cohesin along with components required for the cohesin loading reaction just before introducing the mixture into the flow cell.
2. During imaging and trapping, it is especially important to have fresh and active glucose oxidase/catalase system to remove oxygen as quickly as possible to prevent photobleaching and laser damage to cohesin and DNA. Add the fresh mixture (which can be aliquoted and stored at −80 °C) to the buffer just before the experiment. During intermediate washing steps when no laser light exposure is present, glucose, glucose oxidase, and catalase may be omitted.
3. We use JPK Nanotracker 2.
4. We use a custom-made setup, but a commercial module can also be used.
5. We use the Nikon product.
6. We use Andor iXon Ultra EMCCD camera.
7. We do not describe here expression and purification of cohesin and loader, which were described elsewhere (e.g., [12, 14]).
8. For the single-molecule visualization, it is important to use bright and photostable fluorophores that can be imaged without photodamage for a significant period of time. In our

experience, TMR and LD655 performed best and provided excellent signal-to-noise ratio.

9. Controlled laminar flow is especially important after beads are added to the flow cell and tethered to DNA. Uncontrolled flow may apply strong forces to the beads via drag force, which could disrupt the interaction between the bead and cohesin or DNA. We calculated that at the flow rates of 5 μL/min, drag forces applied to the beads do not exceed 1 pN, which is significantly smaller than the interaction forces we measured. For other reaction steps, flow rates used never exceeded 100 μL/min, with slower flow rates used for key reactions (DNA immobilization, cohesin loading onto DNA, second DNA capture).

10. The glass coverslip affects refraction of the optical trapping beam, meaning that variations in the cover glass thickness may lead to variations in the trapping stiffness. For better reproducibility between the experiments, we recommend using calibrated coverslips with a variation in thickness smaller than 5 μm.

11. Cohesin concentration should typically be ~0.5 nM and the loader concentration should be 2 nM. The exact cohesin concentration required to achieve the single-molecule loading needs to be determined experimentally. We find that the best conditions are when, at the end of all reactions and washes, ~ 80% of DNA molecules are loaded with cohesin and most of them contain just a single cohesin as verified by the single-molecule imaging.

12. If cohesin is to be cross-linked on DNA after the loading, add 30 nM of the cross-linker (i.e., SpyCatcher-linker-SpyCatcher) [12] and incubate for a further 15 min.

13. High salt washes eliminate cohesin bound to DNA only through electrostatic contacts by destabilizing the cohesin–DNA interaction. By increasing the concentration of salt to 2 M, the threshold concentration can be determined that removes all non-salt-resistant cohesin. Exact amount of salt may depend on the specific construct used.

14. The second DNA can be added at a higher concentration to bias the reaction toward capture of the second DNA by cohesin. This reaction is inefficient because of the small number of cohesin molecules on the first DNA. After the incubation, excess of the second DNA is washed out.

15. Due to the very low number of single molecules, binding of the beads is slow and inefficient. To bias the reaction toward formation of complexes with beads, high concentration of beads is added, and unbound beads are washed out after the incubation.

16. Because of the multiple steps required to assemble the correct construct and extremely small overall number of molecules, there should be few beads left bound to DNA. One must move around through multiple fields of view to find a bead attached to DNA. If many beads are bound, this most likely indicates that too much cohesin remained on DNA, potentially hindering the measurement of single molecules. Pulling experiments are not useful in this case as many beads are likely attached by more than one cohesin.
17. Optical traps measure the force as the product of the stiffness of the optical trap, which depends on the laser intensity, and the displacement of the bead from the center of the optical trap, as measured by the quadrant photodetector. This displacement is limited by the size of the beam and is typically less than 250 nm. Thus, displacements smaller than ~20 nm cannot be measured accurately (unless special measures are taken to improve the signal-to-noise ratio), and displacements larger than 250 nm cannot be measured at all (because the bead moves out of the detection area of the laser beam). Therefore, the laser intensity must be chosen as to set the stiffness such that the desired force is achieved at the reliably measured displacement (e.g., 100 nm). For instance, if the expected force to be measured is 25 pN, the stiffness needs to be set such that 25 pN force is reached at the ~100 nm displacement. Therefore, adjust the laser intensity such that the stiffness of the optical trap for this experiment is $k = 0.25$ pN/nm. Use smaller value to measure smaller forces and higher value for higher forces.
18. In the absence of DNA stain, correct attachment of the bead to the molecule of interest can be identified through movement. Freely floating beads will be washed off, whereas beads that may bind to the surface nonspecifically will have no movement at all. It is possible that some of the first DNA will remain single tethered and therefore some beads may bind to the ends of the single tethered DNA. However, there is a clear difference in the tethered motion of the beads attached to the second DNA or cohesin and the beads attached directly to the first DNA, which can be seen by naked eye. In the latter case, the tethered motion is much less restricted and more symmetrical.
19. Photobleaching of the fluorophore on cohesin prior to the force measurement can, in principle, damage the protein. However, we performed experiments in the same conditions and verified that the average force of rupture does not change. This suggests that if there is some damage to cohesin due to photobleaching, it is likely insignificant.

20. Photobleaching in one step is not a guarantee that the molecule contains only one fluorophore and therefore a single cohesin. Often, several fluorophores may bleach in what appears as one step, but this step would be larger than the elementary step corresponding to the single fluorophore. To make sure that one photobleaching step does correspond to a single fluorophore, sufficient photobleaching statistics must be collected in advance from many molecules to determine the minimum photobleaching step. This step needs to be compared to the photobleaching step of the molecule to ensure it does indeed contain only one complex. All single photobleaching steps must be the same and should not vary more than the signal-to-noise ratio allows.
21. Positioning of the bead not far from the coverslip is important to ensure the force is not applied axially. Optical traps are much softer in z and pulling the bead axially may significantly change the calibration, leading to an error in measurements. To position the bead at precise distance above the coverslip, move the stage or the trap until the bead hits the surface. The contact point will be visible at the z-signal of the quadrant photodetector and as an image change of the bead on the camera. Once the contact point is established, move 50–100 nm just above the coverslip.
22. We recommend the speed of movement to be 0.16 μm/s during the force–distance recording. In principle, this rate can be varied and if sufficient statistics is collected at different rates, this may provide additional information into the properties of the interaction [15].
23. When cohesin ruptures, the bead pulled by the optical trap jumps back to the center of the trap. On the force–distance curve, this corresponds to the jump of the force to zero. Some force–distance curves may have multiple jumps, which indicate ruptures of multiple connections. While those could be interesting on their own, a single rupture on the FD curve is typically a signature of the single cohesin complex being detached from the DNA and serves as an additional confirmation that the bead was indeed attached by the single cohesin.

References

1. Can S, Lacey S, Gur M et al (2019) Directionality of dynein is controlled by the angle and length of its stalk. Nature 566(7744): 407–410. https://doi.org/10.1038/s41586-019-0914-z
2. Guydosh NR, Block SM (2009) Direct observation of the binding state of the kinesin head to the microtubule. Nature 461(7260): 125–128. https://doi.org/10.1038/nature08259
3. Yildiz A, Tomishige M, Gennerich A et al (2008) Intramolecular strain coordinates kinesin stepping behavior along microtubules. Cell 134(6):1030–1041. https://doi.org/10.1016/j.cell.2008.07.018

4. Qiu W, Derr ND, Goodman BS et al (2012) Dynein achieves processive motion using both stochastic and coordinated stepping. Nat Struct Mol Biol 19(2):193–200. https://doi.org/10.1038/nsmb.2205
5. Capitanio M, Canepari M, Maffei M et al (2012) Ultrafast force-clamp spectroscopy of single molecules reveals load dependence of myosin working stroke. Nat Methods 9(10): 1013–1019. https://doi.org/10.1038/nmeth.2152
6. Woody MS, Winkelmann DA, Capitanio M et al (2019) Single molecule mechanics resolves the earliest events in force generation by cardiac myosin. eLife 8. https://doi.org/10.7554/eLife.49266
7. Davidson IF, Peters JM (2021) Genome folding through loop extrusion by SMC complexes. Nat Rev Mol Cell Biol 22(7): 445–464. https://doi.org/10.1038/s41580-021-00349-7
8. Uhlmann F (2025) A unified model for cohesin function in sister chromatid cohesion and chromatin loop formation. Mol Cell 85(6): 1058–1071. https://doi.org/10.1016/j.molcel.2025.02.005
9. Gruber S, Haering CH, Nasmyth K (2003) Chromosomal cohesin forms a ring. Cell 112(6):765–777. https://doi.org/10.1016/s0092-8674(03)00162-4
10. Haering CH, Farcas AM, Arumugam P et al (2008) The cohesin ring concatenates sister DNA molecules. Nature 454(7202):297–301. https://doi.org/10.1038/nature07098
11. Murayama Y, Samora CP, Kurokawa Y et al (2018) Establishment of DNA-DNA interactions by the Cohesin ring. Cell 172(3): 465–477 e415. https://doi.org/10.1016/j.cell.2017.12.021
12. Richeldi M, Pobegalov G, Higashi TL et al (2024) Mechanical disengagement of the cohesin ring. Nat Struct Mol Biol 31(1): 23–31. https://doi.org/10.1038/s41594-023-01122-4
13. Pobegalov G, Chu LY, Peters JM et al (2023) Single cohesin molecules generate force by two distinct mechanisms. Nat Commun 14(1): 3946. https://doi.org/10.1038/s41467-023-39696-8
14. Guerin TM, Barrington C, Pobegalov G et al (2024) An extrinsic motor directs chromatin loop formation by cohesin. EMBO J 43(19): 4173–4196. https://doi.org/10.1038/s44318-024-00202-5
15. Arya G (2016) Models for recovering the energy landscape of conformational transitions from single-molecule pulling experiments. Mol Simul 42(13):1102–1115. https://doi.org/10.1080/08927022.2015.1123257

Chapter 3

Visualizing Binding of Single SMC Complexes to DNA Substrates Using Combinational Optical Tweezers and Fluorescence Microscopy

Anno I. Koetje, Gemma L. M. Fisher, Menelaos Voulgaris, Sanjana Saravanan, Richard J. Bartlett, Stephen J. Cross, and Luis Aragon

Abstract

Structural Maintenance of Chromosomes (SMC) complexes undertake a variety of DNA-centric processes to both organize and maintain genomes throughout the cell cycle. Their multiple DNA binding sites and conformational dynamics have made fully characterizing their interactions with DNA challenging, and indeed, fundamental questions remain around DNA substrate preferences, the stability of their interaction, and requirements of complex loading and unloading. Recent advances in combinational single-molecule techniques now allow for the correlation of force spectroscopy and real-time analysis of protein binding dynamics on DNA at high spatial and temporal resolution with the complementary use of microfluidics for rapid exchange of experimental conditions. This is a powerful toolkit that allows for the direct visualization of individual SMC complexes binding to DNA substrates while probing the contributions of associated factors including ATP turnover and force dependencies of SMC action. Additionally, the single-molecule approach grants the ability to probe complex stoichiometry and subunit kinetics while providing access to heterogeneous and transient behaviors otherwise averaged in ensemble assays. Here, we describe how to recombinantly prepare eukaryotic SMC complexes—cohesin and Smc5/6—with fluorophore labelling and a variety of DNA substrates for single-molecule experimentation using correlative optical tweezers and fluorescence microscopy. We highlight practical considerations for their study using the LUMICKS C-Trap system and provide custom data analysis scripts for automated detection and localization of DNA binding events.

Key words Optical tweezers, Fluorescence microscopy, Microfluidics, Single molecule, DNA substrates

1 Introduction

Structural Maintenance of Chromosomes (SMC) complexes play essential roles in genome organization and maintenance across all domains of life. Eukaryotes possess three main SMC complexes:

Yasuto Murayama and Yasutaka Kakui (eds.), *SMC Proteins: Methods and Protocols*, Methods in Molecular Biology, vol. 2991, https://doi.org/10.1007/978-1-0716-5005-9_3,

cohesin, condensin, and the Smc5/6 complex. Cohesin provides sister chromatid cohesion and organizes interphase chromatin into functional domains, thus impacting gene expression [1]. Condensin is essential for mitotic chromosome condensation and the proper segregation of sister chromatids [2]. Smc5/6 is involved in chromosomal replication and repair processes and serves as a viral restriction factor [3].

SMC complexes are characterized by their conserved architecture, consisting of two SMC proteins with two long coiled-coil arms connected at one end by a hinge domain and at the other end by an ATPase-containing head domain, which is bridged by a kleisin subunit. This architectural arrangement forms rod-shaped complexes with ~50 nm arms that separate the "hinge" dimerization domain from the ATPase "head" domain, which engages/disengages during the ATPase cycle. Additional subunits are recruited to complete SMC holocomplexes; in the case of condensin and cohesin, HEAT repeat proteins associate with the kleisin (HAWKs), and in the case of Smc5/6, kleisin-interacting tandem winged-helix elements (KITEs) [1].

A widely accepted view to explain the various functions of SMC complexes in genome organization is that they operate as molecular motors that bind chromatin and extrude progressively larger loops of DNA through an ATP-dependent mechanism [4]. This activity, referred to as loop extrusion, results in the formation of stable chromatin loops, which are essential for processes like sister chromatid cohesion, mitotic chromosome condensation, and genome organization [1].

Total Internal Reflection Fluorescence (TIRF) microscopy has facilitated single-molecule studies characterizing SMC binding, diffusion, and translocation on DNA. In these experiments, purified fluorescently labelled SMC complexes are bound to DNA molecules that are immobilized on a surface, allowing real-time observation of individual complexes and their interactions. Recent studies using condensin, cohesin, and Smc5/6 have demonstrated that these complexes can extrude DNA loops, with the loops growing progressively over time in an ATP-dependent manner [5–8]. While TIRF microscopy, and other similar approaches, have provided invaluable information, these methods typically do not allow for the controlled application of mechanical forces, which might be an important parameter affecting SMC function.

Optical tweezers systems, such as the LUMICKS C-Trap, offer a powerful alternative to surface-based assays for studying SMC complex dynamics. The system combines dual-trap optical tweezers with fluorescence microscopy and microfluidics, enabling high-resolution manipulation and visualization of single molecules in solution [9, 10] (Fig. 1). This platform allows precise force application, facilitating the study of force-dependent molecular

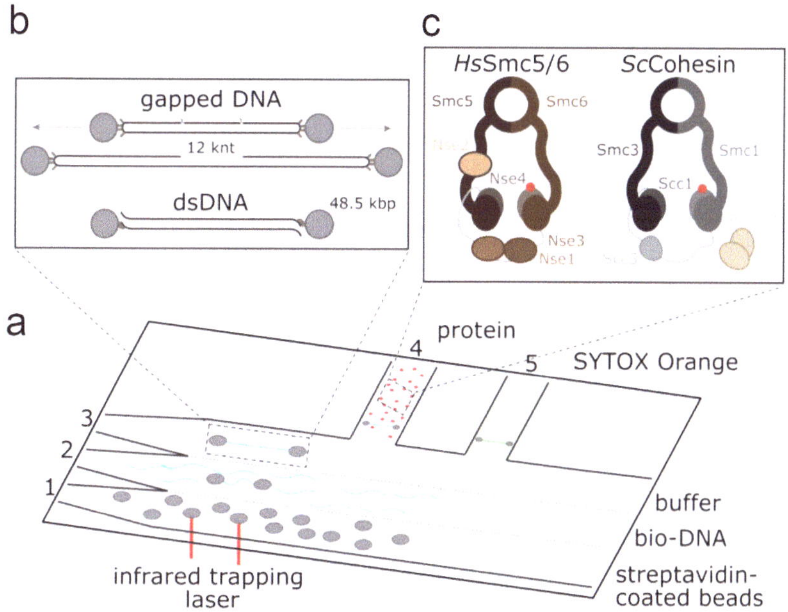

Fig. 1 A single-molecule assay for visualizing SMC binding to different DNA substrates. (**a**) Schematic of the 5-channel microfluidics chip compatible with the LUMICKS C-Trap system. Channels 1–3 converge to create a single laminar flow channel. Channel 1 contains streptavidin-coated polystyrene beads with a diameter of 4.0–4.9 μm. Channel 2 contains the DNA substrate of choice and channel 3 contains buffer. Channels 4–5 can be used for protein-containing solutions or SYTOX Orange (SxO) to visualize the DNA substrates. (**b**) The double-stranded (ds) DNA and gapped DNA substrates described here. dsDNA is λ-DNA modified to have biotinylated 3′ ends. Gapped DNA is generated by stretching end-capped λ-DNA that has been nicked in two locations by Cas9. (**c**) Illustration of the SMC complexes described here

behaviors such as DNA compaction. Additionally, the integrated microfluidics system enables rapid buffer exchange, providing unparalleled control over experimental conditions and permitting the study of dynamic molecular responses under multiple biochemical environments in real time. This capacity to finely control experimental parameters makes the C-Trap system particularly well suited for dissecting the molecular mechanisms of SMC complexes.

Here, we provide a detailed experimental protocol for visualizing and localizing DNA binding by individual cohesin and Smc5/6 complexes using the LUMICKS C-Trap system. We describe the optimization of experimental conditions, calibration procedures, and considerations for data acquisition and analysis. By leveraging the unique advantages of optical tweezers, this approach offers a powerful platform for uncovering the molecular mechanisms underlying SMC complex function and expanding our understanding of how these essential protein complexes contribute to genome organization.

2 Materials

Use ultrapure MilliQ water (18.2 MΩ.cm resistivity at 25 °C) and analytical grade reagents to prepare all solutions. Access to standard molecular biology apparatus is required.

2.1 SMC Holocomplex Expression and Purification

2.1.1 Protein Expression in Insect Cells

1. Dedicated 27 °C incubators with 120 rpm shaking.
2. Sterile Erlenmeyer flasks.
3. Microbiological safety cabinet.
4. Automated brightfield and fluorescence cell counter and slides (e.g., Luna Technologies).
5. Sf9 (Spodoptera frugiperda IPLB-Sf21-AE) cells.
6. Liquid medium for insect cell culture (*see* **Note 1**).
7. 10,000 U/mL penicillin–streptomycin.
8. Viral bacmids and subsequent P2 virus generated by transposition of pACEbac1 *SMC1 SMC3 SCC1*-ybbR-HRV3C-Twin-StrepII and pACEbac1 TwinStrepII-*SCC3* into DH10EMBacY (Geneva Biotech) according to ref. [11].
9. Viral bacmid and subsequent P2 virus generated by transposition of pBIG2 *SMC5*, *SMC6*-HRV3C-10xHis-3xHA, *NSE4*-HRV3C-ybbR-TEV-TwinStrepII, *NSE1*, *NSE2*, and *NSE3* into DH10EMBacY (Geneva Biotech).
10. 0.4% (w/v) Trypan Blue solution.
11. PBS buffer: 137 mM NaCl, 2.7 mM KCl, 10 mM Na_2HPO_4, and 1.8 mM KH_2PO_4. Sterilized by filtration.

2.1.2 Protein Expression in S. cerevisiae

1. pRS426 *SCC2*-3myc-3XTwinStrepII-ybbR-*SCC4* described in ref. [12].
2. Strain W303-1a: *MATa Lys2::pGAL1-GAL4::LYS2 pep4::HIS3 bar1::hisG ade2-1 trp1D2 can1-100 leu 2-3112 ura3-52*.
3. Selective drop-out media: 6.9 g/L yeast nitrogen base without amino acids, 590 mg/L CSM multiple drop-out -adenine, -histidine, -leucine, -tryptophane, -uracil, 40 mg/L adenine, 20 mg/L histidine, 100 mg/L leucine, 20 mg/L tryptophane, 2% (w/v) raffinose, and 0.05% (w/v) glucose. Sterilize by filtration. 2% (w/v) agar added for agar plates.
4. 20% (w/v) galactose: Prepared in selective drop-out media minus glucose and sterilized by filtration.
5. Buffer A: 25 mM HEPES-NaOH (pH 7.5), 5% (v/v) glycerol, and 5 mM β-mercaptoethanol (BME).
6. Protease inhibitor cocktail tablets (*see* **Note 2**).

2.1.3 Protein Purification

1. ÄKTA pure protein purification system with multiwavelength detection (Cytiva).
2. Cryogenic grinder (*see* **Note 3**).
3. Liquid nitrogen.
4. 40 mL KIMBLE dounce homogenizer.
5. Bronson digital sonifier with a 12.5 mm flat tip.
6. High-speed centrifuge.
7. Chromatography columns: 5 mL StrepTrap XT (Cytiva), 5 mL HiTrap Heparin HP (Cytiva), 5 mL HiTrap Q HP (Cytiva), and Superose 6 Increase 10/300 GL (Cytiva) or Superose 6 Increase 5/150 GL (Cytiva).
8. Buffer A: 25 mM HEPES-NaOH (pH 7.5), 5% (v/v) glycerol, and 5 mM β-mercaptoethanol (BME).
9. Buffer B: 50 mM HEPES-NaOH (pH 7.5), 5% (v/v) glycerol, and 1 mM tris(2-carboxyethyl)phosphine (TCEP).
10. Buffer C: 25 mM Tris-HCl (pH 7.5), 200 mM potassium glutamate, 5% (v/v) glycerol, and 5 mM BME.
11. Standard polyacrylamide gel electrophoresis (PAGE) equipment.
12. 0.45 μm pore syringe filters.
13. Protease inhibitor cocktail tablets (*see* **Note 2**).
14. 250 U/μL Benzonase.
15. 10 mg/mL RNAse A.
16. 100 mM phenylmethylsulfonyl fluoride (PMSF) solution prepared in isopropanol.
17. 0.5 M ethylenediaminetetraacetic acid disodium salt solution (EDTA).
18. 2 U/μL HRV 3C protease.
19. 10 U/μL AcTEV protease.
20. Biotin (≥99% lyophilized powder).

2.1.4 Conjugation and HPLC Purification of Fluorophore-CoA

1. High-performance liquid chromatography (HPLC) system with UV/vis detector and fraction collector.
2. LC-18 (5 μm) HPLC column (SUPELCOSIL).
3. TEAA buffer: 0.1 M triethylamine and 5% (v/v) acetonitrile, adjusted to pH 7.0 using acetic acid (*see* **Note 4**).
4. HPLC buffer B: HPLC-grade acetonitrile.
5. ATTO 647 N-maleimide (ATTO-TEC).
6. Anhydrous dimethyl sulfoxide (DMSO).
7. Coenzyme A (CoA) trilithium salt.

8. 0.5 M TCEP (pH 7.0).
9. 1 M dithiothreitol (DTT).
10. Buffer A: 25 mM HEPES-NaOH (pH 7.5), 5% (v/v) glycerol, and 5 mM β-mercaptoethanol (BME).
11. 100 μM Sfp enzyme (prepared as described in ref. [13]).
12. 1 M $MgCl_2$.
13. Strep-Tactin resin (*see* **Note 5**).
14. 0.8 mL micro spin chromatography columns.

2.2 Mass photometry of Purified SMC Holocomplexes

1. TwoMP Mass Photometer with isolation table and AcquireMP and DiscoverMP software (Refeyn).
2. MP uncoated and clean sample carrier slides (Refeyn).
3. CultureWell gaskets (Grace Bio-Labs).
4. Buffer D: 50 mM Tris-HCl (pH 7.5), 125 mM potassium glutamate, 2.5 mM $MgCl_2$, and 1 mM DTT.
5. 20 mg/mL thyroglobulin prepared in Buffer D.
6. 1 mg/mL bovine serum albumin (BSA) prepared in Buffer D.
7. Isopropanol.

2.3 Generation of DNA Substrates for Optical tweezer Experiments

2.3.1 Production of Torsionally Unconstrained Biotinylated λ-DNA

1. 0.3 mg/mL λ-DNA.
2. 0.4 mM Biotin-14-dATP solution.
3. 0.4 mM Biotin-14-dCTP solution.
4. 10 mM dGTP solution.
5. 10 mM dTTP solution.
6. 50 U/μL Klenow Fragment (3′ to 5′ exo-).
7. 10× NEB2 buffer (NEB).
8. DNA purification kit (*see* **Note 6**).

Biotinylated end-capped λ-DNA with a 12 knt single-stranded gap produced using Cas9-mediated nicking according to ref. [19].

2.4 LUMICKS C-Trap Microfluidics System

1. 5-channel u-Flux microfluidics flow cell (LUMICKS).
2. Inlet tubing: 1/16″ × 0.004″ FEP tubing, cut to 5× 25 cm and 5× 50 cm (LUMICKS).
3. Outlet tubing: 1/16″ × 0.010″ FEP tubing (LUMICKS).
4. 2.5 mL Luer lock syringes.
5. 50 mL centrifuge tube (any polypropylene type suitable).
6. 0.2 μm pore size syringe filters.
7. 5% (v/v) sodium hypochlorite (bleach).
8. 0.025 M sodium thiosulfate.

9. 1:50 diluted Hellmanex III cleaning concentrate.
10. 0.5% (w/v) Pluronic-F127 solution.

2.5 C-Trap Experiments

1. 0.5% (w/v) streptavidin-coated polystyrene beads, 4.0–4.9 μm diameter.
2. High BSA Passivation Buffer: 50 mM Tris-HCl (pH 7.5), 125 mM potassium glutamate, 2.5 mM $MgCl_2$, and 2 mg/mL BSA.
3. Working Buffer: 50 mM Tris-HCl (pH 7.5), 125 mM potassium glutamate, 2.5 mM $MgCl_2$, 0.5 mg/mL BSA, 1 mM ATP, and 1 mM DTT.
4. 5 mM SYTOX Orange (SxO) in DMSO.

2.6 Data Acquisition

1. LUMICKS Bluelake version 2.5 or above.

2.7 Data Analysis

1. Anaconda software for management of Python environments: https://www.anaconda.com/download/success.
2. Anaconda environment configuration file: smc-binding-analysis.yml, can be downloaded from https://zenodo.org/records/15065253.
3. Jupyter Notebook script files: https://zenodo.org/records/15065253.

3 Methods

3.1 SMC Holocomplex Expression and Purification

3.1.1 Expression of S. cerevisiae and H. sapiens SMC Complexes

Recombinant expression of homogeneous *S. cerevisiae* (*Sc*) and *H. sapiens* (*Hs*) SMC complexes is readily achieved using the baculovirus (BEVS) expression system. We will not provide a guide to plasmid and bacmid construction or P1 and P2 baculovirus production as detailed protocols are described elsewhere [11]. For purification of human hexameric Smc5/6 complex as a holocomplex, we use the following arrangement of cleavable protein tags: *SMC5*, *SMC6*-HRV3C-10xHis-3xHA, *NSE4*-HRV3C-ybbR-TEV-TwinStrepII, *NSE1*, *NSE2*, and *NSE3*. In our experience, co-expression of *S. cerevisiae* cohesin tetramer subunits does not yield homogeneous holocomplex and instead is better produced by expression of cohesin trimer (*SMC1, SMC3, SCC1*-ybbR-HRV3C-TwinStrepII) and TwinStrepII-*SCC3* separately for subsequent reconstitution.

1. Prewarm insect cell culture media to 27 °C and add penicillin–streptomycin to a final concentration of 50 U/mL.
2. For μg yields of purified protein, expression is advisable on a scale of 500–1500 mL. The virus titer required for peak expression should be determined empirically; we typically use 800 μL

P2 per 200 mL of cells at a density of 2×10^6 cells/mL. Grow cultures at 27 °C in the dark with 120 rpm shaking.

3. Critical to optimal expression is the point at which cells are harvested following infection (*see* **Note 8**). Periodically review cell viability so to harvest within a viability window of 85–90% (*see* **Note 9**). This typically takes 48–72 h.
4. To harvest cells, pellet cells gently by centrifugation (22 °C, 200 g, 15 min), wash with PBS buffer (*see* **Note 10**), repeat centrifugation, and store at −80 °C following snap freezing.

3.1.2 Expression of the Scc2-Scc4 Loader Complex

In the case of cohesin, loading onto DNA requires the loader complex Scc2-4 [14]. Full-length *S. cerevisiae* Scc2-Scc4 co-expresses well in its native organism under the control of a galactose-inducible promoter using a multicopy episomal 2-micron vector: pRS426 *SCC2*-3myc-3XTwinStrepII-ybbR-*SCC4*.

1. Transform W303-1a with the expression vector using the LiAc/single-stranded carrier DNA/PEG method [15] and plate onto selective drop-out media (lacking uracil and supplemented with 2% (w/v) raffinose, 0.05% (v/v) glucose, and 2% (w/v) agar) (*see* **Note 11**).
2. Grow subsequent strain at 30 °C to an OD_{600} of 1.1 in 6 L of selective drop-out media.
3. Induce protein expression by the addition of 20% (w/v) galactose solution to a final concentration of 2% (v/v) and continue induction for 16 h at 20 °C.
4. Harvest cells by centrifugation (3500 g, 4 °C) and resuspend pellets in Buffer A supplemented with 200 mM NaCl and an EDTA-free protease inhibitor cocktail. Repeat centrifugation and resuspension using 2/3rd the volume of pellet mass.
5. Snap freeze as "popcorn" by dropwise freezing the yeast suspension in liquid nitrogen and subsequently store at −80 °C.

3.1.3 Purification of SMC Holocomplexes

Purification of all SMC holocomplexes uses a 3-step purification methodology: affinity chromatography, anion exchange chromatography, and size exclusion chromatography.

1. For lysis of insect-cell expressed complexes, thaw the Sf9 cell pellets gently at 4 °C and resuspend in Buffer B (~40 mL per 500 mL cell culture) supplemented with 150 mM NaCl, 1 cOmplete EDTA-free Protease Inhibitor Cocktail Tablet per 50 mL buffer, 70 μg RNAse A, 1 mM $MgCl_2$, and 75 U/mL of Benzonase.
2. Lyse the Sf9 cells using a dounce homogenizer chilled to 4 °C—complete ~20 strokes in ~30 mL volumes followed by

2 rounds of gentle sonication (20% amplitude, 20 s pulse with 50% duty for 3 min).

3. Add 100 mM PMSF to a final concentration of 1 mM to the lysate and clarify by centrifugation (45,000 g, 4 °C, 45 min).
4. In the case of yeast-expressed Scc2-4, cryogenically grind the yeast popcorn using a freezer/mill with liquid nitrogen.
5. Thaw the resulting yeast cell powder at 4 °C with an additional volume of Buffer A supplemented with 200 mM NaCl, EDTA-free protease inhibitor cocktail, 1 mM $MgCl_2$, and 75 U/mL of Benzonase.
6. Clarify the yeast lysate by centrifugation (45,000 g, 4 °C, 1 h) and filter supernatant through 0.45 μm pore syringe filter.
7. Supplement lysate (from Subheading 3.1.3, **step 3** or **step 6**) with 0.5 M EDTA to a final concentration of 2 mM and load at 2–5 mL/min onto a 5 mL HiTrap StrepTrap XT column equilibrated in Buffer A with 200 mM NaCl for Scc2-4 or Buffer B with 150 mM NaCl for all other SMC complexes.
8. Elute using a step of Buffer A/B supplemented with 50 mM biotin (*see* **Note 12**) and collect 1 mL fractions. For optimal yield and to reduce the eluate volume, pause flow over the column for 10 min after 1.5 column volumes of elution buffer.
9. For cohesin and Scc3, load the eluate onto a 5 mL HiTrap Q HP column equilibrated in Buffer B with 100 mM NaCl.
10. In the case of Scc2-4 and Smc5/6, load the eluate onto a 5 mL HiTrap Heparin HP column equilibrated in Buffer A with 150 mM NaCl. For Scc2-4, [NaCl] must be reduced to 150 mM prior to loading by gradual dilution.
11. Complete a linear gradient in both instances to 1 M NaCl over 20 column volumes and collect 300 μL fractions.
12. Assess peak fractions by SDS-PAGE, pool accordingly (*see* **Note 13**), and concentrate by centrifugal ultrafiltration (where required) ahead of setup of fluorophore labelling and affinity tag cleavage. Details of these processes are described in Subheading 3.3.

3.2 HPLC Purification of Fluorophore-CoA

Site selective fluorophore labelling of SMC complexes can be undertaken with high efficiency using the ybbR-Sfp system [13, 16]. It is optimal to complete conjugation of the desired fluorophore to CoA prior to reaction with the ybbR-tagged protein so to remove unreacted fluorophore–maleimide that may react with native surface-accessible cysteine residues, and this also circumvents the need to have DMSO present (required for dissolving most fluorophore–maleimide conjugates) which are destabilizing to proteins even at low concentrations. Selection of the appropriate fluorophore will depend on dual–/triple-color experimental plans and

Table 1
Fluorophore options for conjugation to SMC proteins for use in a LUMICKS C-Trap system

Laser (nm)	Emission filter (nm)	Fluorophores conjugated to SMCs
488	512/25	Alexa Fluor 488
532	582/75	Alexa Fluor 555, Cy3B, Janelia Fluor 549
639	700/100	Alexa Fluor 647, ATTO 647 N

the lasers/filters installed for microscopy. Table 1 lists our system specifications and the preferred fluorophores used in each instance. Here, we will describe the production and purification of ATTO 647 N-CoA conjugate using reverse-phase high-performance liquid chromatography (RP-HPLC).

1. Prepare 200 nmol of Coenzyme A sodium salt hydrate (CoA) in 50 μL of 100 mM sodium phosphate pH 7.0 (*see* **Note 14**) and 200 nmol aliquot of ATTO 647 N maleimide in 50 μL of anhydrous DMSO (*see* **Note 15**).
2. Mix them together and supplement 400 nmol TCEP from 0.5 M TCEP stock solution (pH 7.0). React the mixture for 1 h at 22 °C with end-over-end mixing. Quench reaction by the addition of 2000 nmol DTT and incubate for a further 15 min.
3. Dilute the conjugated sample two-fold with 0.1 M TEAA buffer (pH 7.0).
4. Apply the diluted sample to a SUPELCOSIL LC-18 (5 μm) HPLC column and run the column with a 5–95% acetonitrile gradient in 0.1 M TEAA (pH 7.0). Run at 1 mL/min while collecting 100 μL fractions to separate ATTO 647 N-CoA conjugate from unreacted ATTO 647 N-maleimide and CoA. Typical chromatograms for ATTO 647 N-CoA and Alexa Fluor 647-CoA are shown in Fig. 2.
5. Pool all relevant fractions and split into aliquots for removing solvent by vacuum concentration. Store at −80 °C with silica gel desiccant.
6. Quantify ATTO 647 N-CoA conjugate yield by spectrophotometry using A_{260} and A_{646} absorbances; extinction coefficients of $1.4 \times 10^4\ M^{-1}\ cm^{-1}$ and $1.5 \times 10^5\ M^{-1}\ cm^{-1}$ for CoA and ATTO647N, respectively; and a correction factor of 0.06 for ATTO 647 N absorbance at 260 nm (*see* **Note 16**).

3.3 Protein Labelling with Fluorophore-CoA

Reaction of ybbR-tagged SMC complexes with fluorophore-CoA can be completed after the second column purification step to minimize the duration for which protein samples are being processed.

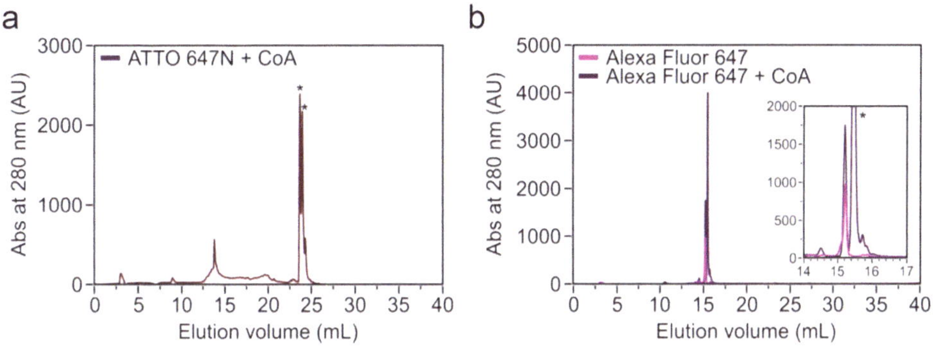

Fig. 2 Purification of fluorophore-CoA conjugates by HPLC. (**a**) RP-HPLC chromatogram of a preparative run (50 nmol) following reaction of ATTO 647 N-maleimide and CoA at a 1:1 molar ratio. ATTO 647 N is a relatively hydrophobic dye and elutes distinctively to CoA and ATTO 647 N-maleimide alone. Asterisks (*) denote fluorophore-CoA peaks. Note ATTO 647 N is supplied as a mixture of two isomers (absorption/fluorescence properties are identical in practice) hence its two peaks. (**b**) RP-HPLC chromatograms for an analytical run of Alexa Fluor 647-maleimide alone and a preparative run (50 nmol) following reaction of Alexa Fluor 647-maleimide and CoA at a 1:1 molar ratio. Alexa647 is hydrophilic and less well resolved when conjugated to CoA (see inset of elution during 14–17 mL)

1. For optimal labelling efficiency, the SMC sample for labelling should be at a minimal concentration of 1 μM (*see* **Note 17**). Labelling reactions in the volume range of 100–1000 μL are practical for subsequent processing.
2. Fully resuspend the dried fluorophore-CoA aliquot(s) (from Subheading 3.2) in Buffer A with 150 mM NaCl to a final concentration of 1 nmol/μL. Add 5-fold molar excess of the fluorophore-CoA to the protein sample. Also add 1 M $MgCl_2$ and 100 μM Sfp enzyme to a final concentration of 10 mM and 2.5 μM, respectively.
3. Proteolytic cleavage of the TwinStrepII tag can be completed simultaneously using HRV 3C or AcTEV protease in the case of cohesin complexes and *Hs*Smc5/6, respectively. Add according to the vendor-supplied protocols and leave both reactions to proceed at 4 °C for 16 h while protected from light.
4. Remove any uncleaved protein still retaining a TwinStrepII tag by incubation with 100 μL Strep-Tactin resin. Equilibrate resin by diluting with 900 μL Buffer A with 150 mM NaCl and gentle centrifugation (200 g, 4 °C) three times. Then add to proteins sample and incubate at 4 °C for 30 min on a roller mixer. Collect the unbound fraction by passing sample-resin mixture over a micro spin chromatography column.
5. Centrifuge the eluate (10,000 g, 4 °C, 15 min) to remove any aggregates or particulates.

6. Immediately load the supernatant onto an appropriate size exclusion chromatography column (Superose 6 Increase 10/300 or 5/150 GL run at 0.4 mL/min or 0.03 mL/min, respectively) equilibrated in Buffer A with 300 mM NaCl, Buffer B with 150 mM NaCl or Buffer C for Scc2-4, cohesin/Scc3, and Smc5/6, respectively.
7. Collect fractions and use a multiwavelength detector to monitor A_{280}, A_{260}, and A_{645}. When using a Superose 6 Increase 10/300 GL column, *Hs*Smc5/6 hexamer, *Sc*Cohesin trimer, and Scc2-4, elute as monomers at ~12.5 mL, ~12 mL, and ~14.5 mL, respectively.
8. Analyze appropriate fractions in SDS-PAGE. Fluorescently image gel, prior to Coomassie staining, using an excitation wavelength appropriate to the fluorophore now conjugated (Fig. 3).
9. Pool fractions corresponding to the labelled and purified complex. Quantify your protein concentration using spectrophotometry (A_{280}). Determine the degree of fluorophore labelling achieved by measuring absorbance at the wavelength corresponding to the fluorophore's maximal absorption and using a correction factor for any absorbance at 280 nm (see Table 2).
10. Snap freeze aliquots and store at −80 °C.

3.4 Mass photometry of Purified SMC Complexes

Single-molecule assays inherently necessitate the use of protein concentrations in the range of pM to low nM. In turn, it is best practice to ensure multi-subunit complexes, such as SMCs, are stable at such concentrations and in the assay buffer to be used. Mass photometry offers a convenient means with which to test this rapidly and with minimal use of samples (*see* **Note 18**) [17, 18] (Fig. 4).

3.4.1 Instrument Setup

1. Turn on the instrument and open the AcquireMP software at least 2 h before use to allow for thermal equilibrium. Also turn on the isolation table.
2. Prepare sample slide with 6-well gasket. Use a clean blade to cut a 6 well (3 by 2) gasket and use tweezers to position it onto the Refeyn mounting guide. Place a pre-cleaned coverslip on top of this and gently apply pressure using tweezers to ensure all edges of the gaskets are adhered to the coverslip.
3. Apply a drop of immersion oil onto the objective and then place the coverslip onto the sample holder ensuring the gaskets are facing upward. Secure coverslip using a pair of magnets on either side.
4. Position the objective near center of the first well (upper left typically) using the lateral movement objective controls at 50× speed and set this as "well position."

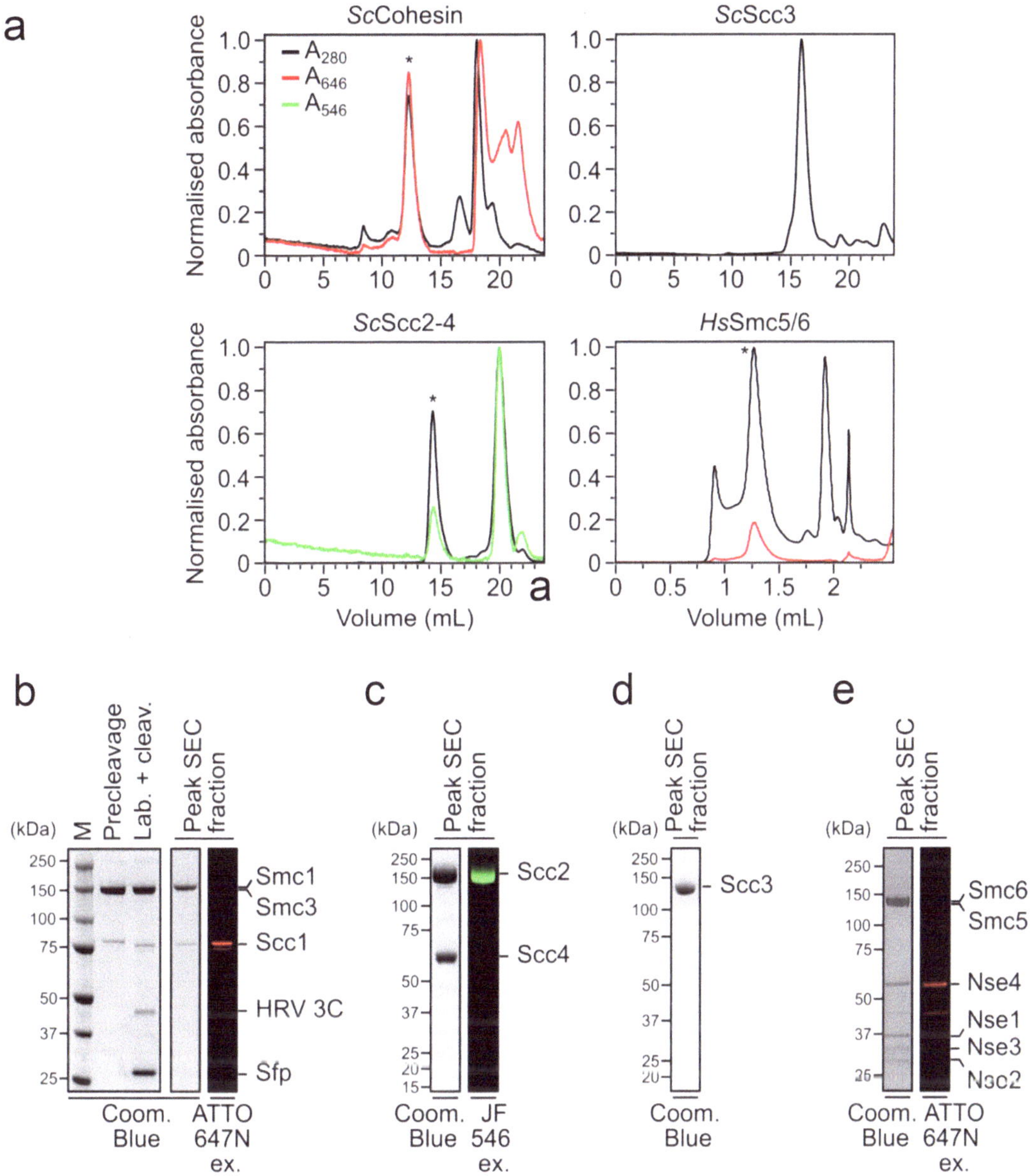

Fig. 3 Recombinant purification and fluorophore labelling of eukaryotic SMC complexes. (**a**) SEC chromatograms of *Sc*Cohesin, *Sc*Scc2-4, *Sc*Scc3, and *Hs*Smc5/6 following fluorophore labelling (where indicated) and TwinStrepII cleavage. Peaks corresponding to the fluorophore labelled SMC species of interest are identified with an asterisk (*). (**b**–**e**) SDS-PAGE sampling of peak fractions from A. In the case of fluorophore labelling for *Sc*Cohesin, *Sc*Scc2-4, and *Hs*Smc5/6, each gel was exposed to excitation of an appropriate wavelength before Coomassie Blue staining

Table 2
Correction factors for absorbance at 280 nm of fluorophores used in single-molecule assays with SMC complexes

Fluorophores conjugated to SMC	Correction factor for A_{280}
Alexa Fluor 488	0.11
Alexa Fluor 555	0.08
Janelia Fluor 549	0.169
Alexa Fluor 647	0.03
ATTO 647 N	0.03

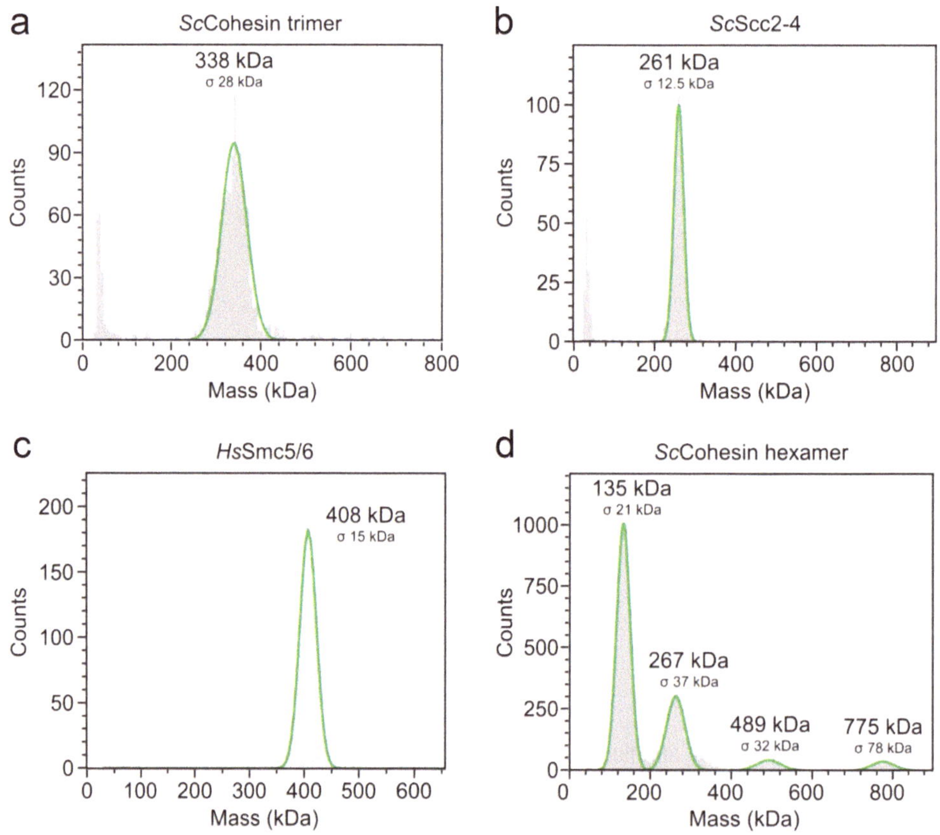

Fig. 4 Mass photometry can be used to confirm the homogeneity of purified SMC complexes and the tractability of single-molecule assay conditions. (**a**–**c**) Mass photometry analysis of purified *Sc*Cohesin trimer (expected mass 348 kDa) (**a**), *Sc*Scc2-4 (expected mass 255 kDa) (**b**), and the hexameric *Hs*Smc5/6 complex (expected mass 415 kDa) (**c**). (**d**) Reconstitution conditions of the hexameric *Sc*Cohesin complex (expected mass 736 kDa) can be readily assessed by mass photometry. All measurements were collected in 125 mM KGlu-containing buffer

3.4.2 Calibration

Mass standards must be sampled during each data acquisition session and in the buffer to be used for your samples. We recommend using the same buffer to be used in the single-molecule assays, which in this case is Buffer D (50 mM Tris-HCl (pH 7.5), 125 mM potassium glutamate, 2.5 mM $MgCl_2$, and 1 mM DTT) (*see* **Note 19**). SMC complexes are typically ~300–500 kDa and so BSA (bovine serum albumin) and bovine thyroglobulin are suitable protein standards to use for calibration for a size range of 66 kDa to 660 kDa.

1. Filter Buffer D through 0.22 μm filter and keep at room temperature.
2. Prepare 1 mg/mL BSA and 20 mg/mL thyroglobulin stocks using Buffer D (stable at 4 °C for <12 months) and dilute 1 in 200 into Buffer D.
3. Select "Regular" as the field of view size in AcquireMP software.
4. First check buffer contributions to noise and the cleanliness of the imaging surface by adding 16 μL Buffer D into the first well. Click "droplet dilution" to automatically focus the objective at the glass-water interface and imaging surface. The focus ring shown in the AcquireMP software should be solid and the readout of sharpness should not exceed 6%. If this is not the case, bubbles are likely present between the objective and the coverslip. Remove coverslip and reposition in the sample holder. To verify the surface is suitably clean, first check the native image of the surface for bright spots—there should be none. Toggle to ratiometric imaging mode as well to check for any apparent binding events. If either imaging mode indicates the surface is not clean, unlock focus and move laterally at 1× speed. Repeat droplet dilution focus following each lateral movement of the objective. To sample buffer-based noise, record a 60 s video and check for any detected binding events in real time.
5. With the imaging surface and buffer preparation confirmed to be sufficiently clean, sample BSA as mass standards 1 and 2 (BSA will be in a monomer-dimer equilibrium and will have binding events of 66 and 132 kDa). Add 4 μL of the 1:200 diluted BSA solution to the well containing 16 μL Buffer D (final concentration 15 nM).
6. Mix well contents with a pipette twice without contacting the coverslip. Record a 60 s movie and assess that binding events are occurring at a suitable density and rate using the native imaging (binding events will be evident as dark spots) and real-time histogram generation of contrast distributions—aim for 1000–2000 events in a 60 s window (*see* **Note 20**).

7. Use software controls to reposition objective at the center of the second well and add 16 μL buffer. Repeat the process of automatically focusing and checking the imaging surface. Collect a 60 s dataset for the third mass standard with thyroglobulin: Add 2 μL of the 1:200 diluted solution (final concentration 5 nM). Acquire dataset as previously in **steps 5** and **6**.
8. Open the BSA and thyroglobulin datasets in the DiscoverMP software to visualize their contrast distributions. Process with standard analysis settings. Assign Gaussian function fits manually to the two peaks in the BSA histogram and the single peak of the thyroglobulin dataset. Generate calibration file with these three populations by assigning them the expected MWs (66 kDa, 132 kDa, and 660 kDa) as listed in the predefined mass standards listings. The calibration curve output should have a mass-to-contrast fit with a R^2 above 0.95 (repeat acquisition if not the case). Save calibration file.
9. If acquiring data over a timescale of more than 1 h, repeat measurement of the BSA and thyroglobulin standards at the end of data acquisition to verify the initial calibration remains valid (*see* **Note 21**).

3.4.3 Data Acquisition

1. Prepare 10 μL SMC samples at 50 nM in Buffer D. If the complex is to be reconstituted, that is, cohesin trimer with Scc3, do this at 500 nM and incubate for 30 min before diluting to 50 nM. Remove samples from ice 10 min before sampling.
2. Use software controls to reposition objective at the center of the next well. Repeat the process of automatically focusing and checking the imaging surface. Dilute each 50 nM SMC sample 1:10 in 18 μL of Buffer D in the well (final concentration 5 nM). Assess the density of binding events and adjust dilution accordingly.
3. Collect each sample in triplicate.

3.4.4 Data Processing and Visualization

1. Process datasets in DiscoverMP software to identify and quantify binding events. Plot each dataset as a histogram distribution and convert the contrast values to molecular masses using the calibration file previously generated. Use the same analysis settings as used for the calibration standards (*see* **Note 22**).
2. Save analysis results and record the total number of binding events, the standard deviation of Gaussian fit (sigma) to each distribution component.

3.5 Generation of DNA Substrates for C-Trap Assays: Biotinylated Double-Stranded λ DNA with Unconstrained Ends

1. Mix the following for the biotinylation reaction (final concentrations for a 100 μL reaction): 4 nM λ-DNA, 100 μM dGTP, 100 μM dTTP, 80 μM Biotin-14-dATP, 80 μM Biotin-14-dCTP, and 0.05 U/μL Klenow Fragment (3′ to 5′ exo-) in 1× buffer NEB2 adjusted with nuclease-free water to 100 μL.
2. Incubate the reaction in a thermocycler with the following program: 37 °C for 30 min, 70 °C for 15 min, and then 4 °C hold.
3. Purify the reaction using a DNA purification kit following the vendor-provided instructions.

3.6 Cleaning and Passivation of the Microfluidics System Channels

1. Filter all solutions using a 0.2 μm pore size syringe filter.
2. Flush ~1 mL MilliQ water through at 2 bar.
3. If the system was previously left in bleach, flush ~1 mL 0.025 M sodium thiosulfate solution through at 2 bar.
4. Flush ~1 mL MilliQ water through at 2 bar.
5. Flush ~1 mL 0.5% (w/v) Pluronic-F127 solution through at 2 bar.
6. Flush ~1 mL of the High BSA Passivation Buffer for 8 min at 2 bar and 15 min at 0.5 bar through all channels.
7. Flush ~0.5 mL of the Working Buffer through at 2 bar.

3.7 Setup of LUMICKS Instrumentation and Bluelake for Force Spectroscopy Measurements and Confocal Imaging

1. Place a 60–70 μL drop of water on the objective and a drop of immersion oil on top of the flow cell.
2. Set the trapping laser to 100% and the overall power to 30%.
3. Set the objective focus position to the middle of the flow cell in the z-finder tab.
4. Lower the condenser to focus using the automated program. Set the trapping laser to 30% and make sure the moon looks similar to the manufacturer's defined "good moon" example.
5. Return the trapping laser to 100% and turn on the flow to ~0.2 bar and wait to catch a single bead in Trap 1 and Trap 2 in channel 1. Configure traps so that Trap 2 is on the left and Trap 1 on the right.
6. Once a single bead has been caught in each trap, move to channel 3 and turn the flow off. Check to ensure that DNA is not tethered between the beads by moving the beads >12 μm apart and checking for any force increase.
7. Create a new bead tracking template and input the diameter of the beads (variable for each manufacture). Future bead match scores should be >85%.

8. Zero the force and run a force calibration with the beads 12 μm apart. Verify that the trap stiffness is between 0.2 and 0.3 pN/nm. Apply the calibration.
9. For the worm-like chain (WLC) model parameters of non-end-capped biotinylated λ-DNA, set the following parameters in the FD viewer reference model: contour length: 48.50 kbp (16.49 μm), persistence length: 56.00 nm, stretching modulus: 1500.00 pN, Boltzmann * temp: 4.11 pN*nm, range (min): 0.01 pN, range (max): 100.00 pN.
10. Input appropriate 2D scan and kymograph settings for imaging λ-DNA held in a dual trap: 50–100 nm pixel size, 0.1–0.2 ms pixel time, and 90–130 ms line time.
11. Turn on lasers necessary for fluorescent imaging.
12. For confocal imaging, set laser powers for use in the range of 4–10 μW.
13. Minimally select the following to be included in .h5 file exports: All dimensions of Force LF, Distance 1, and Distance 2.

3.8 Sample Preparation and Channel Layout for dsDNA and ssDNA Experiments

3.8.1 Channel 1 (Passivated Streptavidin-Coated Polystyrene Beads)

1. Dilute 10 μL of streptavidin-coated polystyrene beads with 10 μL of Working Buffer and 20 μL of High BSA Passivation Buffer and then shake at 1500 rpm for 10 min at 22 °C.
2. Add 160 μL of Working Buffer and shake for a further 2 min.
3. Add 800 μL Working Buffer to a total of 1 mL and keep on ice until use.

3.8.2 Channel 2 (Biotinylated ss/dsDNA)

1. For dsDNA experiments, add 20 ng of biotinylated DNA stock per 1 mL of Working Buffer. For ssDNA experiments, dilute 1 μL of Cas9-nicked DNA in 1 mL of Working Buffer. Do not mix by pipetting up and down to avoid DNA shearing.

3.8.3 Channel 3 (Buffer)

1. Prepare 1 mL of Working Buffer (add 10 μg biotin when using non-cleaved TwinStrepII-tagged proteins).

3.8.4 Channel 4 (Protein)

1. Dilute fluorescently labelled SMC complex in 1 mL of Working Buffer (with 10 ug biotin if required) to a final concentration of 1–2.5 nM.
2. In the case of the *Sc*Cohesin complex, reconstitute the tetramer before dilution by mixing 500 μM trimer with 2.5 molar excess of Scc3 and incubating for 30 min on ice. Just before use, filter sample through a 0.22 μm pore size spin filter to remove aggregates and particulates.

3.8.5 Channel 5 (Other)

1. Prepare 1 mL of Working Buffer supplemented with any other reagent(s) that might be required for the desired experiment, such as SYTOX ORANGE (SxO) for DNA visualization, different ionic strength buffer washes, SDS, or protease treatments. For ssDNA experiments, use 12.5 nM SxO in this channel.

3.9 Data Analysis Software Installation and Usage

3.9.1 Installation of smc-Binding-Analysis Anaconda Environment

1. Download and install Anaconda software from https://www.anaconda.com/download/success.
2. Download and unzip "smc-binding-analysis" code repository from https://zenodo.org/records/15065253 by clicking "Download all."
3. Open an Anaconda-enabled terminal ("Anaconda Powershell Prompt" on Windows or the standard "Terminal" on Linux or Mac).
4. Type the following command to set up Anaconda environment, replacing "[path to environment.yml]" with the full path to the relevant file (e.g., "/home/ab123/Downloads/environment-linux.yml" on Linux, "/Users/ab123/Downloads/environment-mac.yml" on Mac, or "C:\Users\ab123\Downloads\environment-windows.yml" on Windows):

```
conda env create -f [path to environment.yml]
```

3.9.2 Running Data Analysis Scripts

1. Open Anaconda-enabled terminal.
2. Type the following to activate *smc-binding-analysis* environment:

```
conda activate smc-binding-analysis
```

3. Type the following to start JupyterLab:

```
jupyter lab
```

4. Select Jupyter notebook file (those with ".ipynb" extensions) using file browser on the left.

3.10 Visualizing SMC Binding to dsDNA

3.10.1 Experiment and Data Acquisition

1. Move trapped beads, following force calibration as described in Subheading 3.8, to channel 2 to capture a single dsDNA tether. Place beads ~15 μm apart and turn on flow for all channels to 0.2 bar. Move Trap 1 relative to Trap 2 to decrease the distance between beads from 15 μm to 5 μm before extending the DNA back to 15 μm. Monitor the measured force on bead 1 during this movement. If the force starts to increase around or before 15 μm, one or more DNAs has been tethered, and the beads should be moved to channel 3. Otherwise, repeat these actions until a tether is detected (*see* **Note 23**).

2. In channel 3, ensure only one DNA is tethered between the beads by stretching the DNA from 5 μm to 15 μm with no flow on. If the force upon extension follows the WLC model, the force will plateau at 60 pN once the DNA is stretched beyond 16 μm. If more than one DNA is tethered, the tension on the DNA will plateau above 100 pN. If more than one DNA is tethered, stretch the DNA until the force falls to 60 pN, indicating that only one DNA remains. Collect a force–distance (FD) curve using the automated program in Bluelake to extend the DNA from 5 to 16 μm at 1 μm/s.
3. Position traps 12 μm apart (~5 pN force). Open all channels and turn on flow at 0.3 bar for 30 s to refresh contents of all channels (*see* **Note 24**).
4. Turn off flow and immediately move beads with DNA tether to channel 4 to load SMC proteins while keeping DNA in a 12 μm distance clamp. Incubate here for 30–60 s (*see* **Note 25**).
5. Return bead pair and DNA with bound protein to channel 3 for a single 2D scan. Set 2D scan area to encompass desired imaging region including both beads (*see* **Note 26**; use bright-field image as guide). Ensure the confocal laser box and photon detector are switched on and use the red (639 nm) laser (in the case of ATTO 647 N labelled *Sc*Cohesin) at 6 μW and/or the green (532 nm) laser (in the case of Janelia Fluor 546 labelled *Sc*Scc2-4) at 4.5 μW. Export image as a scan.h5 file.

3.10.2 Data Analysis: Foci Detection and Quantification of Binding Events

1. Start JupyterLab as described in Subheading 3.9.2, **step 2**.
2. Open "Foci_counting.ipynb" Jupyter notebook file.
3. Set the relevant parameters at the start of the third cell: The "data_folder" parameter defines the full path to the folder containing scan.h5 files for use in foci detection. The "scan_-name" parameter defines the full name of the scan.h5 file to process. The "channel" parameter defines the channel to extract from multichannel images. Indexing starts at 0, so for example, the third channel would use "channel = 2." Channels 0, 1, and 2 refer to imaging using red, green, and blue wavelength excitation and corresponding emission filters. The "bead_size" parameter defines the physical diameter of beads used (specified in μm).
4. Run all three cells. This will automatically crop the image to remove the beads, detect the y-position of the tether, and crop a 5 px high region centered on this tether, which will be averaged to a single 1D intensity profile along the tether (Fig. 5). Multiple Gaussians will be fit to this intensity profile. The quality of fit, reported as RMSE, is stated above the displayed plot.

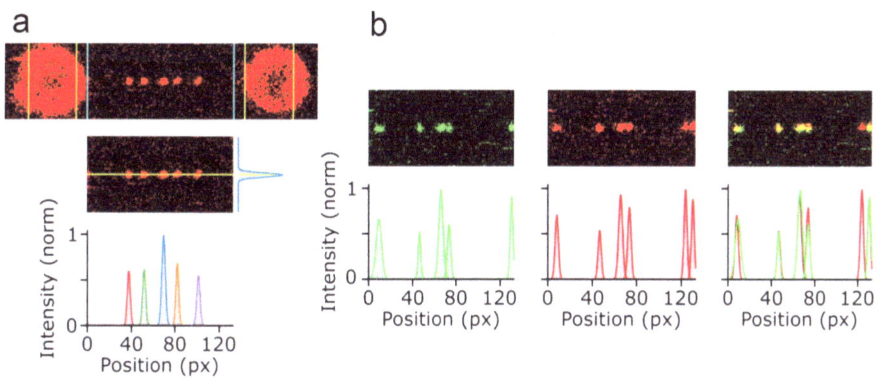

Fig. 5 Visualization of dsDNA binding by *Sc*Cohesin and detection of binding events. (**a**) Example processing of a confocal image of individual *Sc*Cohesin-ATTO 647 N complexes bound to a dsDNA tether. *Top*: Automated detection of beads (yellow lines) and bead proximal margin (blue lines). *Middle*: Detection of DNA tether position along y-axis using single peak Gaussian fitting. *Bottom*: Multi-Gaussian fitting of intensity profile along DNA tether. (**b**) Example foci detection for 2-channel image of *Sc*Cohesin-ATTO 647 N (red) and *Sc*Scc2-4-Janelia Fluor 546 (green)

5. Once complete, this will create a subfolder called "foci_counts" within the folder containing the input .h5 file. Within this folder, the script will save a TIFF image showing the 5 px high cropped region, a PNG image showing the detected Gaussian profiles (each peak assigned a random color; Fig. 5), and a CSV file containing the fitting parameters for each profile.

3.11 Visualizing SMC Binding to ssDNA

3.11.1 Experiment and Data Acquisition

1. Move the trapped beads to channel 2 containing Cas9-nicked DNA and capture a single DNA tether as described in Subheading 3.11.1.
2. Stretch the singly tethered DNA to 25 μm and hold for 10 s to unpeel a complementary strand between the defined nick sites.
3. Relax the DNA to 5 μm and then collect a FD curve stretching the DNA at 1 μm/s from 5 μm to 23 μm, before also collecting the reverse by relaxing the DNA at 1 μm/s to 5 μm. If an ssDNA gap has been made successfully, the FD curve will have shifted to the right compared to the WLC model, with the force plateauing only around 20 μm (Fig. 6a).
4. With the DNA held at 15 μm distance, move to channel 5 containing SxO. Ensure the confocal laser box and photon detector are switched on as well as the green and red lasers.
5. Take a 2D scan, ensuring that the scan window encompasses both beads. The scan should show an unstained ssDNA region flanked by green dsDNA between the beads. If the gap is not located in the center or if multiple DNAs are present, eject the beads and start again at **step 1**.

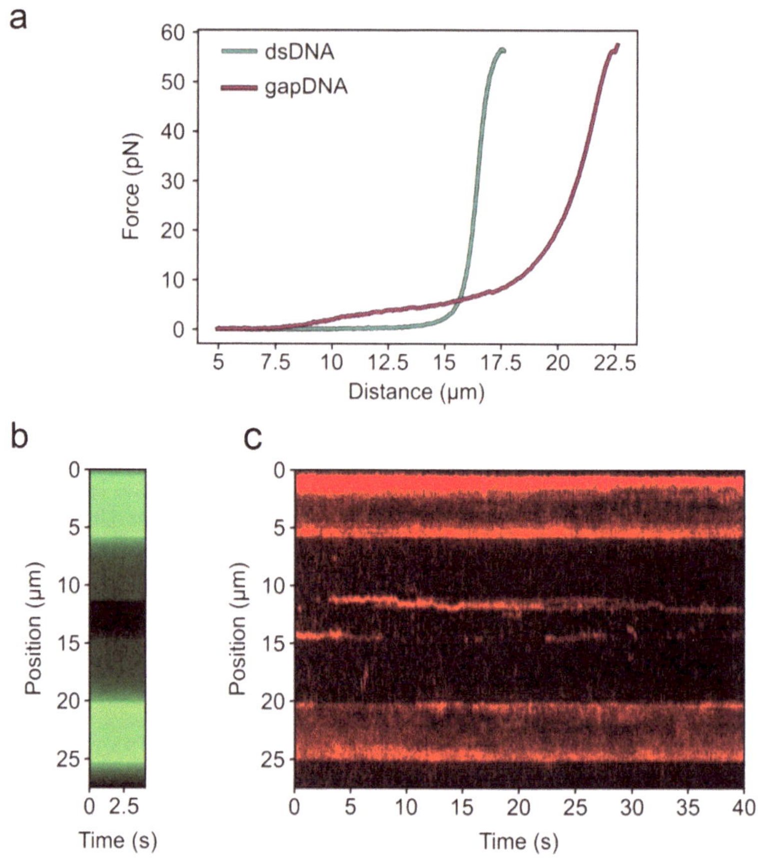

Fig. 6 Gapped DNA generation and visualization of ssDNA binding by *Hs*Smc5/6. (**a**) Force–distance (FD) curves recorded for λ-DNA before gap formation (cyan) and after the formation of a 12 knt gap (magenta). (**b**) A kymograph of a λ-DNA containing a central 12 knt ssDNA gap recorded in a 12.5 nM SYTOX Orange containing channel. The gap is evident as a dark region flanked by dsDNA stained in green, between the two beads. (**c**) A kymograph of the same λ-DNA as in (**b**) recorded in a channel containing 2.5 nM *Hs*Smc5/6-ATTO 647 N

6. Line up the kymograph window along the DNA visible in the 2D scan to ensure that kymographs are properly aligned. It may be necessary to move the y-position of the right bead to ensure that the DNA is exactly horizontal.
7. Record a kymograph for 5 s, which should show a clear unstained ssDNA flanked by dsDNA and beads (Fig. 6b).
8. Return to channel 3 and hold the DNA at 15 μm stretch and flow for 30 s at 0.2 bar.
9. Switch off the flow in all channels and move the beads and DNA to channel 4 to load SMC proteins while still at 15 μm apart. Record a kymograph for 40 s. You will see traces appearing on the DNA, localized primarily in the central ssDNA region (Fig. 6c).

10. Once the kymographs have been completed, all FD curves, scans, and kymographs from the same DNA molecule are saved within one marker and exported as a .h5 file for later analysis.

3.11.2 Data Analysis: Localization of SMC Complex Binding

Optical tweezers data collected in Subheading 3.11.1 can be analyzed using the following python scripts. These scripts cover functions such as detecting the bead positions in kymographs or tracking SMC protein binding events in kymographs. The scripts can either be used all together as described below (Fig. 7) or can be employed individually as part of a custom analysis pipeline. Scripts should be opened in JupyterLab running within an environment setup as in Subheading 3.9.

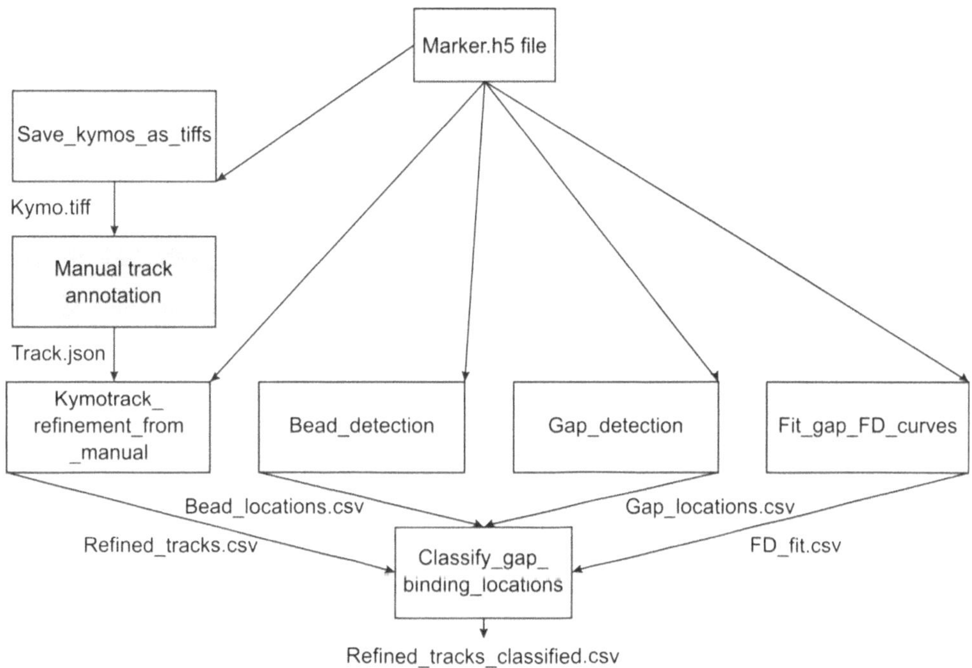

Fig. 7 Data analysis pipeline for localizing *Hs*Smc5/6 binding events to hybrid ss/dsDNA substrates. Schematic outlining how optical tweezers data saved in marker ".h5" files (green box) is processed through six python scripts (blue boxes) and a manual annotation step (red box) to analyze the collected data. The marker ".h5" file is used as an input file into all the python scripts other than the final "Classify_gap_binding_locations. ipynb" script. The "Save_kymos _as_tiffs.ipynb" saves the kymographs inside the inputted marker file as ".tiff" files which are then used for manual labelling of SMC binding tracks. The manual annotations are saved as ".json" files which are then used as an input into the "Kymotrack_refinement_ from_manual.ipynb," which refines the tracks and saves them as CSV files. The "Gap detection.ipynb," "Bead_detection .ipynb," and "Fit_gap_FD_-curves.ipynb" all use the marker file and create CSV files with the gap location, bead locations, and ssDNA gap length, respectively. These files are then all inputted into the final "Classify_gap_binding_ locations.ipynb" script which classifies whether each track is bound in ssDNA, dsDNA, or the junctions between them and adds this information to the previously created refined track CSV files

Run "Gap_detection.ipynb" Script

This script detects the position of ssDNA–dsDNA junctions in kymographs recorded of gapped DNA in SxO.

1. To use the script, change the "data_folder" parameter to match the path to the folder in which C-Trap marker .h5 files are saved and the "marker_name" parameter to the name of the marker file you wish to analyze, before running the script.
2. This will generate a folder inside the data folder called "sytox_ssDNA_classification" containing the pixel positions of the ssDNA gap edges and the average inter-bead distance in μm saved in a CSV file with the same name as the original marker file (Fig. 8a).

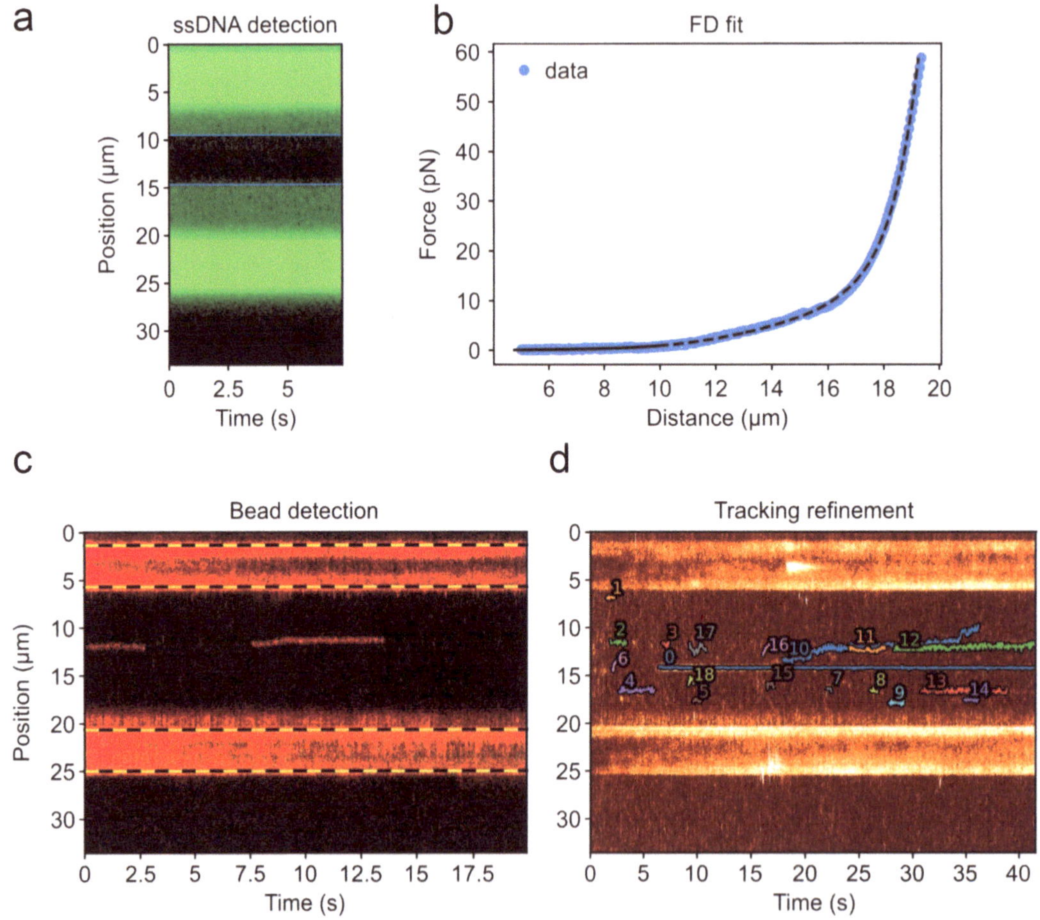

Fig. 8 Example output of scripts analyzing the binding locations of *Hs*Smc5/6 on gapped DNA. (**a**) Kymograph recorded of gapped DNA in SYTOX Orange channel with blue lines overlayed showing the positions of the ssDNA–dsDNA boundaries as detected by the "Gap_detection.ipynb" script. (**b**) Force–distance curve recorded of gapped DNA fitted to a model governing the stretching dynamics of gapped DNA. The fitted curve is overlayed as a dashed black line. (**c**) Kymograph recorded of gapped DNA with dashed black and yellow lines denoting the positions of the beads as detected by the "Bead_detection.ipynb" script. (**d**) Kymograph of *Hs*Smc5/6 binding to gapped DNA, with refined tracks overlayed on the image

3. The script works by going through the specified marker and selecting the kymographs recorded in the SxO channel, by choosing those less than 30 s long with a mean green intensity greater than 8 Au. The selected kymographs are averaged along the y-axis to generate an array with the average green intensity along the kymograph plane. The DNA region within this array is found by detecting the high signal intensity caused by the beads and cropping these out. The ssDNA region is then found by looking for an area within the DNA with a lower signal intensity due to the lack of SxO staining.
4. The "process_files" function in the script has an optional "max_ss_intensity" parameter which governs the maximum allowed green signal intensity of the ssDNA. If no ssDNA is detected in kymographs, the background signal may be interfering with the ability to detect ssDNA, and the "max_ss_intensity" should be increased.
5. Alternatively, the gap edges and distance can also be manually inputted by making a CSV file called "{input marker file name}_kymo{input kymograph id}.csv" containing the gap locations and average inter-bead distance under the headings of "Distance", "ssDNA start," and "ssDNA end."

Run "Fit_gap_FD_curves.ipynb" Script

This script fits gapped DNA FD curves to calculate the size of the gap in kilonucleotide (knt).

1. To run the script, change the data_folder parameter to match the path to the folder in which C-Trap marker .h5 files are saved and the "marker_name" parameter to the name of the marker file you wish to analyze, before running the script. All FD curves within the marker are plotted, and you will be asked to input the ID of the FD curve you wish to be fitted. The FD curve to be fitted should be the extension curve and not the retraction as this typically shows hysteresis which will affect the quality of the fit.
2. This will generate a folder inside the data folder called "gap_fd_fit," containing a CSV file with the length of the dsDNA and ssDNA in kbp and knt respectively, as well as the R2 of the curve fit and the ID of the FD curve fitted (Fig. 8b).
3. The script works by going through all the FD curves within the marker and asking you to select the one for fitting. The selected FD curve is fitted to a model which combines the freely jointed chain (FJC) [19] governing the stretching of ssDNA, and the twistable worm-like chain (tWLC) [20] governing the stretching of dsDNA. The model can then estimate the length of the ssDNA and dsDNA regions within the gapped DNA molecule. These values are saved in a CSV file for later analysis.

Run "Bead_detection.ipynb" Script

The bead detection script analyzes all the kymographs inside a marker file and outputs the position of the beads, the inter-bead distance, and force in the kymograph for each timepoint in a CSV file.

1. To run the script, change the "data_folder" parameter to match the path to your marker .h5 files and the "marker_name" parameter to the name of the marker file you wish to analyze, before running the script.
2. The script will create a folder called "Bead_locations" in your data folder, containing a CSV file with six columns; "Left_bead_start," "Left_bead_end," "Right_bead_start," "Right_bead_end," "Force," and "Kymo_time." These first five contain the locations of the bead edges in pixels and the force on the DNA in pN for every timepoint of the kymograph, while the "Kymo_time" contains the length of the kymograph in s (Fig. 8c).
3. The script works by averaging the green intensity in the kymograph along the plane of the DNA to produce an array of the mean green intensity and then applying a moving average across this array with a window equal to the bead size. The beads have sufficient autofluorescence in the green channel for the beads to be visible even without any green dyes present. The script assumes that the left bead is not moved during the kymograph and therefore uses the inter-bead distance saved in the marker file to get an accurate location of the right bead even if it is moved. The force during the kymograph is also extracted from the marker file at the same time.
4. The "process_kymographs" function has an optional parameter called "bead_size" which represents the size of the beads used in the experiment in μm. If beads of varied sizes were used to collect data, this parameter is set by default to 4.35 μm and should be updated to the actual bead size to ensure accurate bead detection.

Run "Save_kymos_as_tiffs.ipynb" Script

This script saves all kymographs found in a marker which are longer than 35 s as 8-bit tiffs which can be manually annotated to label protein tracks.

1. To run the script, change the data_folder parameter to match the path to the folder in which C-Trap marker .h5 files are saved and the "marker_name" parameter to the name of the marker file you wish to analyze, before running the script.
2. This will generate a folder inside the data folder called "tiffs," containing the kymographs in an 8-bit tiff format.

3. The script reads all kymographs in the specified marker and choses those longer than 35 s. These are then converted to 8-bit format and saved as tiff files.

Manually Annotate Kymographs Recorded in the Protein Channel to Approximately Denote the Individual Protein Tracks

The tiffs generated in Subheading "Run "Save_kymos_as_tiffs.ipynb" Script" must be manually annotated to give very rough coordinates of individual protein tracks in recorded kymographs. The tracks should be saved in a json file so that they can be refined in later steps.

1. Manual annotation software used in this case is Supervisely (https://supervisely.com/), although any platform which generates json files containing track information is suitable.
2. Before starting manual tracking, it is important that the kymographs to be analyzed have good signal to noise, with protein tracks clearly visible and differentiated from the background signal. Additionally, kymographs must not be too crowded with many overlapping tracks, as this will make it difficult to distinguish individual molecules.
3. When annotating the tracks, enough points of the track should be chosen so that the track is fully encompassed by the manual annotation. For non-moving tracks, this may mean that simply selecting the start and end of the track is sufficient, while highly diffusive molecules will require more individual points. All tracks however short should be annotated as very short traces can be excluded from analysis in subsequent steps if desired.

Run "Kymotrack_refinement_from_manual.ipynb" Script

This script takes the manually annotated tracks from Subheading "Manually Annotate Kymographs Recorded in the Protein Channel to Approximately Denote the Individual Protein Tracks" and uses Gaussian fitting to refine the tracks and give the position of each track for each time frame. The track positions are saved in a CSV file for further analysis.

1. To run the script, change the "data_folder" parameter to match the path to your marker .h5 files.
2. The script will create a folder called "refined tracks" containing three subfolders: "images," "long_csv," and "short_csv." A png file showing the kymograph with the refined tracks overlayed on it is saved in the images folder. A CSV file for each track is saved in the "long_csv" subfolder with five columns: "track_time_pix," "track_time_s," "track_pos_pix," "track_pos_um," and "track_photon_counts" containing the track time in pixels, the track time in s, the track x-position (DNA axis) in pixels, the track x-position (DNA axis) in µm, and the track photon counts, respectively. In the "short_csv" subfolder, a CSV file with the track ID and dwell time in s is saved.

3. The script works by taking the json files containing manually annotated tracks and loading the corresponding kymograph from the relevant marker .h5 file. Then using the LUMICKS pylake.refine_tracks_gaussian function, the manual tracks are fitted to the kymograph image to generate a refined track, with an x-position for every pixel along the time axis (Fig. 8d).

Run "Classify_gap_binding_locations.ipynb" Script

This is the final script which combines the earlier created output files to classify the refined protein tracks into binding to either ssDNA, dsDNA, or the junctions between them.

1. To run the script, change the "data_folder" parameter to match the folder where your marker .h5 files are saved and the "marker_name" parameter to the name of the marker file you wish to analyze, before running the script.
2. The script will create a folder called "corrected_ssDNA_location" containing a CSV file with the corrected pixel positions of the ssDNA region in the DNA for each time frame in the kymograph. The script will also add information to the CSV files in the "long_csv" subfolder of the "refined_tracks" folder to class every position in the refined tracks as being bound to ssDNA, dsDNA, or junctions. In the "short_csv" subfolder, the CSV files are appended to add the proportion of the time each track spends bound to each class of DNA.
3. The script performs two main functions: First, it applies a correction to the gap edge positions detected earlier in SxO images, and secondly, it uses the revised positions to class every timepoint in a molecule track as being in ssDNA, dsDNA, or junction DNA. The gap edge correction is done by taking the length of ssDNA in knt calculated in Subheading "Run "Fit_gap_FD_curves.ipynb" Script" and assuming that the interbead distance is still the same as when the gapped DNA was imaged in the SxO channel, allowing us to calculate the expected extension of the ssDNA in μm in the protein channel. The increase in gap size in the protein channel is then used to adjust the positions of the ssDNA gap edges used for classification. All the kymograph track positions in the marker are then compared to the bead and gap edge positions to determine what type of DNA they are bound to at each timepoint. If a track is within two pixels of a predicted gap edge, it is classed as being bound to junction DNA.

3.12 Cleaning and Maintenance of the C-Trap

1. Best practices and troubleshooting for even flow across channels and tubing maintenance.
2. Ensure that all tubing is of similar length to ensure that the channels flow at the same rate.

3. Use clean and sharp razor blades to cut the tubing to avoid jagged or bent edges as this can cause shearing of DNA and blockages.
4. Bubbles and blockages can often be eliminated by flicking the tubing gently near the valves or at the output.
5. If bubbles and blockages persist, first check that the connections are properly tightened. Methodically cut small fragments from the ends of the different segments of tubing. Check that the apertures of the valves are visibly free of debris and clean them periodically by removing them and soaking in 1:50 diluted Hellmanex III solution or water to avoid buildups.
6. If the C-Trap is to be used in a subsequent session for experiments using the same channel contents, all channels can be flushed with ~1 mL Working Buffer and stored in that overnight.
7. The microfluidics system can be cleaned with steps below for day-to-day use and can be left in 0.5% (v/v) sodium hypochlorite (bleach) solution long term to inactivate proteins and fluorophores and eliminate bacterial growth. For more thorough cleaning that will help strip the surface of the chip, continue to a Hellmanex wash (*see* **Note 27**): Flush channels with ~1 mL MilliQ water at 2 bar. Flush channels with ~1 mL 5% (v/v) bleach at 2 bar. Flush channels with ~1 mL MilliQ water at 2 bar. Flush channels with ~1 mL 0.025 M sodium thiosulfate solution through at 2 bar. Flush channels with ~1 mL MilliQ water at 2 bar. Flush channels with ~1 mL 1: 50 dilution of Hellmanex III cleaning concentrate. The microfluidics system can be left in this until the next use.

4 Notes

1. We use Sf-900 III media (Thermo Fisher Scientific).
2. We use cOmplete EDTA-free Protease Inhibitor Cocktail Tablets (Roche).
3. We use a SPEX CertiPrep 6850 Freezer/Mill (SPEX).
4. Prepare 2 L of 0.1 M TEAA buffer by mixing 100 mL acetonitrile with 28 mL of triethylamine in 1.9 L of water. Use ~10 mL of acetic acid to adjust to pH 7.0 (starting pH should be ~9). Filter through a 0.22 μm pore size polyvinylidene difluoride (PVDF) membrane.
5. We use Strep-Tactin Superflow Plus resin (QIAGEN).
6. We use a QIAquick PCR Purification Kit (QIAGEN).
7. Any adverse effects of fluorophore conjugation can be first assessed by an ATPase assay carried out according to ref.

[21]. While absolute reported rates for each SMC complex vary based on the methodology undertaken, stimulatory and inhibitory effects of associated factors can be used to assess the activity and identification of the complex under study.

8. Peak protein expression is sensitive to a multitude of factors. It is critical to only use healthy log phase cells for virus infection. We observe a drop in expression after storage of a working stock of baculovirus (P2) at 4 °C for 9–12 months.
9. Viral infection and protein production can also be accessed by YFP signal if using DH10EMBacY bacmids where *YFP* is also under control of a polH promoter.
10. Biotin is a common ingredient in cell culture media and must be removed prior to loading onto a Strep-Tactin resin as biotin will bind and reduce the binding capacity of this matrix for StrepII-tagged proteins.
11. The W303-1A strain carries the ade2-1 mutation which disrupts adenine biosynthesis and leads to a buildup of an intermediate in this pathway—an oxidized form of ribosylaminoimidazole. Accumulation of this results in a red coloring visible at the level of colonies. Media should be supplemented with at least 40 mg/L adenine to circumvent this affecting optimal growth [22].
12. The addition of biotin will increase the buffer pH; however, it will only fully dissolve under more alkaline conditions. To aid this, add 10 M NaOH dropwise to the solution until the initial buffer pH is regained.
13. In the case of the cohesin trimer complex, the HiTrap Q column is essential for resolving Smc1–3 dimers and the trimeric complex. The leading shoulder of the trimer peak will contain dimeric species so be selective in pooling fractions here.
14. CoA is not stable in aqueous solutions so prepare fresh in degassed buffer.
15. ATTO 647 N maleimide is available in 1 mg quantities. Ideally store long-term dried. Alternatively, snap freeze surplus sample in DMSO and store at −80 °C for future use.
16. In practice, the nmol yield calculated for CoA and the fluorophore for the conjugate will not be exactly equal, as you would expect. Use the fluorophore readout as the final fluorophore-CoA conjugate quantity.
17. *Sc*Cohesin has an increased propensity to form aggregates when left at higher concentrations (>5 μM) for extended periods of time such as during the 16 h incubation for fluorophore labelling.
18. The Refeyn mass photometry instrument is compatible with measurement of most fluorophore-conjugated proteins.

19. Tris and HEPES-based buffers are described to be noisy in mass photometry; however, we find in practice they work as all components of SMC complexes are greater than 50 kDa and in turn resolved away from buffer-related noise.
20. Collecting data at a suitable density of binding events is critical for good-quality data. If binding event density is too high, the point spread function (PSF) of species will overlap and will compromise accurate detection. If binding density is too low, the resulting contrast distribution will contain disproportionate noise. In the case of the former, decrease sample concentration and in the case of the latter, either increase sample concentration or combine data from several acquisitions sets at a low concentration.
21. If repeat measurement of the BSA and thyroglobulin standards results in shifted masses, use these measurements to create at second calibration file to use for analyzing more recently collected samples.
22. Default analysis settings typically work well for the analysis of SMC-sized complexes. If smaller species are expected and not observed, modify the integration time (or number of averaged frames) and/or noise rejection.
23. Difficulty in catching DNA tethers is usually indicative of disrupted flow. Check for bubbles by inspecting the chip area using brightfield imaging and remove any bubbles found using pulses of 2 bar flow. It may be necessary to raise the condenser to inspect other regions of the flow cell and connected tubing.
24. Even with BSA and Pluronics-based passivation of the microfluidics system, the effective concentration of protein in solution will decrease over time with adsorption to the channel surface. To circumvent this, flush fresh reagent into channels immediately before protein loading.
25. Determining the incubation time required for loading SMC complexes onto DNA tethers to a density suitable for analysis, that is, able to resolve and fit Gaussians to single PSFs, must be empirically tested. This will be dependent on the effectiveness of surface passivation.
26. It is critical that the whole of the leftmost bead is included in the 2D scan region for automatic bead detection during analysis.
27. If using the Hellmanex cleaning solution after bleach (or vice versa), the system must be completely cleared of any residual bleach as Hellmanex contains an alkaline concentrate, which, when mixed with bleach, can produce chlorine gas.

Acknowledgments

We thank James Collier for supplying constructs for *Sc*Cohesin trimer and *Sc*Scc3 expression and Erin Cutts for designing the construct for *Hs*Smc5/6 expression. This work was funded by the Medical Research Council (UKRI MC-A652-5PY00).

References

1. Hoencamp C, Rowland BD (2023) Genome control by SMC complexes. Nat Rev Mol Cell Biol 24:633–650
2. Hirano T (2016) Condensin-based chromosome organization from bacteria to vertebrates. Cell 164:847–857
3. Aragón L (2018) The Smc5/6 complex: new and old functions of the enigmatic long-distance relative. Annu Rev Genet 52:89–107
4. Davidson IF, Peters JM (2021) Genome folding through loop extrusion by SMC complexes. Nat Rev Mol Cell Biol 22:445–464
5. Ganji M, Shaltiel IA, Bisht S et al (2018) Real-time imaging of DNA loop extrusion by condensin. Science 360:102–105
6. Davidson IF, Bauer B, Goetz D et al (2019) DNA loop extrusion by human cohesin. Science 366:1338–1345
7. Kim Y, Shi Z, Zhang H et al (2019) Human cohesin compacts DNA by loop extrusion. Science 366:1345–1349
8. Pradhan B, Kanno T, Umeda Igarashi M et al (2023) The Smc5/6 complex is a DNA loop-extruding motor. Nature 616:843–848
9. Gross P, Farge G, Peterman EJG, Wuite GJL (2010) Combining optical tweezers, single-molecule fluorescence microscopy, and microfluidics for studies of DNA–protein interactions. In: Methods in Enzymology, pp 427–453
10. Candelli A, Wuite GJL, Peterman EJG (2011) Combining optical trapping, fluorescence microscopy and micro-fluidics for single molecule studies of DNA-protein interactions. Phys Chem Chem Phys 13:7263–7272
11. Collier JE, Lee BG, Roig MB et al (2020) Transport of DNA within cohesin involves clamping on top of engaged heads by SCC2 and entrapment within the ring by SCC3. elife 9:1–36
12. Gutierrez-Escribano P, Newton MD, Llauró A et al (2019) A conserved ATP- and Scc2/4-dependent activity for cohesin in tethering DNA molecules. Sci Adv 5:1–16
13. Yin J, Lin AJ, Golan DE, Walsh CT (2006) Site-specific protein labeling by Sfp phosphopantetheinyl transferase. Nat Protoc 1:280–285
14. Ciosk R, Shirayama M, Shevchenko A et al (2000) Cohesin's binding to chromosomes depends on a separate complex consisting of Scc2 and Scc4 proteins. Mol Cell 5:243–254
15. Gietz RD, Schiestl RH (2007) High-efficiency yeast transformation using the LiAc/SS carrier DNA/PEG method. Nat Protoc 2:31–34
16. Yin J, Straight PD, McLoughlin SM et al (2005) Genetically encoded short peptide tag for versatile protein labeling by Sfp phosphopantetheinyl transferase. Proc Natl Acad Sci USA 102:15815–15820
17. Young G, Hundt N, Cole D et al (2018) Quantitative mass imaging of single biological macromolecules. Science 360:423–427
18. Sonn-Segev A, Belacic K, Bodrug T et al (2020) Quantifying the heterogeneity of macromolecular machines by mass photometry. Nat Commun 11:1–10
19. Smith SB, Cui Y, Bustamante C (1996) Overstretching B-DNA: the elastic response of individual double-stranded and single-stranded DNA molecules. Science 271:795–799
20. Gross P, Laurens N, Oddershede LB et al (2011) Quantifying how DNA stretches, melts and changes twist under tension. Nat Phys 7:731–736
21. Voulgaris M, Gligoris TG (2019) A protocol for assaying the ATPase activity of recombinant cohesin holocomplexes. In: Methods in Molecular Biology, pp 197–208
22. Kokina A, Kibilds J, Liepins J (2014) Adenine auxotrophy - be aware: some effects of adenine auxotrophy in Saccharomyces cerevisiae strain W303-1A. FEMS Yeast Res 14:697–707

Chapter 4

Atomic Force Microscopy Analysis of Bridging-Induced Phase Separation Mediated by SMC Proteins

Sara Illodo, Hyeong Ku Kim, and Je-Kyung Ryu

Abstract

Structural maintenance of chromosome (SMC) protein is essential for chromosomal architecture and genome stability across all domains of life. SMC proteins, including cohesin, condensin, and the Smc5/6 complex, are known to organize chromosomes through DNA loop extrusion. However, recent atomic force microscopy (AFM) studies showed ATP-independent DNA-bridging-dependent cohesin/DNA condensate formation, suggesting bridging-induced phase separation (BIPS) for genome organization. Here, we present a comprehensive protocol for AFM image acquisition and data analysis to explore the BIPS induced by cohesin SMC complexes. Since the BIPS mechanism can potentially apply to other multivalent DNA-binding proteins involved in DNA/protein condensate formation, this protocol provides valuable insights into the molecular basis of DNA/protein co-condensation.

Key words Bridging-induced phase separation (BIPS), Structural maintenance of chromosome (SMC), Cohesin, DNA-cohesin cluster, AFM

1 Introduction

Structural maintenance of chromosome (SMC) protein is highly conserved molecular machines present in all domains of life. They are central to the regulation of chromosomal architecture, genome stability, and dynamic processes such as DNA condensation [1–6]. Among the SMC family, cohesin, condensin, and the Smc5/6 complex are particularly important for orchestrating the spatial and temporal organization of chromosomes [7, 8].

A well-established function of these complexes is DNA loop extrusion, a mechanism that facilitates higher-order chromatin structure. Recent studies have shown that cohesin, condensin, and Smc5/6 are all capable of mediating this activity [2, 4, 7–10]. However, additional mechanisms beyond loop extrusion are increasingly being recognized [11–17]. One such mechanism is

Yasuto Murayama and Yasutaka Kakui (eds.), *SMC Proteins: Methods and Protocols*, Methods in Molecular Biology, vol. 2991,
https://doi.org/10.1007/978-1-0716-5005-9_4,

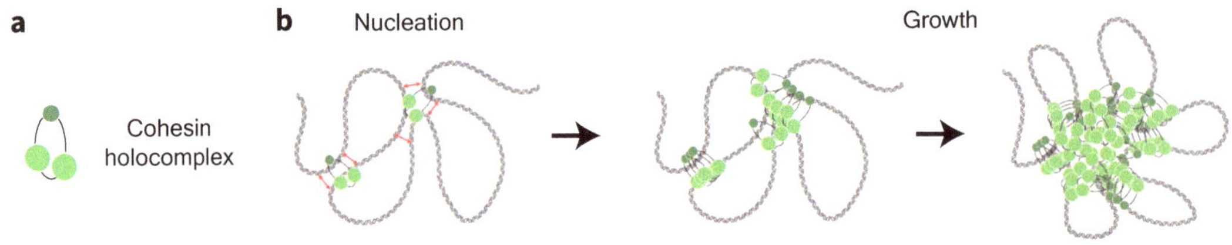

Fig. 1 Cohesin-driven bridging-induced phase separation. (**a**) A representation of the cohesin holocomplex. (**b**) Nucleation: A cohesin holocomplex binds to DNA through a multivalent DNA–protein interaction forming a DNA loop that works as nucleation point. Growth: This bridging increases the local concentration of DNA in the specific region, promoting the recruitment of additional cohesin molecules. Also, entropic penalty minimization through targeting the bridged region stimulates the growth of condensates. The accumulation of multiple cohesin molecules results in the formation of a large DNA/cohesin holocomplex droplet

phase separation, which may play a complementary or independent role in genome organization.

We have recently demonstrated that cohesin can form DNA-associated clusters via bridging-induced phase separation (BIPS), a specific form of phase separation driven by protein-mediated bridging between distal DNA segments [9, 18–20]. This process stands in contrast to the more commonly discussed self-interaction-induced phase separation (SIPS), which arises from multivalent protein–protein interactions [3, 18]. In BIPS, the multivalency arises from protein–DNA interactions that connect distant regions of the same DNA molecule, resulting in the formation of DNA loops. These loops locally concentrate both DNA and protein, promoting further cohesin binding and ultimately leading to the emergence of a DNA–cohesin condensate (Fig. 1).

Importantly, because DNA exhibits intrinsic stiffness over several persistence lengths, a minimum DNA length threshold is required for BIPS to occur (Fig. 2). Our findings suggest that DNA length must be at least ~3 kbp in length to support this process, with larger condensates forming as DNA length increases beyond 3 kbp.

To investigate these condensates at the molecular level, we employ atomic force microscopy (AFM). Since its development in 1986 [21], AFM has become a powerful tool for visualizing the structure of biological molecules at nanometer resolution with high contrast [21, 22]. Unlike fluorescence microscopy, AFM enables direct imaging of biomolecules without labeling, making it ideally suited for analyzing protein–DNA condensates. Recent advances, particularly peak-force tapping mode, allow for high-resolution imaging with minimal sample perturbation [22–25].

In this chapter, we present a comprehensive protocol for studying cohesin-mediated DNA condensation using AFM. We highlight critical steps for sample preparation, imaging, and data interpretation to investigate the molecular mechanisms underlying BIPS.

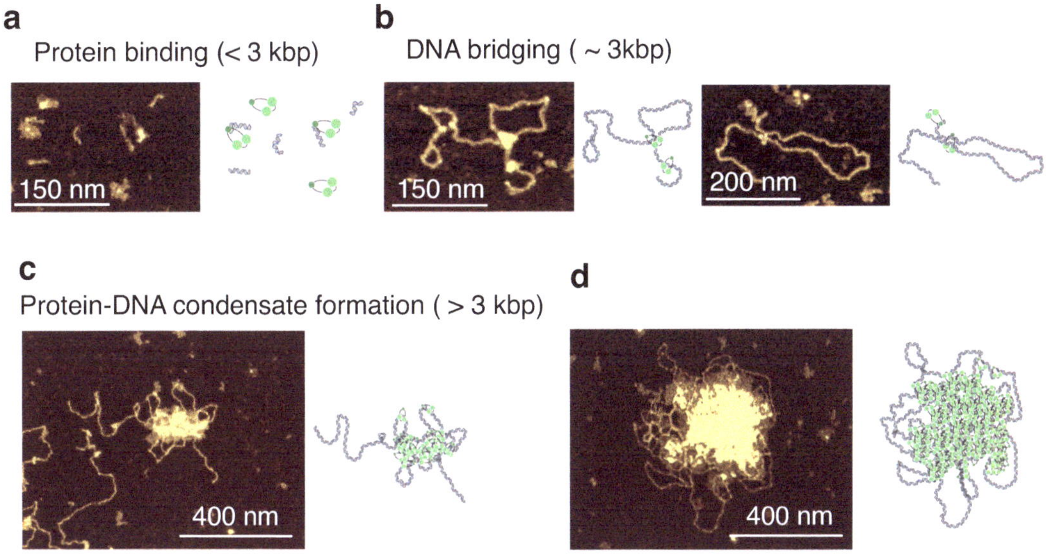

Fig. 2 Length dependency of cluster formation. AFM image and model figure of (**a**) Yeast cohesin binding to DNA with length < 3 kbp. (**b**) DNA–cohesin–DNA bridging with length around 3 kbp (**c**) Cohesin–DNA condensates (DNA length > 3 kbp) (**d**) A large cohesin/DNA cluster with 48.5 kbp DNA length

2 Materials

2.1 Yeast Strains for Protein Purification

1. Cohesin tetramer: The budding yeast strain (*Saccharomyces cerevisiae*) overexpressing the *S. cerevisiae* Smc1 (fused with a 3x Pk tag at the C-terminus), Smc3, Scc1 (fused with a 3C protease cleavage sequence-2x protein A tag at the C-terminus), Scc3 (fused with a 1x Myc tag at the C-terminus), and Gal4 transcription factor under the control of *GAL1/GAL10* promoters [14] (*see* **Note 1** for the strain construction).
2. The Scc2-Scc4 loader complex: The budding yeast strain (*Saccharomyces cerevisiae*) overexpressing the *S. cerevisiae* Scc2 (fused with a 3C protease cleavage sequence-2x protein A tag at the C-terminus), Scc4 (fused with a 3x HA tag at the C-terminus), and Gal4 transcription factor under the control of *GAL1/GAL10* promoters [14] (*see* **Note 2** for the strain construction).

2.2 Yeast Medium

1. YP-raffinose medium: 2% (w/v) peptone, 1% (w/v) yeast extract, 2% (w/v) raffinose. Prepare in distilled water and sterileze by autoclaving.
2. 20% (w/v) galactose: Prepare in distilled water and sterileze by autoclaving.

2.3 Protein Purification

1. Buffer A for cohesin tetramer: 50 mM HEPES-NaOH (pH 7.5), 2 mM $MgCl_2$, 0.5 mM Tris(2-carboxyethyl) phosphine (TCEP), 0.5 mM Pefabloc, 20% glycerol (v/v), protease inhibitor cocktail. Prepare Buffer A containing 0, 0.3, and 1 M NaCl, respectively.
2. Buffer B for cohesin loader: 50 mM Tris-HCl (pH 7.5), 2.5 mM $MgCl_2$, 0.5 mM phenylmethylsulfonyl fluoride (PMSF), 1 mM dithiothreitol (DTT), 10% glycerol (v/v), protease inhibitor cocktail. Prepare Buffer B containing 0, 0.3, and 1 M NaCl, respectively.
3. Buffer R: 20 mM Tris-HCl (pH 7.5), 150 mM NaCl, 0.5 mM TCEP, 10% (v/v) glycerol.
4. Benzonase.
5. RNase A.
6. PreScission protease.
7. Human IgG agarose (Merck).
8. 1 mL HiTrap Heparin HP column (Cytiva).
9. Superose 6 10/300 GL gel filtration column (Cytiva).
10. Superdex 200 10/300 GL gel filtration column (Cytiva).
11. Shaking incubator.
12. Freezer mill.
13. Centrifuge: Large centrifuge for harvesting yeast cells, ultracentrifuge.
14. FPLC system.

2.4 Dry AFM Analysis

1. DNA fragments (0.1–20 kbp, Thermo Scientifics NoLimits).
2. Unmethylated lambda DNA.
3. Reaction buffer: 50 mM Tris-HCl (pH 7.5), 50 mM NaCl, 2.5 mM $MgCl_2$, 2 mM TCEP.
4. 0.1 μg/mL Poly-L-lysine: Prepared in ultrapure water.
5. Ultrapure water.
6. 1.5 mL tubes.
7. Mica (6–9 mm diameter).
8. Magnet metal AFM discs.
9. Tweezers.
10. Epoxy glue.
11. Compressed nitrogen gas contourable with nitrogen gun.
12. MultiMode 8 AFM (Bruker).
13. NanoScope VI controller (Bruker).

14. SCANASYST-AIR-HR cantilevers (nominal stiffness and tip radius of 0.4 N/m and 2 nm, respectively).
15. Gwyddion software.
16. Python.

3 Methods

3.1 Cohesin Tetramer and the Loader Purifications

Both the cohesin tetramer (Fig. 3) and the Scc2-Scc4 cohesin loader are prepared by the same protocol described by Minamino et al. [14] with slight modifications as noted below:

1. Culture the budding yeast cells (cohesin or Scc2-Scc4, Materials Subheading 2.1) at 30 °C in YP-raffinose medium until the OD_{600} reaches to 1.0.
2. To induce protein expression, add 1/10 volume of 20% (w/v) galactose (2% (w/v) final) and incubate for 2 h for the cohesin tetramer, or 1.5 hours for Scc2-Scc4.
3. Collect the yeast cells by centrifugation and wash once with deionized water.
4. Resuspend cell pellets in Buffer A (for cohesin) or Buffer B (for Scc2-Scc4) containing 300 mM NaCl. Freeze the cell suspension with liquid nitrogen.
5. Ground the frozen cells with a freezer mill.
6. Thaw the cell powder on ice.
7. Add two volumes of Buffer A (for cohesin) or Buffer B (for Scc2-Scc4) containing 300 mM NaCl and 0.3 μg/mL RNase A.
8. Clarify lysates by two rounds of centrifugation (30,000 g for 30 min at 4 °C and 142,000 g for 1 h at 4 °C).

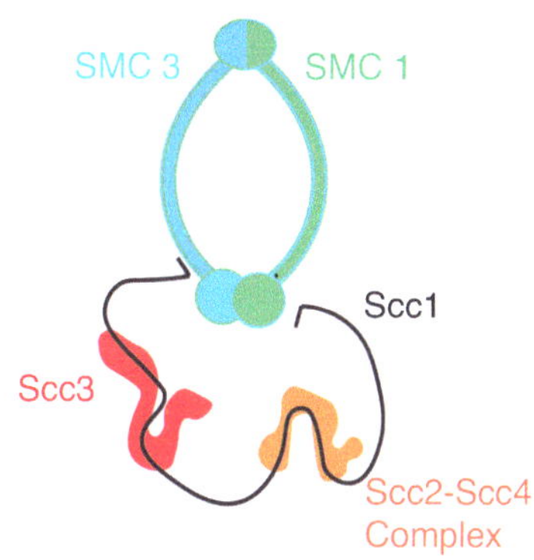

Fig. 3 Model of a yeast cohesin architecture. Each unit is represented by a different color

9. Transfer the clarified lysates to a new tube and add Benzonase to a final concentration of 1.25 U/mL.
10. Incubate the clarified lysate with 2 mL of IgG agarose beads (pre-equilibrated with Buffer A or B) for 2 hours at 4 °C.
11. Wash the resins with Buffer A (for cohesin) or Buffer B (for Scc2-Scc4) containing 300 mM NaCl.
12. Resuspend the beads with 5 mL of Buffer A (for cohesin) or Buffer B (for Scc2-Scc4) containing 300 mM NaCl.
13. Add PreScission protease to a final concentration of 10 μg/mL, and incubate overnight (~15 h) at 4 °C.
14. Load eluate onto a 1 mL HiTrap Heparin HP column and develop it with a linear gradient of 300 mM to 1 M NaCl in Buffer A (cohesin) or Buffer B (Scc2-Scc4). Collect the peak fractions.
15. Load the eluate onto a gel filtration column equilibrated with Buffer R. Use a Superose 6 10/300 GL column for cohesin purification, and a Superdex 200 10/300 GL column for Scc2-Scc4 purification.

3.2 Dry AFM Imaging

For dry AFM observation, cohesin/DNA cluster is formed in a tube and then mounted on a poly-L-lysine-coated mica (Fig. 4). We use a MultiMode 8 AFM equipped with a NanoScope VI controller. All dry AFM experiments are performed by the peak-force tapping mode (*see* **Note 3**, Fig. 5) with a peak force set point lower than 70 pN and an oscillation frequency of 8 kHz [7, 9]. The peak force value is important to minimize any sample disruption that could be potentially caused by the constant interaction with the tip. Before AFM measurement, calibration and laser alignment must be performed (*see* **Notes 4** and **5**).

3.2.1 Mica Stage Assembly

1. Place a magnet metal AFM disc on a horizonal bench.
2. Apply a small amount of epoxy glue to the center of the magnet disc.
3. Carefully place the circular mica onto the epoxy using tweezers.

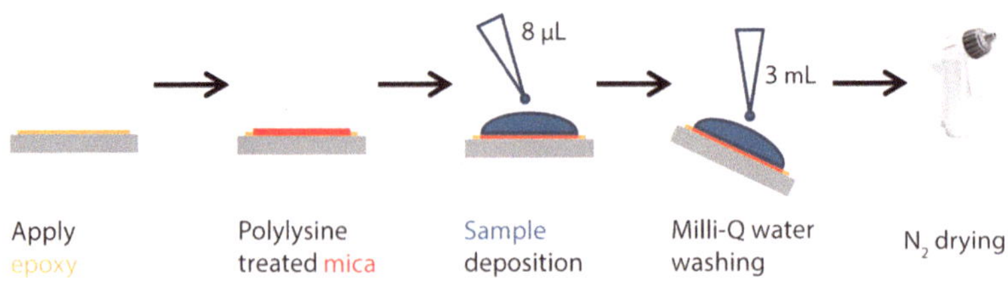

Fig. 4 Schematic representation of sample deposition into previously treated mica substrate for AFM imaging

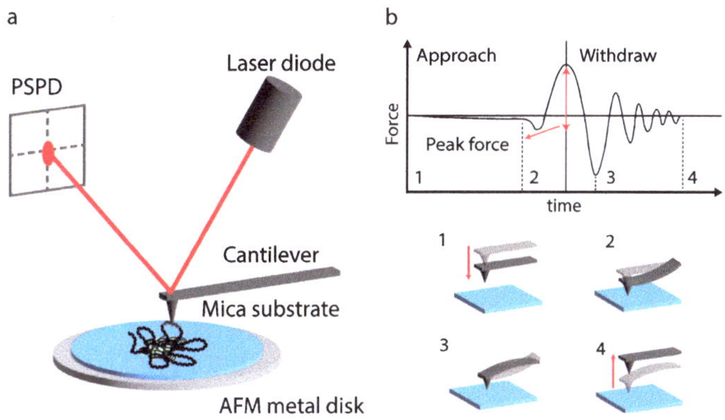

Fig. 5 AFM peak-force tapping imaging acquisition mode. (**a**) The laser is focused on the cantilever and reflected into the detector (position-sensitive photodetector, PSPD). The cantilever taps the DNA/cohesin sample that has been loaded into the mica substrate. (**b**) Graphical representation of the force behavior with time in a single tapping cycle when the cantilever is (1) approaching the sample, (2) and (3) interacting with the sample, and 4) withdrawing from the sample

4. Press mica with the back of a pipette tip for better adherence. To remove air bubbles, press the mica onto the epoxy while rotating the magnetic disc.
5. Wait until epoxy is fully dried.
6. Soak a circular mica in 0.1 μg/mL poly-L-lysine for 90 seconds to make the mica surface positively charged (*see* **Note 6**).
7. Dry the mica by nitrogen stream.

3.2.2 Preparation of Cohesin/DNA Sample for Dry AFM Observation

1. Mix 10 nM cohesin, 10 nM Scc2-Scc4, and 4 ng/μL DNA with the desired length in reaction buffer.
2. Incubate the reaction mixture at room temperature for the indicated times:
   (a) Incubate for 5 min to observe cohesin/DNA cluster.
   (b) Incubate 10 s to observe a single cohesin molecule bound to the DNA.
3. Cleave the surface of the assembled mica with tape (*see* **Note 7**).
4. Spot 8 μL of the cohesin/DNA mixture on the mica surface and incubate for 30 seconds (*see* **Notes 6** and **8**).
5. Wash the mica assembly by flowing 3 mL of ultrapure water.
6. Dry the mica assembly by nitrogen stream (*see* **Note 9**).

3.2.3 AFM Image Acquisition

1. Set the sample-mounted mica assembly on an AFM sample holder.

2. Carefully approach the cantilever to the mica surface according to the instruction of the AFM.
3. Scan the sample using the peak-force tapping mode with the following settings:
   (a) A peak force set point lower than 70 pN.
   (b) An oscillation frequency of 8 kHz.
   (c) A scan area of 10 μm x 10 μm (5120 × 5120 pixels).
   (d) AFM Analysis of Cohesin-mediated BIPSA scanning speed of 0.7 kHz.
4. Save the images.

3.3 AFM Image Processing

All AFM data are processed using the Gwyddion software, which is freely downloadable, using the following protocol:

1. Open images in Gwyddion.
2. Apply level data by mean plane subtraction.
3. Open align rows using various methods tool. Select median option.
4. Remove polynomial background. Both horizontal and vertical polynomial degree should be 0 at this point.
5. Using the mark grains by threshold tool mask all the molecules. Proteins and DNA usually present a height of around 0.2 nm.
6. Open align rows using various methods tool again. Select polynomial option. The polynomial degree should be set to 5. It is important that the "exclude masked region" option is on.
7. Deconvolute tip effect for further improvement of image quality by reducing the broadened effects caused by tip convolution.
   (a) SPM Modes and on Tip select "blind estimation." This step will obtain the tip image.
   (b) Using the obtained tip image, select the "surface reconstruction" option in Tip.
8. Save image in gwy, tiff, or other desired file types.

3.4 Cluster Characterization

Physical parameters such as cluster volume (Subheading 3.4.1), area, size, and radius (Subheading 3.4.2) can be used to characterize the clusters. For defining the radius, several approaches may be used, including the volume-derived radius, $R_{\mathrm{V}} = \sqrt[3]{3V/4\pi}$, the area-derived radius, $R_{\mathrm{A}} = \sqrt{A/\pi}$, and the radius of gyration (Subheading 3.4.2).

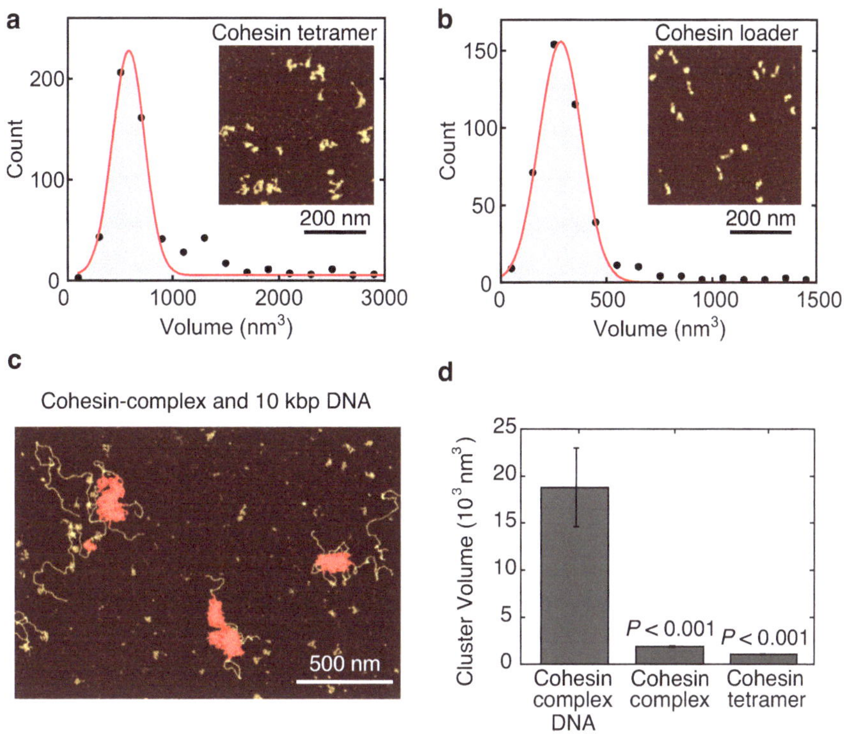

Fig. 6 AFM images cluster volume analysis. (**a**, **b**) Volume distribution from the cohesin tetramer (**a**) and the cohesin loader (**b**) obtained from AFM images and representative AFM images (inset) ($N = 602$ and 419, for cohesin tetramer and cohesion loader, respectively). (**c**) AFM images of cohesin/DNA cluster, the red area is used for volume calculation. (**d**) Volume comparison of cohesin complex/DNA cluster, cohesin complex, and cohesin tetramer ($N = 21$, 11 and 57, median ± SEM). Permissioned from ref. [9]

3.4.1 Cluster Volume Analysis

Volume analysis was done using Gwyddion as follows [9]:

1. Manually define the cluster area in Gwyddion by masking the desired clusters (Fig. 6c).
2. Using the mark grains by threshold tool, apply the desired threshold to the mask (usually 0.2 nm is accurate for DNA and proteins). Before applying, make sure the "combined with existing mask" option is on.
3. Open filter grain tool. Set the lower and higher threshold parameter as needed depending on the size of the sample. The threshold parameter should be defined previously by analyzing the volume distribution of the protein (Fig. 6a, b).
4. Open distribution tool and calculate the volume using the Laplacian background-basis volume option (*see* **Note 10**). Export raw data and save.
5. Repeat for all images of the same sample.
6. Calculate the average volume. Figure 6d shows a representative result of the volume of the cohesin/DNA cluster.

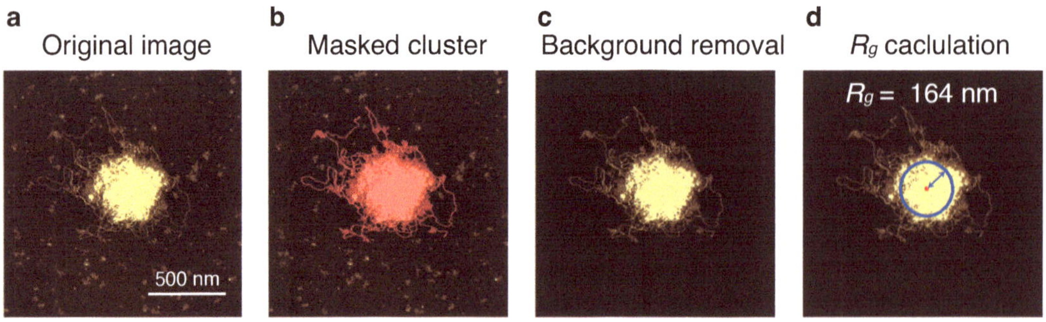

Fig. 7 Processing images using the analysis protocol of the radius of gyration of the cohesin/DNA cluster. Each image corresponds to one of the steps of the analysis. (**a**, **b**) On the original AFM image (**a**), the cluster molecule is first masked (**b**). (**c**) After masking, the cluster background subtraction is done. (**d**) As final step, the *Rg* of the previously masked cluster is calculated

3.4.2 Cluster Radius of Gyration Analysis

Radius of gyration (R_g) is calculated using Gwyddion and python. Python code first calculates the center of mass of each molecule, which is defined as in Eq. (1). Then, it calculated the R_g, which we defined as shown in Eq. (2) (where h_{xy} is the height at a given pixel (x,y), and h_{total} is the sum of all pixel's height included in the masked region) [26].

1. Open image in Gwyddion.
2. Crop image to show an individual cluster within the region of interest (ROI) as shown in Fig. 7a.
3. In the cropped image, manually mask the molecule (Fig. 7b).
4. Remove background excluding masked region (Fig. 7c).
5. Save image.
6. Open image in python.
7. Calculate R_g using the following formula (Fig. 7d):

$$(X_{\text{CM}}, Y_{\text{CM}}) = \left(\sum_{x,y} \frac{x\, h_{xy}}{h_{\text{total}}}, \sum_{x,y} \frac{y\, h_{xy}}{h_{\text{total}}}\right) \quad (1)$$

$$R_g = \sqrt{\sum_{x,y} \frac{h_{xy}\left((x - X_{\text{CM}})^2 - (y - Y_{\text{CM}})^2\right)}{h_{\text{total}}}} \quad (2)$$

4 Notes

1. The cohesin expression strain was constructed as follows: The complementary DNA (cDNA) encoding Scc1 fused with two consecutive protein A tags at the C-terminal end, separated by a 3C protease cleavage site and SMC3 cDNA, is cloned to the YIplac204 (TRP1) vector under the GAL1/10 promoter,

generating the plasmid YIpScc1-Smc3. The cDNA encoding Scc3 adding a myc epitope tag at its C-terminal is inserted into the YIplac211 (URA3) vector under the GAL10 promoter, resulting in the plasmid YIpScc3. Smc1-encoding cDNA was amplified using PCR and linked to three tandem Pk epitopes at its C-terminus and was cloned into the shuttle vector pRSI1402 under the control in the direction of GAL1 of the GAL1 promoter. This plasmid also includes the GAL4 gene of the budding yeast under the control of the GAL10 promoter to enhance galactose-induced expression, resulting in the plasmid pRSIISmc1-Gal4. These three DNA constructs were linearized by restriction digestion and integrated into the eco1Δ rad61Δ *S. cerevisiae* strain at TRP1, URA3, and LEU2 genomic loci, respectively, as reported previously by Minamino et al. [14].

2. The Scc2-Scc4 loader expression strain was constructed as follows: cDNA encoding Scc2 linked a 3C protease cleavage site-2x protein A tag at its C-terminus was cloned into pRSI1402 under the GAL1 promoter into pRSI1402, which also contained the GAL4 gene under the control of the GAL10 promoter, yielding the plasmid pRSIIScc2-Gal4. cDNA encoding Scc4 was modified by adding a triple HA epitope tag at its C-terminus. This modified DNA is cloned into YIplac204, resulting in the plasmid YIpScc4. Following the same procedure as for cohesin tetramer, the loader constructs were integrated into wild-type budding yeast cell at ADE2 and TRP1 genomic loci, respectively.
3. Although conventional tapping mode can also be used, peak-force tapping mode, introduced in 2009, is preferred due to improved force control and reduced tip–sample interaction. It allows for more precise force control, quantitative property mapping, and the reduction of the sample-tip interaction, thus obtaining more consistent images with less artifacts. The reduced interaction also broadens the range of samples and reduces both tip and sample damage [27–29].
4. To calibrate the AFM tip, we first calibrated the deflection sensitivity using a PDMS sample with a known Young's modulus of 1.2 MPa by performing force–distance ramping in the Nanoscope 6 software. The deflection sensitivity, typically ranging from 20 to 40 nm/V, was obtained and used for thermal tuning. Using the calibrated deflection sensitivity, we then performed thermal tuning to determine the spring constant of the cantilever, completing the tip calibration process. For piezo scanner calibration, we used a grating standard sample (VGRP-15 M) with a known step height and pitch, following the manufacturer's instructions.
5. When loading the sample and cantilever into the AFM system, tip calibration and laser alignment must be completed prior to

image acquisition. After carefully approaching the cantilever toward the sample until the tip shadow becomes visible, the X and Y positions of the cantilever holder should be adjusted to maximize the laser reflection signal. The sum signal on the photodetector should typically fall within the range of 5.99–6.05 V. Improper alignment may result in laser misalignment, preventing signal detection. Once the laser is properly aligned, the cantilever is engaged with the sample surface. Following engagement, thermal tuning is performed to determine the cantilever's spring constant, as per the manufacturer's instructions.

6. Poly-L-lysine coating gives the mica surface a positive charge, which mediates the bond between mica and DNA. This reduces the time taken for the sample to deposit onto the mica. However, if poly-L-lysine is used, the incubation time for the sample deposition must be adhered to strictly.
7. As AFM is a label-free imaging technique and we cannot guarantee that our molecule of interest can be distinguished from other unwanted molecules, it is crucial that the samples are clean and free of contaminants.
8. Cohesin concentration is critical for successful imaging. AFM can only be used to visualize condensates under specific conditions. Because cohesin presents a high binding affinity (less than 1 nM), using low concentrations helps minimize background proteins, allowing clear imaging of cohesin–DNA interactions. Without this control, high background noise makes AFM imaging not possible.
9. When drying the sample with the nitrogen gun, it may not dry completely. It is important to ensure that any remaining water does not reach the central region of the mica again, as this can disturb the sensitive structures within the sample.
10. To determine whether the cluster is not a single cohesin complex but instead composed of multiple cohesin complexes, a confidence level higher than 99.9% was used, as well as seven standard deviations (SDs) and the average volume of a single cohesin complex (2134 nm^3 in the case of yeast cohesin holocomplex). Any defined area whose cluster volume was higher than the threshold was considered a cluster.

Acknowledgments

J.-K.R. acknowledges the Suh Kyoungbae foundation [SUHF23010081], Institute of Applied Physics of Seoul National University, Creative-Pioneering Researchers Program, and AI-Bio Research Grant through Seoul National University, the Brain Korea

21 Four Project grant funded by the Korean Ministry of Education, Samsung Electronics Co., Ltd. [Project Number IO220811-0196401], and the National Research Foundation of Korea [Project Number RS-2023-00212694, RS-2023-00265412, RS-2023-00218318, RS-2022-NR066591, RS-2024-00410733]. J.-K.R. and S.I. acknowledges the Global-LAMP Program of the National Research Foundation of Korea (NRF) grant funded by the Ministry of Education [No. RS-2023-00301976].

References

1. Moon KW, Kim D, Ryu JK (2024) Anisotropic scrunching of SMC with a baton-pass mechanism. Commun Biol 7(1):881
2. Moon KW, Ryu JK (2023) Current working models of SMC-driven DNA-loop extrusion. Biochem Soc Trans 51(5):1801–1810
3. Ryu JK, Hwang D, Choi JM (2021) Current understanding of molecular phase separation in chromosomes. Int J Mol Sci 22(19):10736
4. Uhlmann F (2016) SMC complexes: from DNA to chromosomes. Nat Rev Mol Cell Biol 7(7):399–412
5. Hirano T (2016) Condensin-based chromosome organization from bacteria to vertebrates. Cell 164(5):847–857
6. Nasmyth K, Haering CH (2005) The structure and function of SMC and Kleisin complexes. Annu Rev Biochem 74:595–648
7. Ryu JK, Katan AJ, van der Sluis EO et al (2020) The condensin holocomplex cycles dynamically between open and collapsed states. Nat Struct Mol Biol 27(12):1134–1141
8. Ryu JK, Rah SH, Janissen R et al (2022) Condensin extrudes DNA loops in steps up to hundreds of base pairs that are generated by ATP binding events. Nucleic Acids Res 50(2): 820–832
9. Ryu JK, Bouchoux C, Liu HW et al (2021) Bridging-induced phase separation induced by cohesin SMC protein complexes. Sci Adv 7(7): eabe5905
10. Fudenberg G, Imakaev M, Lu C et al (2016) Formation of chromosomal domains by loop extrusion. Cell Rep 15(9):2038–3049
11. Forte G, Boteva L, Conforto F et al (2023) Bridging condensins mediate compaction of mitotic chromosomes. J Cell Biol 223(1): e202209113
12. Hirano T, Kinoshita K (2025) SMC-mediated chromosome organization: does loop extrusion explain it all? Curr Opin Cell Biol 92: 102447
13. Kinoshita K, Tsubota Y, Tane S et al (2022) A loop extrusion-independent mechanism contributes to condensin I-mediated chromosome shaping. J Cell Biol 221(3):e202109016
14. Minamino M, Higashi TL, Bouchoux C et al (2018) Topological in vitro loading of the budding yeast cohesin ring onto DNA. Life Sci Alliance 1(5):e201800143
15. Murayama Y, Uhlmann F (2014) Biochemical reconstitution of topological DNA binding by the cohesin ring. Nature 505(7483):367–371
16. Pastic A, Nosella ML, Kochhar A et al (2024) Chromosome compaction is triggered by an autonomous DNA-binding module within condensin. Cell Rep 43(7):114419
17. Uhlmann F (2025) A unified model for cohesin function in sister chromatid cohesion and chromatin loop formation. Mol Cell 85(6): 1058–1071
18. Park J, Kim JJ, Ryu JK (2024) Mechanism of phase condensation for chromosome architecture and function. Exp Mol Med 56(4): 809–819
19. Erdel F, Rippe K (2018) Formation of chromatin subcompartments by phase separation. Biophys J 114(10):2262–2270
20. Brackley CA, Taylor S, Papantonis A et al (2013) Nonspecific bridging-induced attraction drives clustering of DNA-binding proteins and genome organization. P Natl Acad Sci U S A 110(38):3605–3611
21. Binnig G, Quate CF, Gerber C (1986) Atomic Force Microscope. Phys Rev Lett 56(9): 930–933
22. Krieg M, Fläschner G, Alsteens D et al (2019) Atomic force microscopy-based mechanobiology. Nat Rev Phys 1(1):41–57
23. Hu J, Gao MY, Wang ZB et al (2022) Review on the applications of atomic force microscopy imaging in proteins. Micron 159:103293
24. Müller DJ, Janovjak H, Lehto T et al (2002) Observing structure, function and assembly of

single proteins by AFM. Prog Biophys Mol Bio 79(1–3):1–43

25. Zhai ZY, Schmid SY, Lin ZX et al (2024) Unveiling the nanoscale architectures and dynamics of protein assembly with in situ atomic force microscopy. Aggregate 5(5):e604
26. Brennan L, Kim HK, Colmenares S et al (2025) HP1a promotes chromatin liquidity and drives spontaneous heterochromatin compartmentalization. bioRxiv 2024.10.18.618981
27. Schillers H, Medalsy I, Hu SQ et al (2016) PeakForce tapping resolves individual microvilli on living cells. J Mol Recognit 29(2):95–101
28. Li M, Xi N, Liu LQ (2021) Peak force tapping atomic force microscopy for advancing cell and molecular biology. Nanoscale 13(18): 8358–8375
29. Slattery AD, Shearer CJ, Shapter JG et al (2018) Improved application of carbon nanotube atomic force microscopy probes using PeakForce tapping mode. Nanomaterials (Basel) 8(10):807

Chapter 5

Monitoring Local Chromatin Dynamics Regulated by SMC Complexes in Fission Yeast

Yasutaka Kakui

Abstract

Structural maintenance of chromosomes (SMC) complexes organize genome architecture throughout the cell cycle. Condensin, a member of the SMC complex family, engages chromatin loci by embracing chromatin within its ring and constrains the dynamics of chromatin fiber to shape mitotic chromosomes. To understand how condensin constrains chromatin dynamics, it is crucial to monitor the mobility of chromatin fiber with high temporal and spatial resolution. Here, I describe a step-by-step protocol to record the dynamics of chromatin locus visualized using the LacO/LacI system in fission yeast, followed by the determination of its physical properties. This method is applicable not only for monitoring the dynamics of chromatin fiber but also for tracking various types of particles in living cells.

Key words Chromatin, Chromosome condensation, Condensin, Brownian motion, Particle tracking, Mean squared displacement (MSD)

1 Introduction

Structural maintenance of chromosomes (SMC) complexes are the heart of controlling the spatial chromatin organization in living organisms, ranging from bacteria to higher eukaryotes [1]. The SMC complexes form proteinaceous rings that topologically entrap DNA strands or chromatin fibers, thereby organizing the spatial chromatin structure. One of the SMC complexes, condensin, is indispensable to form mechanistically stable condensed chromosomes in mitosis [2]. The genome-wide chromosome conformation capture technique, Hi-C, has revealed condensing-mediated longer-range chromatin interactions within mitotic chromosomes in fission yeast and human cells [3, 4]. Another SMC complex, cohesin, holds sister chromatids together to ensure the faithful chromosome segregation and also mediates the formation of chromatin loops and topologically associating domains (TADs) for controlling gene expressions in interphase [5, 6].

Yasuto Murayama and Yasutaka Kakui (eds.), *SMC Proteins: Methods and Protocols*, Methods in Molecular Biology, vol. 2991, https://doi.org/10.1007/978-1-0716-5005-9_5,

Chromatin is a chain-like structure composed of nucleosomes and dynamically moves within the nucleus. The dynamics of chromatin has been monitored using fluorescently labeled individual nucleosomes [7], or through the LacO/LacI system, where LacI protein tagged with a fluorescent protein binds to Lac operator (LacO) repeats inserted at certain genomic loci [4, 8–10]. To obtain physical properties of chromatin, the mean squared displacement (MSD) is calculated from the trajectories of individual nucleosomes or genomic loci visualized by the LacO/LacI system. MSD represents an average squared distance that a particle travels over a given time interval and is commonly used to characterize diffusive motion. MSDs derived from these trajectories have revealed subdiffusive motion of chromatin in the interphase nucleus. When DNA damage, such as DNA double-strand break, is introduced, chromatin increases its mobility, which in turn enhances efficient homologous search and subsequent DNA repair [8, 9]. During the cell cycle, two SMC complexes, cohesin and condensin, contribute to the regulation of chromatin mobility. Cohesin constrains chromatin mobility throughout the interphase in human cells [7]. Condensin strongly confines chromatin mobility within condensed chromosomes in mitosis [4]. In the interphase fission yeast nuclei, chromatin mobility is partially constrained by condensin, protecting chromatin from transcription-induced DNA damage [10].

This chapter documents an experimental protocol for tracking fluorescently labeled particles with high temporal resolution, followed by the calculation of mean squared displacement (MSD) to determine their physical properties. By analyzing MSDs, this method determines the Brownian motion of fluorescent beads and reveals the confined motion of chromatin fiber mediated by SMC complex condensin. This experimental approach provides insights into the physical properties of various types of objects in living cells and facilitates our understanding of their dynamic behavior.

2 Materials

2.1 Cell Culture

1. Pombe Glutamate Medium (PMG): 14.7 mM potassium hydrogen phthalate, 15.5 mM Na_2HPO_4, 3.75 g/L L-glutamic acid monosodium salt, 2% (w/v) glucose, 5.2 mM $MgCl_2$, 0.1 mM $CaCl_2$, 13.4 mM KCl, 0.28 mM Na_2SO_4, 4.2 μM pantothenic acid, 81.2 μM nicotinic acid, 55.5 μM inositol, 40.8 nM biotin, 8.09 μM boric acid, 2.37 μM $MnSO_4$, 1.39 μM $ZnSO_4$, 0.74 μM $FeCl_2$, 0.247 μM molybdic acid, 0.6 μM KI, 0.16 μM $CuSO_4$, 4.76 μM citric acid. Autoclave after complete dissolution. 150 μg/mL adenine, leucine, uracil, lysine, and histidine are added as required.

2. Thiamine solution: Dissolve 10 mg/mL (w/v) thiamine in deionized water, then filter-sterilized.
3. 0.5 M 3-indoleacetic acid (IAA): Dissolved in methanol. Prepare this freshly before use.
4. Methanol.
5. Fission yeast strains: YUK799 (*h*[−] P_{nmt41}*-kan-slp1 TetR-tdTom::leu1*[+] *LacI-eGFP::his7*[+] *ChrII-3.6 Mb::TetO-hphMX ChrI-1.95 Mb::LacO-natMX6 nat-*P_{nmt81}*-cut14-IAA-ura4*[+] *ura4.D18 ade6::ade6*[+]*-*P_{adh15}*-skp1-OsTIR1-bsd-*P_{adh15}*-skp1-AtTIR1-2NLS*), YUK808 (*h*[−] P_{nmt41}*-kan-slp1 TetR-tdTom::leu1*[+] *LacI-eGFP::his7*[+] *ChrII-3.6 Mb::TetO-hphMX ChrI-1.95 Mb::LacO-natMX6 lys1 ade6-M210*).

2.2 Microscope

1. 35 mm glass bottom dish.
2. 2 mg/mL (w/v) lectin solution: Dissolve lectin in deionized water, then dispense as small aliquots. Store at −20 °C and avoid repeated freeze/thaw cycles. Dilute 10x for use (0.2 mg/mL (w/v) final). Diluted 0.2 mg/mL (w/v) lectin solution can be stored at 4 °C for a few weeks.
3. Fluorescent beads (*see* **Note 1**).
4. Immersion oil.
5. An inverted fluorescence microscope (*see* **Note 2**).

2.3 Particle tracking Software

1. ImageJ (NIH).
2. CALM package: Activate "CALM" in ImageJ updater before starting the analysis.
3. A standard software for statistic application (e.g., R, Microsoft Office Excel).

3 Methods

3.1 Cell Culture

1. Grow cells in PMG media to 4–6 × 10^6 cells/mL.
2. Add 1/2000 culture volume of thiamine solution to arrest cells in mitosis (*see* **Note 3**).
3. Incubate at 25 °C for 90 min.
4. Add 1/1000 culture volume of 0.5 M IAA for *cut14* shutoff or the same amount of methanol for the control (*see* **Note 4**).
5. Incubate at 25 °C for 90 min.
6. Cells are ready for mounting on the 35 mm glass bottom dish (*see* **Note 5**).

3.2 Dish Preparation

3.2.1 For Monitoring Brownian motion of Fluorescent Beads

1. Suspend 1 μL of fluorescent beads in 100 μL of PMG media (*see* **Note 1**).
2. Mount the suspension on the glass bottom dish (*see* **Note 6**).

3.2.2 For Imaging Cells

1. Mount 100 μL of 0.2 mg/mL lectin solution on the glass bottom dish.
2. Leave it for 5 minutes.
3. Remove the excess amount of the lectin solution by pipetting.
4. Mount 10^6 cells suspended in 100 μL of PMG media on the lectin-coated area of the glass bottom dish (*see* **Notes** 7 and **8**).
5. Allow the cells to sit for 5 min.
6. Remove the unattached cells from the glass bottom dish with a micropipette (*see* **Note 9**).
7. Slowly pour 5 mL of PMG media at the rim of the glass bottom dish with 1 mL micropipette (*see* **Note 10**).

3.3 Fluorescence Microscope Imaging

1. Place the prepared glass bottom dish on an inverted fluorescence microscope.
2. Acquire a time series of single focal plane images (*see* **Notes 1**, **2**, and **11**).
3. Export the image as a single file for particle tracking (*see* **Note 12**).

3.4 Particle Tracking

1. Open the series of images in Fiji ImageJ.
2. Split channels to create a single channel image if necessary.
3. Go Plugin > CALM > Tracking > Virus Tracker.
4. Set parameters for the particle detection and their tracking (*see* **Note 13**, Fig. 1).
5. Press "Run" and select a location to save the results.
6. Individual particles are automatically detected using the detection parameters you set, and the MSDs of each particle are calculated. Example of output files are summarized in Fig. 2.

3.5 MSD Analysis

1. Open MSDPlotData_Particle_xx.csv files (xx should be automatically assigned as a particle identifier) from Subheading 3.4, **step 6** in a standard software for statistical application (e.g., Microsoft Excel).
2. To calculate the average MSD in each condition, copy MSD column of each MSDPlotData_Particle_xx.csv file into a new table.
3. Calculate averages of MSD at each time interval. Standard error of the mean (s.e.m) can be also calculated from the same data.

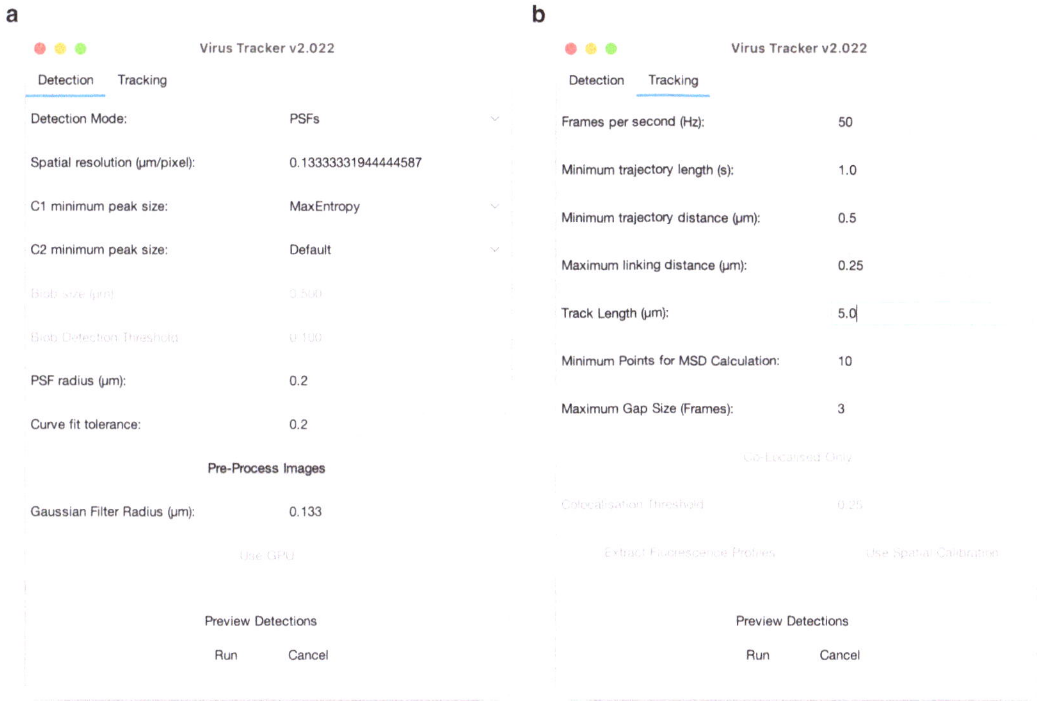

Fig. 1 Parameter settings for CALM Virus Tracker. (**a**) Parameters for detecting fluorescent particles. "Spatial resolution" should match the resolution of the image. (**b**) Parameters for tracking individual particles over time. "Frame per seconds (Hz)" should be calculated from the time interval of the image acquisition

4. Plot the average MSD of the measurements as a function of the time interval (Δt) on a double logarithmic scale (Fig. 3). In the double logarithmic scale, we can estimate the motion of particles from the slope of MDS curve (exponent α). For example, the Brownian motion (random diffusion) gives an exponent $\alpha \sim 1$ and subdiffusion (constrained motion) gives $0 < \alpha < 1$. A lower exponent α represents more constrained motion. In the case of 1 μm beads, the exponent α is nearly 1, indicating that the beads exhibited the Brownian motion. In contrast, fixed cells showed nearly constant MSDs at any given time interval ($\alpha \sim 0$), indicating no chromatin movement. Chromatin in wild type and *cut14* shutoff increased MSDs over time, and their exponents α are 0.28 in wild type and 0.53 in *cut14* shutoff, respectively. These results suggest that chromatin dynamics is confined in mitosis by the function of the SMC complex condensin, as we have reported previously [4, 14].

a

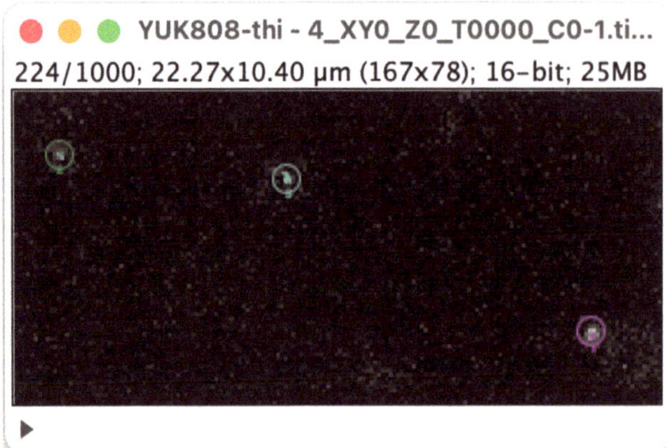

b

Particle	Duration (s)	Displacement (µm)	Velocity (µm/s)	Directionality	Diffusion Coefficient (µm^2/s)
1	9.400	15.683	1.668	0.665	0.000229
2	19.500	40.261	2.065	0.610	0.000371
3	2.520	8.500	3.373	0.563	0.000328
4	1.840	7.400	4.022	0.555	0.000677

c

Particle	Frame	Time (s)	X (µm)	Y (µm)	C1 Intensity	C2 Intensity
1	470.0	9.4	19.887605206441700	8.015899720780540	874.0	NaN
1	469.0	9.38	19.82716429169220	8.117476948979750	858.0	NaN
1	468.0	9.36	19.82348823275490	8.090395554334060	722.0	NaN

d

Time (s)	Mean	SD	N
0.0	0.0	0.0	470.0
0.02	0.0014840304376478700	0.001842536894857870	469.0
0.04	0.0017532143069992800	0.002131448442683680	468.0
0.06	0.002032287900598580	0.002471004314755060	467.0

Fig. 2 Output files obtained by CALM Virus Tracker. (**a**) An example image of particles highlighted by CALM Virus Tracker. The numbers on the images are particle identifiers for the other files. (**b**) Summary of the particle tracking. Number of the particles in the table match to the numbers shown in image a. (**c**) A list of single particle positions for calculating the mean square displacement (MSD). Positions of Particle 1 in a and b are shown. (**d**) A table of MSD (shown as "Mean") and its standard deviation (SD) at a given time interval for Particle 1. Number of data points used for the calculation is shown as N. The table represented in c is used for the calculation

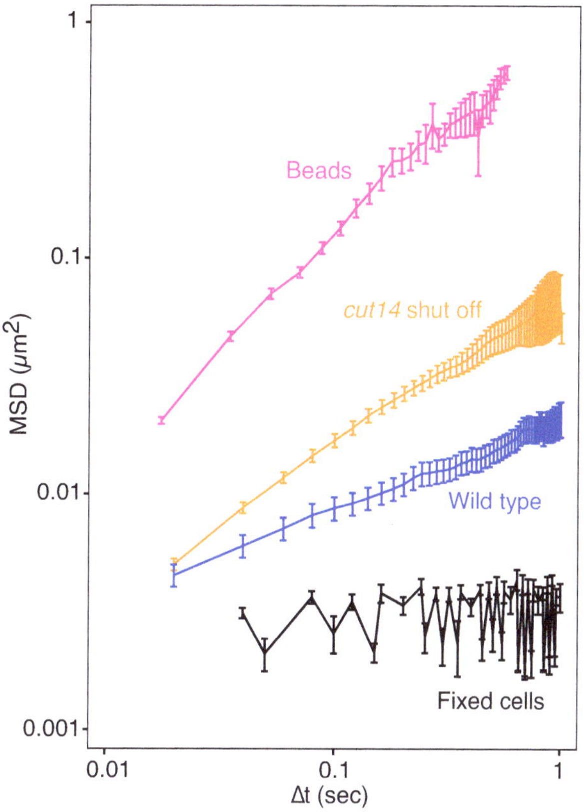

Fig. 3 Average MSDs of chromatin and fluorescent beads. The average MSDs are plotted as a function of the time interval (Δt) in the double logarithmic scales. Mean ± s.e.m. are shown in wild-type mitosis, mitosis in *cut14* shutoff, fixed cells, and 1 μm beads

4 Notes

1. Smaller fluorescent beads move faster. Choose a right size of beads (e.g., FluoSpheres™ Carboxylate-Modified Microspheres (Invitrogen)) for your microscope setting. In our microscope setting, the dynamics of 1 μm beads can be monitored at 57 Hz.
2. Our microscope settings are as follows: An α Plan Apochromat 100×/1.46 NA oil objective (Carl Zeiss) is mounted on an inverted microscope (AxioObserver.Z1, Carl Zeiss) equipped with an Evolve 512 EMCCD camera (Photometrics) and a custom-built spinning-disc confocal system (Intelligent Imaging Innovations) in a temperature-controlled chamber [11]. SlideBook software (Intelligent Imaging Innovations) is used to control the microscope.

3. Fission yeast cells harboring P_{nmt41}-*kan-slp1* halt their cell cycle in mitosis in the presence of thiamine [12]. In our hands, more than 90% of P_{nmt41}-*kan-slp1* strains are arrested in mitosis after 3 h incubation with thiamine at 25 °C.
4. Cut14-AID protein is degraded after IAA addition. For the efficient shut off of *cut14*, endogenous promoter of *cut14* is replaced to thiamine-repressible promoter, P_{nmt81}. The details of *cut14* shutoff in fission yeast are described in ref. [13].
5. If a negative control is required, cells can be fixed at this stage. Follow Subheading 3.2.2 for mounting fixed cells.
6. For monitoring beads, make sure not to mount lectin solution. Lectin adheres fluorescent beads on the surface of the glass bottom dish.
7. Add 5 μg/mL thiamine, 0.5 mM IAA, or methanol if necessary.
8. It is important to adhere cells firmly on the glass surface for accurate measurement of chromatin motion. Any movement of cells would result in the failure of particle tracking. Apply a precise amount of cells to avoid clouding. If too many cells are applied on the glass bottom dish, some of them may not adhere to the glass surface, causing tracking failure.
9. Do not touch the glass surface as this strips the cells. Slightly tilt the glass bottom dish and use 100 μL micropipette to aspirate the media.
10. Do not pour media directly onto the cells. Water flow may strip out the attached cells, as some cells may not adhere firmly to the surface of the glass bottom dish.
11. To monitor chromatin dynamics in fission yeast cells, single focal plane images were acquired up to 1000 times with 20 ms exposure in a streaming mode. To accurately calculate MSD and diffusion coefficient, at least 50 series of time lapse images should be acquired. In the automatic particle detection step (Subheading 3.4), some of the particles would not be traced through the entire time series and would be split into multiple trajectories. Therefore, you should acquire images for as long as possible at this stage.
12. Multilayer tiff or the other appropriate format can be used.
13. The analysis parameters for "spatial resolution" and "frame per sec" should be changed according to your imaging conditions. "Minimal trajectory length" can be calculated from "Frame per sec."

Acknowledgments

I would like to thank M. Way for the fluorescent microscope and D.J. Barry for the particle tracking. Y.K. is supported by the Japan Society for the Promotion of Science (JSPS) for Grant-in-Aid for Scientific Research (C) 22 K06092, Waseda University grants for Special Research Projects 2024C-285, and the Institute for Fermentation, Osaka (IFO).

References

1. Uhlmann F (2016) SMC complexes: from DNA to chromosomes. Nat Rev Mol Cell Biol 17(7):399–412
2. Hirano T (2016) Condensin-based chromosome organization from bacteria to vertebrates. Cell 164(5):847–857
3. Gibcus JH, Saejima K, Goloborodko A et al (2018) A pathway for mitotic chromosome formation. Science 359(6376):eaao6135
4. Kakui Y, Rabinowitz A, Barry DJ et al (2017) Condensin-mediated remodeling of the mitotic chromatin landscape in fission yeast. Nat Genet 49(10):1553–1557
5. Rao SS, Huntley MH, Durand NC et al (2014) A 3D map of the human genome at kilobase resolution reveals principles of chromatin looping. Cell 159(7):1665–1680
6. Uhlmann F (2025) A unified model for cohesin function in sister chromatid cohesion and chromatin loop formation. Mol Cell 85(6): 1058–1071
7. Nozaki T, Imai R, Tanbo M et al (2017) Dynamic organization of chromatin domains revealed by super-resolution live-cell imaging. Mol Cell 67(2):282–293.e7
8. Miné-Hattab J, Rothstein R (2012) Increased chromosome mobility facilitates homology search during recombination. Nat Cell Biol 14(5):510–517
9. Amitai A, Seeber A, Gasser SM et al (2017) Visualization of chromatin decompaction and break site extrusion as predicted by statistical polymer modeling of single-locus trajectories. Cell Rep 18(5):1200–1214
10. Kakui Y, Barrington C, Barry DJ et al (2020) Fission yeast condensin contributes to interphase chromatin organization and prevents transcription-coupled DNA damage. Genome Biol 21(2):272
11. Abella JV, Pernier J, Barry DJ et al (2016) Isoform diversity in the Arp2/3 complex determines actin filament dynamics. Nat Cell Biol 18(1):76–86
12. Petrova B, Dehler S, Kruitwagen T et al (2013) Quantitative analysis of chromosome condensation in fission yeast. Mol Cell Biol 33(5): 984–998
13. Kakui Y, Uhlmann F (2019) Efficient depletion of fission yeast condensin by combined transcriptional repression and auxin-induced degradation. Methods Mol Biol 2004:25–33
14. Gerguri T, Fu X, Kakui Y et al (2021) Comparison of loop extrusion and diffusion capture as mitotic chromosome formation pathways in fission yeast. Nucleic Acids Res 49(3): 1294–1312

Part II

DNA and Protein Topology Assays of SMC Proteins

Chapter 6

DNA Bioconjugation with Polymer Beads Using SNAP-tag® Technology

Hon Wing Liu and Stephan Gruber

Abstract

Polymer beads (such as Dynabeads) are commonly used to capture and isolate DNA and proteins, relevant for many applications in biochemistry and nanobiology. Current approaches to link DNA with beads involve non-covalent attachments such as biotin-streptavidin which pose a risk of spontaneous dissociation, potentially influencing data interpretation. We present a protocol to circumvent this risk by exploiting SNAP-tag technology to establish stable covalent anchoring (bioconjugate) between DNA and Dynabeads. We describe steps for generating benzylguanylated (BG) closed covalent circular DNA (BG-cccDNA), conjugating DNA with SNAP-tag and with beads, and enzymatic treatment and elution of conjugated DNA. We have successfully employed this technique to study the effect of Dynabeads as micrometer-scale DNA roadblocks on the motor activity of bacterial SMC Wadjet complex.

Key words SMC complexes, DNA roadblocks, Wadjet, JetABCD, SNAP-tag, Dynabeads

1 Introduction

Micrometer-scale synthetic polymer beads have been utilized in research for decades. Their widespread use is attributed to their ease of manipulation and capacity to be coated by a vast catalogue of reactive moieties and biomolecules. For the latter, using beads to bind and isolate nucleic acids has become an instrumental technology in biology. Magnetic beads are routinely used to size-selectively purify minute quantities of DNA and RNA for many downstream applications [1], such as library preparation for next-generation sequencing. This technology also facilitates characterization of multistep biological reactions; for example, DNA replication has been reconstituted on immobilized circular DNA templates with purified replication fork components [2]. Linear DNA immobilized onto beads are also utilized as a standard substrate for single-molecule imaging platforms such as optical tweezers.

Yasuto Murayama and Yasutaka Kakui (eds.), *SMC Proteins: Methods and Protocols*, Methods in Molecular Biology, vol. 2991, https://doi.org/10.1007/978-1-0716-5005-9_6,

SMC (structural maintenance of chromosome) complexes are ring-shaped ATPase motors that can extrude large DNA loops. Canonical members, such as eukaryotic cohesin/condensin and bacterial MukBEF/Smc-ScpAB, use loop extrusion to organize DNA into chromosomes that can be faithfully replicated and segregated to daughter cells [3–5]. Other members (bacterial Wadjet and eukaryotic Smc5/6) perform functions in DNA immunity and DNA repair [6–8]. In the SMC field, bead-coupled DNA has been used to understand sister chromatid cohesion establishment by cohesin complexes [9], as well as in single-molecule studies to understand force generation on DNA by SMC complexes [10, 11]. A still-contended question of the field is the capacity of SMC motors to navigate physiological obstacles or "roadblocks" on DNA (such as nucleosomes, replication forks, and transcription bubbles). To this end, beads (and nanoparticles such as colloidal gold) have been exploited as large (ranging from hundreds of nanometers to micrometers) in vitro roadblocks [12].

We and others have recently found that the bacterial SMC complex Wadjet (JetABCD) restricts smaller (<~85 kb), circular plasmids. This is achieved through random DNA cleavage by nuclease subunit JetD, triggered by DNA bending after extrusion of the substrate by JetABC motors [12–15]. We also assessed Wadjet (a type I variant from *E. coli* strain GF4-3) activity upon encountering large roadblocks by using superparamagnetic polymer beads (Dynabeads) as micrometer-scale obstacles—if they hinder loop extrusion by the JetABC motors, cleavage activity and/or site distribution could be affected. We first generated closed covalent biotinylated DNA circles through a plasmid gap filling protocol [16] and attached them onto streptavidin-coated Dynabeads. Although DNA was bound to the beads, we found that it was susceptible to spontaneous dissociation ("DNA leaking") despite the extremely high affinity of biotin-streptavidin linkage ($K_d = 10^{-15}$ M) [17]. DNA leaking was dependent on biotin position and DNA linearity/circularity and was most substantial with circular DNA containing internally positioned biotin, which could influence data interpretation [17].

SNAP is a modified O^6-alkylguanine-DNA-alkyltransferase DNA repair enzyme, which can be conjugated (covalently linked) with ligands containing an O^6-benzylguanine (BG) moiety [18]. Tagging proteins of interest with SNAP allows labelling of BG-containing ligands to them; for example, fluorescent ligands are widely used to track protein behavior in vivo [19]. BG can also be attached on DNA oligonucleotides, and using the aforementioned gap filling procedure, benzylguanylated closed covalent circular DNA (BG-cccDNA) containing one or two BG moieties can be generated, and hence covalent attachment of up to two SNAP-tags. Lysine residues on SNAP-tag can then be reacted with Dynabeads functionalized with epoxy groups, thus establishing (up to two) uninterrupted covalent chains between Dynabeads and defined positions of DNA. After the desired enzymatic treatment (s), conjugated DNA can be eluted from beads after SNAP-tag digestion with Proteinase K (Fig. 1a). We employed this methodology to demonstrate that large roadblocks covalently attached on circular DNA (at least this way) stalls GF4-3 JetABC motor translocation, triggering DNA cleavage exclusively at the Dynabead-anchoring point, allowing us to deduce that Wadjet motors have to converge after loop extrusion for cleavage to occur [17].

Site-specific protein-DNA cross-linking methods have been established, for example, the Flp-nick system [20], but the method presented here has the advantages of studying intact cccDNA (instead of nicked DNA), as well as the flexibility and robustness of SNAP as a protein tag. Indeed, the conjugation protocol can be used to covalently attach any protein of interest (fluorescent proteins, etc.) fused with SNAP onto DNA. The protocol focuses on manipulating and conjugating circular BG-DNA to Dynabeads but is also applicable to conjugating linear BG-DNA derived from PCR (*see* **Note 1**), and in principle any forms of DNA containing BG. The pipeline can be further adapted to study the effect(s) of roadblocks on other DNA-interacting enzymes. It may also be useful for any experimentation requiring stable DNA immobilization onto beads or surface over prolonged periods of time.

2 Materials

2.1 Generation of Benzylguanylated Closed Covalent Circular DNA (BG-cccDNA)

1. Benzylguanylated DNA replacement oligonucleotides (*see* **Note 2**). Oligonucleotides are dissolved in water as a 100 μM stock prior to use.
   BGSiteA:
   5′-[Phos] TCAGCTACACCTCAGCGATCCC [BGdT] CAG CACTAGTCCTCAGCTACGCC -3′ BGsiteB (optional, if two modifications are desired):
   5′-[Phos]TCAGCTCGGCCTCAGCCCCTCC[BGdT]C AGCCTACTTCCTCAGCGAATCC-3′
   [BGdT] denotes O^6-benzylguanine internally attached to a dT nucleobase. [Phos] denotes the phosphate group attached to the 5′ end required for nick ligation to generate closed circles.
2. Starting plasmids pSG6970/pG46 [21] (addgene: 232620) and pSG7084/pG46-46 (addgene: 232621) containing one or two replacement cassettes for chemical modifications, respectively.
3. Nicking enzyme Nt.BbvCI (New England Biolabs).
4. Sterile water.

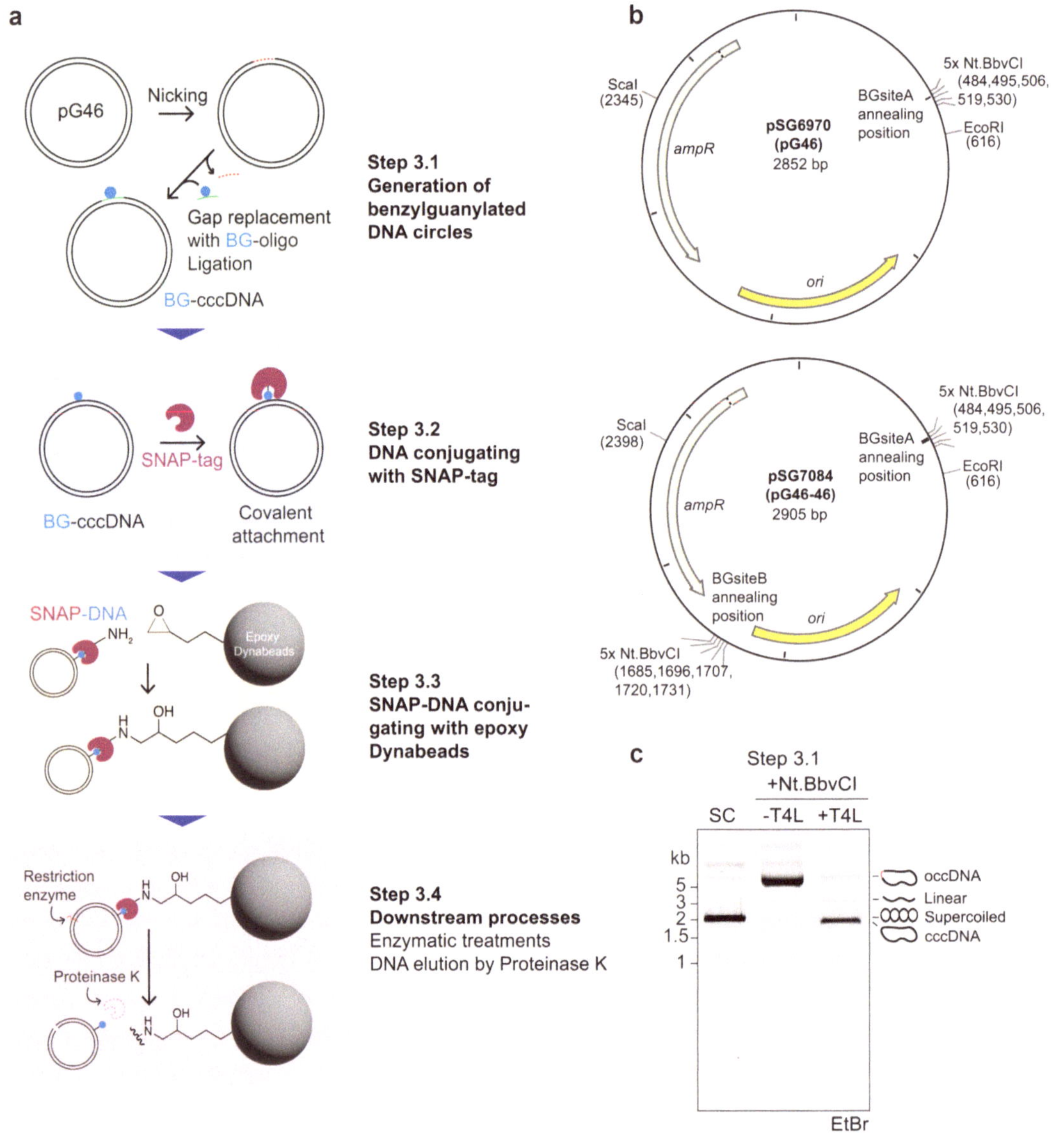

Fig. 1 (**a**) Schematic of the pipeline in four steps as ordered in the methods: First (3.1), pSG6970/pG46 (addgene: 232620) (or pSG7084/pG46-46, addgene: 232621) is nicked at closely spaced Nt.BbvCI sites, and fragments are melted away creating a 46 nt long ssDNA gap, which is replaced by annealing an excess of benzylguanine (BG)-containing oligo (BGsiteA/BGsiteB). After subsequent ligation, BG-cccDNA is generated. The second (3.2) involves covalent attachment of SNAP-tag onto BG-cccDNA. The third (3.3) involves conjugating the DNA-SNAP onto epoxy-coated Dynabeads. Conjugated DNA can then be treated with various enzymes and eluted with Proteinase K in the final step 3.4. (**b**) Maps of starting plasmids pSG6970/pG46 and pSG7084/pG46–46, containing sites for annealing oligos BGsiteA and additional BGsiteB, respectively (to generate either one or two BG modifications). Positions of single cutter enzymes used in this protocol are also highlighted. The depictions are modified from maps displayed on SnapGene (Version 8.0.1) software. (**c**) Agarose gel depicting the expected migration profiles of the intermediates and products of BG-cccDNA generation (step 3.1). After nicking supercoiled pSG6970/pG46 (SC) by Nt.BbvCI and annealing by BGsiteA oligo, the DNA becomes relaxed (as it contains nicks) and thus migrates slower as open covalent circles (occDNA). Nick sealing by T4 ligase (T4L) converts this to closed covalent circles (cccDNA) which migrates slightly faster than supercoiled. The ligated reaction product will always contain a minor fraction of occDNA owing to strand breaks caused by incomplete annealing and/or ligation

5. 10× rCutsmart buffer (New England Biolabs).
6. 200 mM Tris-HCl pH 8.
7. 400 mM EDTA pH 8.
8. 10× T4 ligase buffer (Thermo Fisher Scientific).
9. T4 ligase (5 Weiss U/μL) (Thermo Fisher Scientific).
10. PCR cycler (with temperature ramping function).
11. Thermomixer.
12. Spin columns for DNA purification.
13. Benchtop centrifuge.
14. 1× phosphate-buffered saline (PBS), pH 7.4.
15. DNA loading dye.
16. 1% (w/v) agarose gel for nucleic acid electrophoresis.
17. Ethidium bromide.
18. TBE buffer: 89 mM Tris base, 89 mM borate, 2 mM EDTA, pH 8.2–8.4.
19. Power supply and DNA electrophoresis chamber.

2.2 Conjugating BG-cccDNA with SNAP-tag

1. SNAP-tagged protein or SNAP-tag (*see* **Note 3**).
2. Benchtop centrifuge.
3. BG-cccDNA (from previous step).
4. 1× phosphate-buffered saline (PBS), pH 7.4.
5. 1 M dithiothreitol (DTT) stock.
6. Thermomixer.

2.3 Covalent Attachment of DNA-SNAP Conjugate to Epoxy-Coated Dynabeads

1. Conjugated SNAP-DNA (from previous step).
2. Sterile water.
3. Epoxy A buffer: 19 mM NaH_2PO_4, 81 mM Na$_2$HPO4 dissolved in water.
4. Epoxy B buffer: 3 M $(NH_4)_2SO_4$ dissolved in Epoxy A buffer.
5. Dynabeads™ M-270 Epoxy (Thermo Fisher Scientific).
6. A tube inverter/roller set up in a 37 °C incubator (Fig. 2a).
7. Magnetic rack for 1.5 mL Eppendorf tubes.
8. PicoFuge.
9. PBS-BSA buffer: 1× phosphate-buffered saline (PBS), pH 7.4 supplemented with 0.1% (w/v) bovine serum albumin (BSA).

2.4 Enzymatic Treatment and Elution of Conjugated DNA

1. Suitable buffer for enzyme digest (e.g., rCutsmart, NEB buffers).
2. Wadjet buffer: 10 mM HEPES-KOH pH 7.5, 150 mM KOAc, 5 mM $MgCl_2$ in sterile water.

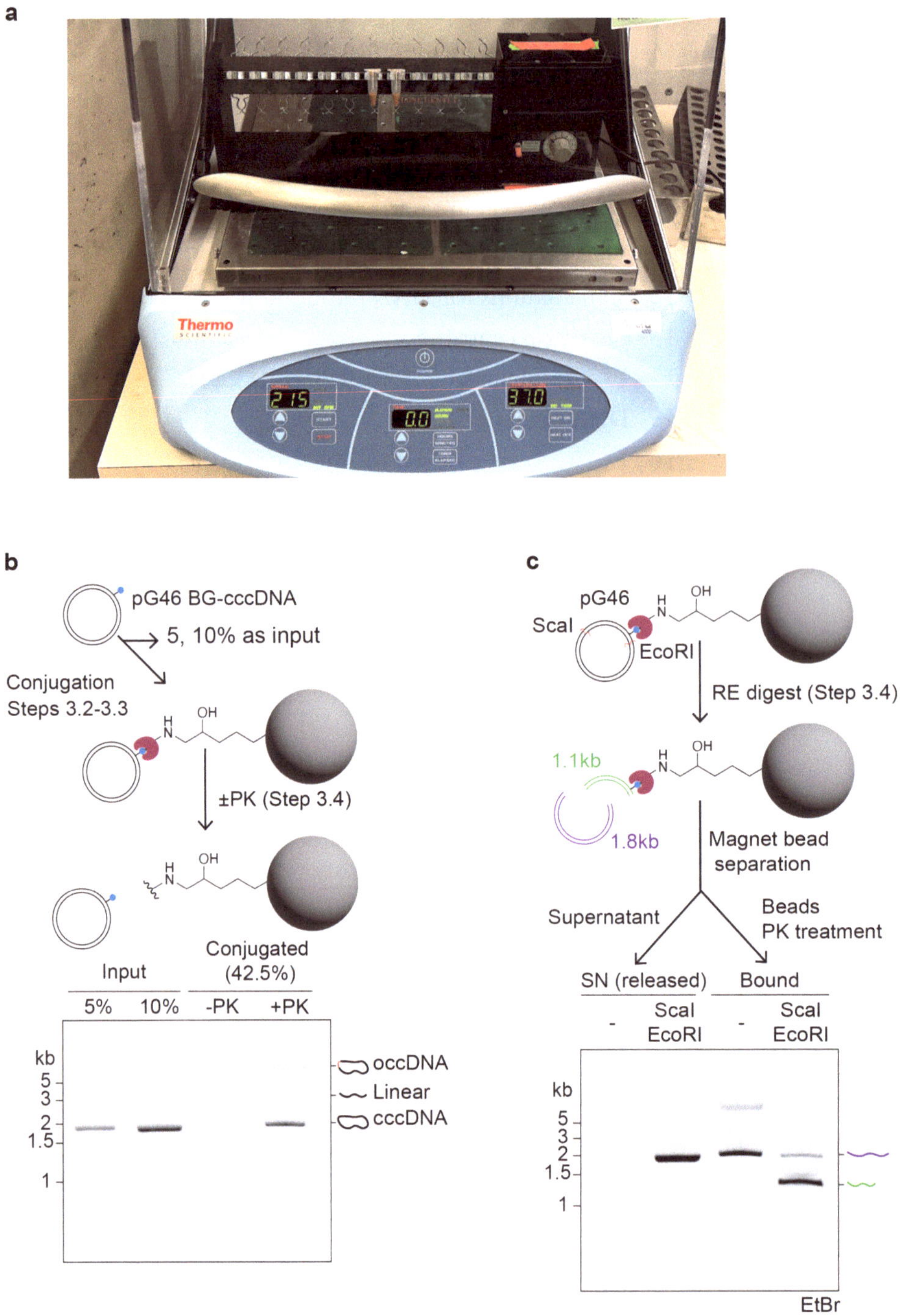

Fig. 2 (**a**) Picture of the conjugating setup used. A tube roller is placed in a suitably sized incubator which is set to 37 °C. Two tubes containing Dynabeads are shown mounted on the roller. Note that the incubator shaking function should not be turned on. Other alternative setups which allow gentle, continuous mixing of beads at 37 °C are possible. (**b**) (Top) Experiment scheme to determine covalent DNA conjugating and its efficiency. After generating 1 μg BG-cccDNA, 5% and 10% of it was taken as input fraction. After taking the rest through the conjugating pipeline, half of the conjugated DNA beads (each constituting 42.5% of the total DNA used) was treated with Proteinase K (+PK) and half without (−PK). The eluate separated from the beads was taken for agarose gel analyses (bottom) as "conjugated" fraction. Eluted DNA circles typically run as two bands. The

3. Suitable single cutter or double cutter restriction enzymes.
4. Magnetic rack for 1.5 mL Eppendorf tubes.
5. 20 mg/mL Proteinase K.
6. Thermomixer.
7. Spin columns for DNA purification (optional).
8. Benchtop centrifuge.
9. DNA loading dye.
10. 1% (w/v) agarose gel for nucleic acid electrophoresis.
11. Ethidium bromide.
12. TBE buffer: 89 mM Tris base, 89 mM borate, 2 mM EDTA, pH 8.2–8.4.
13. Power supply and DNA electrophoresis chamber.

3 Method

3.1 Generation of Benzylguanylated Closed Covalent Circular DNA (BG-cccDNA)

Timing: 4.5 h

The procedure to generate modified DNA circles is adapted from ref. [16], employing starting plasmids (pG46 [21], addgene: 232620) and pSG7084 (pG46–46, addgene: 232621), respectively, containing one or two regions containing closely spaced Nt.BbvCI nicking enzyme sites (Fig. 1b), both 2.9 kb in size.

After nicking by Nt.BbvCI and melting away of resultant short ssDNA fragments by heat, 46 nt ssDNA gaps of defined sequence are generated that can be filled by annealing "replacement oligos" containing the desired modification(s). After replacement, the DNA nicks at the oligo-plasmid junctions are sealed by DNA ligase to form cccDNA (Fig. 1a).

1. Nick the starting plasmid by mixing 15 μL pSG6970/pG46 stock (~100–250 nM final; *see* **Note 4**) with 2 μL 10x rCutsmart buffer and 3 μL of Nt.BbvCI (30 U final). Incubate with gentle shaking on a thermomixer for 1 h at 37 °C. If two modifications are desired, nick pSG7084/pG46–46 instead.

Fig. 2 (continued) faster-migrating band corresponds to cccDNA and a minor slower one occDNA. Comparing the band intensities of the +PK sample with inputs results in an estimate of ~15% conjugating efficiency for this experiment. Without PK, minimal DNA was recovered, indicating that DNA leakage (dissociation) from beads is minimal in this setup. (**c**) (Top) Schematic of experiment to enzymatically treat conjugated DNA circles (step 3.4). DNA circles covalently conjugated to Dynabeads were either treated with or without single cutters EcoRI and ScaI. Treatment with these enzymes should release a 1.8 kb fragment (in purple) to the supernatant, which was isolated from the beads and taken as "SN" fraction (left arrow). The remaining 1.1 kb fragment (in green) should stay bound to the beads and can be eluted by Proteinase K (PK) treatment, which was taken as "bound" fraction after separating from beads (right arrow). Both SN and bound fractions were analyzed with agarose gel electrophoresis (bottom)

2. Quench the reaction by adding 3 μL 200 mM Tris-HCl pH 8.0 (~25 mM final) and 3 μL 400 mM EDTA pH 8.0 (~50 mM final). Transfer reaction into a 200 μL PCR tube.
3. Add 1 μL replacement BG oligo BGsiteA (~3.5 μM final). If two BG modifications are desired, add 1 μL BGsiteA (~3.5 μM final) and 1 μL BGsiteB (~3.5 μM final) into the reaction containing nicked pSG7084/pG46-46.
4. Melt away nicked fragments and anneal replacement oligo by placing the tubes in a thermocycler and performing the following program:

Steps	Temperature	Time	Cycles
Initial denaturation	80 °C	2 min	1
Annealing	ΔT–1 °C	1 min	60
Hold	20 °C	Forever	

5. Column-purify the annealed product. Elute in 52 μL of water. Place on ice.
6. Ligate the modified DNA by adding 6 μL 10× T4 ligase buffer and 2 μL T4 ligase (10 Weiss U final). Incubate for 1 h at 22 °C on a thermomixer with gentle shaking.
7. Check whether ligation was successful by agarose gel electrophoresis. Mix 2 μL of the reaction with 8 μL water and 2 μL 6× DNA loading dye. Perform agarose gel electrophoresis. We typically use 1% (w/v) agarose with 1 μg/mL ethidium bromide. Determine presence of cccDNA band (Fig. 1c; *see* **Note 5**). We note that cccDNA formation is generally slightly less efficient when double gap annealing with pSG7084/pG46-46 is performed.
8. Upon successful ligation, column-purify the DNA and elute in 25 μL 1× PBS. Expected yield (by fluorometric quantification; *see* **Note 6**) ~1 μg total (~40–50 ng/uL, ~25 nM).

We recommend proceeding to the conjugating stages without any interruptions as cccDNA is prone to damage and thus should preferably be freshly prepared for conjugating. If this is unavoidable, flash freeze the DNA in liquid nitrogen and store in a −80 °C freezer.

3.2 Conjugating BG-cccDNA with SNAP-tag

Timing: 2.5 h

Here, BG-cccDNA is conjugated with a molar excess of SNAP-tag protein (or a protein of interest fused with SNAP), as SNAP-tag conjugating efficiency is much lower with DNA than with smaller ligands. We recommend a starting amount of at least 0.5 μg BG-cccDNA, typically 1 μg (~20 nM in a 30 μL reaction), to ensure sufficient recovered material. With a starting material of 1 μg BG-cccDNA, we anticipate 150–250 ng (a conjugating efficiency of 15–25%) DNA can be recovered after the conjugating steps (Fig. 2b; *see* **Note 7**). It is important to plan the number of experimental conditions desired and the amount of conjugated DNA to be used per condition, and calculate the amount of input DNA—the pipeline (including BG-cccDNA preparation) can be scaled up accordingly.

1. Dilute SNAP-tag stock (if homemade, *see* **Note 3**) in PBS to a final concentration of 50 μM. Make 20 μL aliquots. SNAP-tag should not be dissolved or diluted in buffers containing amine components (such as Tris-HCl).
2. Remove protein aggregates by centrifuging SNAP-tag at >16,000 *g* at 4 °C for 10 min with a benchtop centrifuge. Transfer supernatant to a fresh tube.
3. Mix the following in this order:

Reagent	Amount	Final concentration
BG-cccDNA	0.5–1 μg	~20 nM
1× PBS	Up to 26.7 μL	
100 mM DTT diluted in 1× PBS	0.3 μL	1 mM
50 μM SNAP-tag	3 μL	5 μM

4. Incubate for 2 h at 37 °C. Stop at this step if subsequent bead conjugation is not required for your experiment.
5. Add 70 μL PBS, making a final volume of 100 μL.

3.3 Attachment of DNA-SNAP Conjugate to Epoxy-Coated Dynabeads

Timing: 17–20 h

Here, the SNAP-DNA conjugate is conjugated with equilibrated epoxy Dynabeads. These steps are adapted from antibody–Dynabeads coupling guidelines given by Thermo Fisher Scientific. Note that excess, free SNAP from the preceding step will be present at reaction start.

1. Place a rolling tube holder in an incubator (see Fig. 2a for example setup), and preheat the incubator to 37 °C. This can be done during the final 30 min of SNAP-tag conjugating.
2. (Per conjugating reaction) weigh 4 mg M-270 Epoxy Dynabeads in a measuring cup. Do not downscale this step as we found that conjugating efficiency declines with lower quantities of beads. Resuspend beads with 1 mL Epoxy buffer A, and transfer to a 1.5 mL Eppendorf tube.
3. Place tube on a magnetic rack and wait until the beads are separated from the buffer. Then remove buffer carefully with a pipette.

4. Resuspend beads in 100 μL Epoxy buffer A, and vortex thoroughly for 30 s.
5. Repeat **steps 3** and **4** twice.
6. Place in the pre-warmed rolling tube holder and incubate at 37 °C for 10 min with gentle rolling.
7. Add the 100 μL SNAP-tag conjugated DNA to the beads, and flick tube to ensure proper mixing. From this point onward, avoid vortexing the beads as this could cause DNA shearing.
8. Add 100 μL of Epoxy buffer B and flick tube to ensure proper mixing.
9. Place the tube containing the conjugating reaction into the rolling tube holder and incubate overnight with gentle rolling (30 rpm). Do not shorten this step as we found that conjugating SNAP-DNA to beads is a slow reaction.
10. Spin tube briefly (1 s) with a PicoFuge to ensure no suspension is trapped on the tube lid. Place tube on a magnetic rack and wait until the beads are separated from the supernatant.
11. Remove supernatant.
12. Wash beads: Resuspend beads with 1 mL PBS-BSA buffer. Pipette gently up and down three times until all beads are resuspended. Do not vortex. BSA presence is essential here to quench unreacted epoxy groups on the Dynabeads.
13. Place tube on a magnetic rack and wait until the beads are separated from the buffer. Then remove buffer with a pipette ensuring minimal disturbance to the beads.
14. Take tube out of rack and repeat **steps 12** and **13** until a total of three washes.
15. Final wash: Resuspend beads in 1 mL desired reaction buffer (e.g., 1× rCutsmart). Pipette gently up and down three times until all beads are resuspended. The beads conjugated with DNA are now ready for enzymatic treatment.

3.4 Enzymatic Treatment and Elution of Conjugated DNA

Timing: 1–2 h
In this section, we will describe use of restriction enzymes on bead-conjugated DNA, as well as subsequent elution of DNA via SNAP digestion with Proteinase K. In principle, the protocol can be adapted to other DNA-interacting enzymes of interest.

1. Depending on the number of reactions desired and the amount of DNA per reaction (given the conjugating efficiency is ~15–25% of input), distribute the beads accordingly to 1.5 mL Eppendorf tubes (*see* **Note 7**).
2. Place tube on a magnetic rack and wait until the beads are separated from the buffer. Then remove buffer carefully with a pipette, ensuring minimal disturbance to the beads.

3. Take tube out of rack and resuspend in 15 μL 1× rCutsmart buffer. Other buffers and reaction volumes can be used; however, 15 μL was chosen in consideration of the volume of sample an agarose DNA gel well can hold.
4. Add 0.5 μL restriction enzyme(s).
5. Incubate for 15 min at 37 °C with gentle shaking. Flick the tube every few minutes to ensure the beads are well resuspended.
6. After the reaction, add 1 μL Proteinase K (20 mg/mL stock) to the beads per 15 μL reaction. Incubate for 30 min at 55 °C with gentle shaking. Flick the tube every few minutes to ensure the beads are well suspended. The DNA should now be released from the beads as SNAP-tag is digested by Proteinase K (*see* **Note 8**).
7. Place tube on a magnetic rack and wait until the beads are separated from the buffer. Then transfer 15 μL of the supernatant containing eluted DNA to a 1.5 mL Eppendorf tube. The amount of DNA in the eluate can be quantified by fluorometric quantification (*see* **Note 6**). This can be helpful to determine the amount of sample to be used for visualization on an agarose gel.
8. Add 3 μL of 6× DNA loading dye and perform agarose gel electrophoresis (Fig. 2b; *see* **Note 9** if reaction buffer contains high salt concentrations).
9. If DNA is released during the enzymatic treatment (e.g., dual restriction enzyme treatment or a dual cutter), follow the protocol until **step 5**.
10. Isolate the beads by magnet and transfer the supernatant "SN" containing the released DNA to a separate 1.5 mL Eppendorf tube.
11. Wash once with 50 μL 1× rCutsmart buffer.
12. Resuspend beads with 15 μL 1× rCutsmart buffer.
13. Proceed with **steps 6–8**.
14. Perform agarose gel electrophoresis on both SN (released) and eluted (bound) DNA. An example of this is shown in Fig. 2c.

4 Notes

1. The protocol works for linear BG-DNA generated from PCR (we used KOD-one polymerase). Since their generation requires much less BG oligonucleotide, we recommend familiarizing yourself with this protocol using short PCR products (amplifying a 0.5–1 kb product using BGsiteA oligo with a

suitable reverse primer, with pSG6970/pG46 as template). Proceed to step 3.2 once the PCR product is obtained.

2. The BGsiteA/BGsiteB oligos were obtained from biomers.net GmbH.
3. Purified SNAP-tag protein can be obtained commercially as a 50 μM stock (New England Biolabs) or, as in this protocol, purified from expression plasmid pSG7163 (addgene: 232622). Please refer to ref. [17] for the purification method.
4. The starting plasmids should be stocked by maxiprep or miniprep and eluted in water. We find that the final yield of BG-cccDNA is dependent on the amount of starting plasmid material. Ideally, the starting plasmid (2.9 kb) should be at a concentration of ~300–500 ng/μL (~150–250 nM).
5. Some DNA nicking (strand breaks) is inevitable during the long conjugating steps, visualized as the open covalent circle (occDNA) band (Fig. 2b,c). It is important to manipulate the DNA as gently as possible as any excessive mechanical action would cause further damage. We note that occDNA tends to be enriched when the DNA is subject to strand-nicking by freeze-thaw cycles and/or rigorous agitation (such as vortexing) and thus can be mitigated by careful manipulation.
6. If fluorometric quantification (e.g., Qubit, strongly recommended) is not available, use Nanodrop spectrophotometer.
7. After conjugating, the amount of anchored DNA cannot easily be precisely determined (only estimated) without elution and may vary between experiments due to the two-step nature of the conjugating process; therefore, the protocol is less suitable for sensitive assays (e.g., enzyme kinetics). We anticipate that after successful SNAP-tag and epoxy bead conjugating, 15–25% of the input DNA can be recovered after elution (Fig. 2b). This protocol may not be applicable for experiments requiring larger amounts of conjugated DNA (>500 ng). Further, conjugating DNA circles harboring two BG modifications will result in some DNA anchored to beads only on one site (see ref. [17]).
8. Proteinase K is robust in a wide range of buffers [22]. We found that Proteinase K is active in rCutsmart buffer and Wadjet reaction buffer (10 mM HEPES-KOH pH 7.5, 150 mM KOAc, 5 mM $MgCl_2$). Further optimization may be required if Proteinase K is not active in the designated reaction buffer.
9. Reaction buffers that contain high concentrations of salt may hamper band migration and visualization during agarose gel electrophoresis. If so, we recommend column-purifying the eluate after bead separation at this step. After eluting DNA in 15 μL water, proceed with **step 8**. Note that this will result in DNA loss.

Acknowledgments

This work was supported by EMBO Postdoctoral Fellowship (ALTF-490-2021) to H.W.L. and an SNSF project grant (310030_197770) to S.G.

References

1. Berensmeier S (2006) Magnetic particles for the separation and purification of nucleic acids. Appl Microbiol Biotechnol 73:495–504. https://doi.org/10.1007/s00253-006-0675-0
2. Yeeles JTP, Deegan TD, Janska A et al (2015) Regulated eukaryotic DNA replication origin firing with purified proteins. Nature 519:431–435. https://doi.org/10.1038/nature14285
3. Davidson IF, Peters J-M (2021) Genome folding through loop extrusion by SMC complexes. Nat Rev Mol Cell Biol 22:445–464. https://doi.org/10.1038/s41580-021-00349-7
4. Yatskevich S, Rhodes J, Nasmyth K (2019) Organization of chromosomal DNA by SMC complexes. Annu Rev Genet 53:445–482. https://doi.org/10.1146/annurev-genet-112618-043633
5. Uhlmann F (2016) SMC complexes: from DNA to chromosomes. Nat Rev Mol Cell Biol 17:399–412. https://doi.org/10.1038/nrm.2016.30
6. Aragón L (2018) The Smc5/6 complex: new and old functions of the enigmatic long-distance relative. Annu Rev Genet 52:89–107. https://doi.org/10.1146/annurev-genet-120417-031353
7. Liu HW, Roisné-Hamelin F, Gruber S (2023) SMC-based immunity against extrachromosomal DNA elements. Biochem Soc Trans 51:1571–1583. https://doi.org/10.1042/BST20221395
8. Doron S, Melamed S, Ofir G et al (2018) Systematic discovery of antiphage defense systems in the microbial pangenome. Science 359. https://doi.org/10.1126/science.aar4120
9. Minamino M, Bouchoux C, Canal B et al (2023) A replication fork determinant for the establishment of sister chromatid cohesion. Cell 186:837–849.e11. https://doi.org/10.1016/j.cell.2022.12.044
10. Tanasie NL, Gutiérrez-Escribano P, Jaklin S et al (2022) Stabilization of DNA fork junctions by Smc5/6 complexes revealed by single-molecule imaging. Cell Rep 41. https://doi.org/10.1016/j.celrep.2022.111778
11. Pobegalov G, Chu LY, Peters JM et al (2023) Single cohesin molecules generate force by two distinct mechanisms. Nat Commun 14:3946. https://doi.org/10.1038/s41467-023-39696-8
12. Pradhan B, Deep A, König J et al (2024) Loop-extrusion-mediated plasmid DNA cleavage by the bacterial SMC Wadjet complex. Mol Cell 85:107–116. https://doi.org/10.1016/j.molcel.2024.11.002
13. Liu HW, Roisné-Hamelin F, Beckert B et al (2022) DNA-measuring Wadjet SMC ATPases restrict smaller circular plasmids by DNA cleavage. Mol Cell 82:4727–4740.e6. https://doi.org/10.1016/j.molcel.2022.11.015
14. Deep A, Gu Y, Gao YQ et al (2022) The SMC-family Wadjet complex protects bacteria from plasmid transformation by recognition and cleavage of closed-circular DNA. Mol Cell 82:4145–4159.e7. https://doi.org/10.1016/j.molcel.2022.09.008
15. Weiß M, Giacomelli G, Assaya MB et al (2023) The MksG nuclease is the executing part of the bacterial plasmid defense system MksBEFG. Nucleic Acids Res 51:3288–3306. https://doi.org/10.1093/nar/gkad130
16. Luzzietti N, Knappe S, Richter I et al (2012) Nicking enzyme–based internal labeling of DNA at multiple loci. Nat Protoc 7:643–653. https://doi.org/10.1038/nprot.2012.008
17. Roisné-Hamelin F, Liu HW, Taschner M et al (2024) Structural basis for plasmid restriction by SMC JET nuclease. Mol Cell 84:883–896.e7. https://doi.org/10.1016/j.molcel.2024.01.009
18. Juillerat A, Gronemeyer T, Keppler A et al (2003) Directed evolution of O6-Alkylguanine-DNA Alkyltransferase for efficient labeling of fusion proteins with small molecules in vivo. Chem Biol 10:313–317. https://doi.org/10.1016/S1074-5521(03)00068-1
19. Lukinavičius G, Reymonz L, Johnsson K (2015) Fluorescent labeling of SNAP-tagged proteins in cells. In: Gautier A, Hinner M (eds) Site-specific protein labeling. Methods in Molecular Biology, vol 1266. https://doi.org/10.1007/978-1-4939-2272-7_7

20. Nielsen I, Bentsen IB, Lisby M et al (2009) A Flp-nick system to study repair of a single protein-bound nick in vivo. Nat Methods 6: 753–757. https://doi.org/10.1038/nmeth.1372
21. Ralf C, Hickson ID, Wu L (2006) The bloom's syndrome helicase can promote the regression of a model replication fork. J Biol Chem 281: 22839–22846. https://doi.org/10.1074/jbc.M604268200
22. What is the Proteinase K activity in commonly used buffers? | NEB. https://www.neb.com/en/faqs/2012/10/16/what-is-the-proteinase-k-activity-in-commonly-used-buffers

Chapter 7

Gel Electrophoresis Analysis of SMC-Mediated Effects on DNA Topology

Joaquim Roca

Abstract

SMCs might alter the topology of interacting DNA as a result of two processes: stoichiometric binding and dynamic supercoiling. Stoichiometric effects refer to DNA deformations constrained at the DNA interaction sites. Dynamic effects are cumulative alterations of DNA topology generated by a cyclic, energy-dependent activity. This chapter describes basic in vitro ensemble assays to measure the stoichiometric alterations of DNA topology caused by SMC binding and to test the potential DNA supercoiling generated by SMC activity. Lastly, it outlines gel electrophoresis conditions to reveal these changes.

Key words Cohesin, Condensin, Topoisomerase, DNA linking number, DNA twist, DNA supercoiling, DNA loop

1 Introduction

SMCs are multiprotein complexes that contain multiple DNA-binding sites. Conformational changes driven by ATP usage lead SMCs to rearrange their DNA interactions, translocate along DNA, and extrude DNA loops. It is not surprising, therefore, that SMCs have been found to alter the topology of interacting DNA [1–5]. Such alterations in DNA topology might result from two processes: stoichiometric binding and cumulative dynamic effects.

Stoichiometric binding effects refer to DNA deformations constrained by molecular interactions with a given factor (a small molecule, a protein complex, etc.). Binding to DNA might locally alter the DNA double helical twist (*Tw*) and/or the writhe (*Wr*) of the DNA axis (Fig. 1a) [6]. An example of ΔTw is DNA base pair intercalators such as ethidium, which untwists neighboring base pairs by 26 degrees upon intercalation [7]. An example of ΔWr is nucleosome assembly, where 147 bp of DNA wrap around each histone octamer to complete about 1.5 left-handed superhelical turns [8].

Yasuto Murayama and Yasutaka Kakui (eds.), *SMC Proteins: Methods and Protocols*, Methods in Molecular Biology, vol. 2991, https://doi.org/10.1007/978-1-0716-5005-9_7,

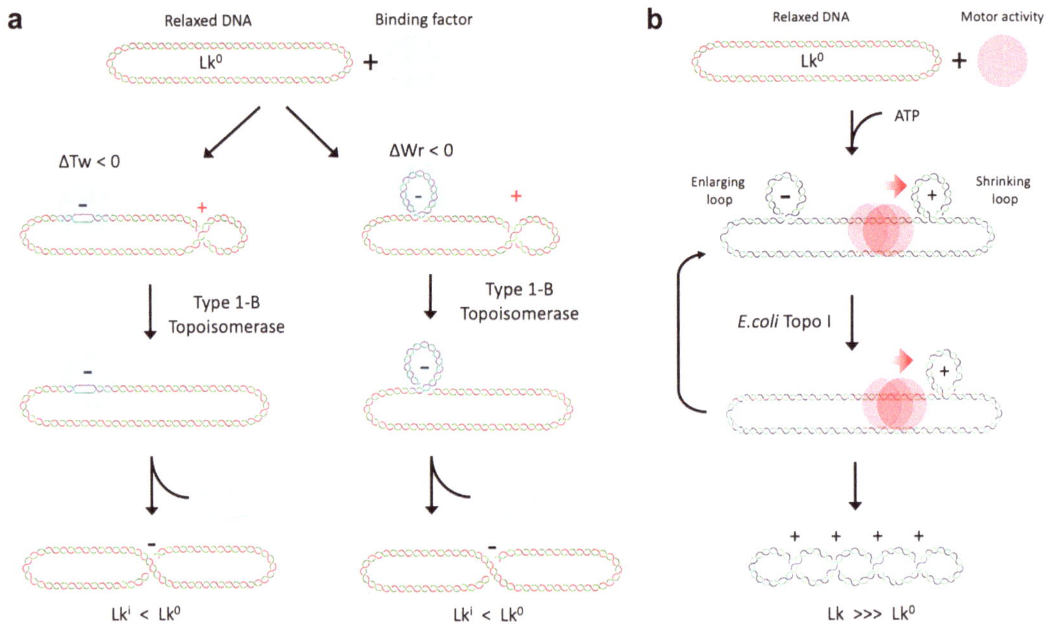

Fig. 1 Alterations of DNA topology by stoichiometric effects and dynamic activities. (**a**) *Stoichiometric effects.* A DNA-binding factor may deform the duplex in terms of twist (*Tw*) and/or writhe (*Wr*). The figure illustrates one example of DNA unwinding ($\Delta Tw < 0$) and another of left-handed bending of DNA ($\Delta Wr < 0$). Since the DNA is covalently closed (fixed *Lk*), a compensatory DNA deformation (depicted as a (+) supercoil) is produced. This unconstrained deformation can be relaxed by a type 1B topoisomerase. The result is a negative ΔLk difference ($Lk^i < Lk^0$), which reflects the extent of deformation stabilized by the DNA-binding factor. (**b**) *Dynamic effects.* The figure depicts an ATP-dependent DNA translocation motor, whose activity generates (+) and (−) twin supercoiling domains in the DNA. If these two domains are permanently or transiently held in separate loops (an enlarging loop and a shrinking loop), selective relaxation of (−) supercoils by *E. coli* Topo I leads to the accumulation of (+) supercoils ($Lk > Lk^0$)

These stoichiometric alterations in DNA topology can be accurately measured based on the following principles. The helical repeat (*h*) of the DNA double helix at 37 °C (under physiological salt concentrations and pH) is approximately 10.5 bp per turn [9]. In a covalently closed circular DNA molecule (e.g., a plasmid), this helical repeat corresponds to a DNA linking number (*Lk*), which represents how many times one strand of the duplex is linked around the other ($Lk =$ number of bp$/h$). Due to thermal motion, DNA molecules typically exhibit a Gaussian distribution of *Lk* values [9]. When DNA is naked and relaxed, the mean of this *Lk* distribution (Lk^0) corresponds to the *Lk* of minimal torsional energy, which is not necessarily an integer.

Interaction between relaxed DNA and a binding factor might locally deform the double helix, altering ΔTw and/or ΔWr. Since *Lk* remains fixed and is defined as $Lk = Tw + Wr$, this deformation will generate compensatory changes in *Tw* and *Wr*, resulting in torsional stress outside the interaction site (Fig. 1a). In a covalently closed circular DNA molecule, this torsional stress can be relieved by a type 1B topoisomerase [10], which transiently cuts one DNA strand, allowing the other to swivel before resealing the duplex. This process resets the former Lk^0 of relaxed DNA into a new distribution of *Lk* values with a mean Lk^i. The difference between Lk^0 and Lk^i (ΔLk) thus captures the extent of deformation stabilized by a DNA-binding factor (Fig. 1a). This ΔLk can be clearly observed via gel electrophoresis of circular DNA molecules because individual Lk values are topological invariants that exhibit distinct electrophoretic mobilities. By adjusting electrophoresis conditions, the distributions of *Lk* topoisomers corresponding to Lk^0 and Lk^i can be resolved side by side, allowing precise measurement of ΔLk [3, 8, 11].

In contrast to stoichiometric effects on DNA topology, dynamic effects refer to cumulative alterations in DNA topology generated by cyclic and energy-dependent activity. A well-known example is the activity of DNA translocation motors (such as DNA-transcribing complexes), which enforce axial rotation of the double helix rather than rotating themselves [12, 13]. Their activity generates DNA overtwisting (positive supercoiling) ahead of the translocating complex and DNA untwisting (negative supercoiling) behind it (Fig. 1b). In a covalently closed circular DNA molecule, these twin supercoiled domains may eventually cancel each other when they merge on the opposite side of the DNA ring. Likewise, any transient or partial dissociation of the motor complex from DNA will allow these supercoiled domains to relax. However, supercoiled domains will persist if the motor encloses these domains in separate DNA loops or if two motors travel in convergent directions. Notably, in all these scenarios, the *Lk* of the circular DNA remains unaltered. Therefore, electrophoresis of such a DNA molecule will not reveal any changes. However, if twin supercoiled domains persist or do not immediately cancel each other, a topoisomerase can be used to selectively relax one type of supercoil while leaving the other intact. Conveniently, *E. coli* topoisomerase I (a type 1A topoisomerase) relaxes negatively supercoiled DNA but does not affect positively supercoiled DNA [10]. Therefore, the presence of *E. coli* topoisomerase I during the formation of twin supercoiled domains will result in an increase in DNA *Lk*, as positive supercoils accumulate while negative supercoils are relaxed (Fig. 1b). This unidirectional change in *Lk* will not occur with a type 1B topoisomerase (used in stoichiometric assays) because it equally relaxes both positive and negative supercoils.

This chapter describes basic in vitro assays to measure the stoichiometric alterations in DNA topology induced by SMC binding and the potential dynamic supercoiling generated by SMC activity. Lastly, it outlines gel electrophoresis conditions for observing the *Lk* changes produced in these assays.

2 Materials

2.1 DNA and Enzymes

1. Purified SMC complexes, such as *S. cerevisiae* condensin [14], that have been snap-frozen and stored at −80 °C should be available in the laboratory at a concentration of approximately 1–5 μM.
2. Since SMCs do not recognize specific base pair sequences to load and manipulate DNA, any bacterial plasmid of 3–8 kb in size can be used as a covalently closed circular DNA substrate (Fig. 2a, lane 1). The plasmid, typically amplified in *E. coli* cells, can be purified either by the classical alkaline lysis method or using commercially available spin columns (*see* **Note 1**).
3. Type 1B topoisomerases, which fully relax both positively and negatively supercoiled DNA, can be purchased from commercial sources such as Gibco BRL (calf thymus Topo I), Promega (wheat germ Topo I), and TopoGEN (human Topo I) (*see* **Note 2**).
4. Type 1A topoisomerases, such as *E. coli* Topo I, which selectively relaxes negatively supercoiled DNA, can be purchased from commercial sources such as New England Biolabs and NZYTech (*see* **Note 3**).

2.2 Buffers

1. SMC dilution buffer: 50 mM Tris–HCl (pH 7.5), 1 mM EDTA, 200 mM NaCl, 1 mM DTT, 500 μg/mL BSA, 50% glycerol.
2. DNA reaction buffer: 25 mM Tris–HCl (pH 7.5), 50 mM KCl or NaCl, 5 mM $MgCl_2$, 1 mM DTT, 1 mM EDTA, 0.1 mg/mL BSA. The buffer can be supplemented with 1 mM ATP or 2 mM of its non-hydrolyzable analog AMPPNP (*see* **Note 4**).
3. 5× stop buffer: 2% (w/v) SDS, 0.5 mg/mL proteinase K, 25 mM EDTA, 25% glycerol, 0.25 mg/mL bromophenol blue.
4. 10× TBE gel electrophoresis buffer: 0.89 M Tris-borate, 20 mM EDTA (pH 8.0).
5. Chloroquine stock solution: Dissolve 10 mg/mL chloroquine diphosphate in distilled water and store it in the dark at 4 °C.

3 Methods

3.1 Assay for Stoichiometric Effects on DNA Topology

Since SMC activities on DNA (loading, translocation, loop extrusion) are typically observed using low protein and DNA concentrations (<10 nM), DNA topology assays should preferably be conducted under similar conditions [3]. High concentrations of SMCs (>100 nM) may promote aggregation of protein–DNA

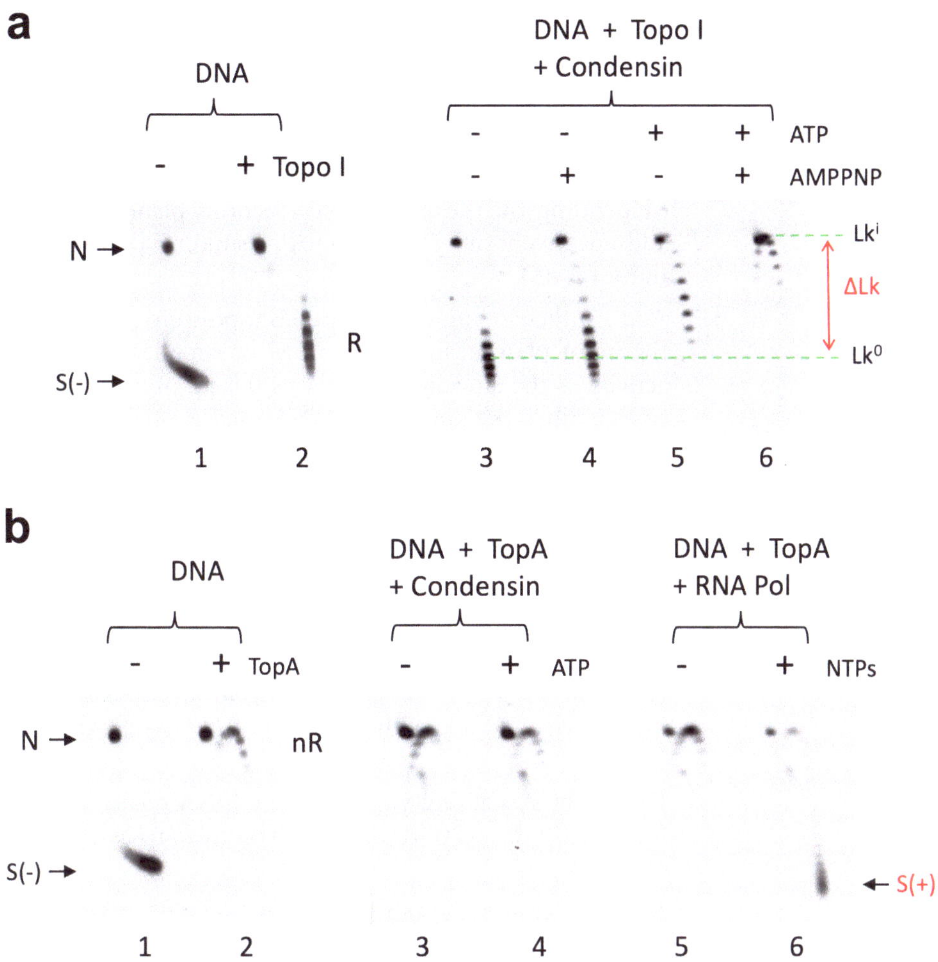

Fig. 2 DNA topology analyses by 2D gel electrophoresis. (**a**) Example of a stoichiometric effect. Lane 1: Negatively supercoiled 4-kb DNA. Lane 2: DNA from lane 1 relaxed with type 1B topoisomerase (Topo I). Lane 3: Relaxed DNA (0.3 nM) from lane 2 incubated for 20 min at 30 °C with *S. cerevisiae* condensin (3 nM) in the presence of Topo I. Lane 4: As in lane 3 but supplemented with AMPPNP (2 mM). Lane 5: As in lane 3 but supplemented with ATP (1 mM). Lane 6: As in Lane 3 but supplemented with ATP (1 mM) for 10 min followed by AMPPNP (2 mM) for 10 min. Chloroquine concentration was 0.4 μg/mL in the first and 1.2 μg/mL in the second gel dimension. S(−) negatively supercoiled topoisomers, (N) nicked circles, R relaxed distribution of *Lk* topoisomers, with a mean of Lk^0. The condensin complex produce *Lk* distributions of mean $Lk^i \neq Lk^0$. The resulting ΔLk in lane 6 is −8 *Lk* units, meaning that each condensin complex stabilizes a deformation (ΔTw and/or ΔWr) equivalent to −0.8 *Lk* units. (**b**) Example of dynamic supercoiling. Lane 1: Negatively supercoiled 4-kb DNA. Lane 2: DNA from lane 1 relaxed with *E. coli* Topo I (TopA). Lane 3: DNA (0.3 nM) from lane 2 incubated for 20 min at 30 °C with *S. cerevisiae* condensin (3 nM) in the presence of TopA. Lane 4: As in lane 3 but supplemented with ATP (1 mM). Lane 5: DNA (0.3 nM) from lane 2 incubated for 20 min at 30 °C with RNA polymerase in the absence of NTPs. Lane 6: As in lane 5 but in the presence of NTPs (1 mM). Chloroquine concentration was 0.1 μg/mL in the first and 1 μg/mL in the second gel dimension. S(−) negatively supercoiled topoisomers, (N) nicked circles, (nR) nearly relaxed distribution of Lk topoisomers produced by TopA, S(+) positively supercoiled topoisomers. The accumulation of (+) supercoiling ($Lk > Lk^0$), which is observed in lane 6 but not in lane 4, evidences that TopA is relaxing the underwound DNA region continuously generated by a DNA translocating complex

complexes, leading to erratic DNA topology effects that are difficult to interpret.

1. Gently dilute the SMC complex (stored at −80 °C) with cold (4 °C) SMC dilution buffer to a final concentration of 100 nM.
2. Prepare an appropriate volume of DNA reaction buffer (depending on the number of reactions planned below) and add negatively supercoiled DNA plasmid to a final concentration of 5 nM.
3. Add a sufficient amount of type 1B topoisomerase to fully relax the input plasmid in <15 min. Most suppliers define 1 U of activity as the amount that relaxes 0.5 μg of plasmid in 1 h at 37 °C. Since the amount of DNA used is small, 1 U of topoisomerase should be sufficient. Importantly, the temperature of this DNA relaxation reaction (25 °C, 30 °C, or 37 °C) should match that used in subsequent SMC reactions (*see* **Note 5**).
4. Take 20 μL of the above relaxation reaction and quench it with 5 μL of stop buffer, followed by incubation for 30 min at the relaxation temperature (25 °C, 30 °C, or 37 °C). Keep these 25 μL at room temperature or 4 °C, ready for gel electrophoresis. This sample serves as the control for proper DNA relaxation activity and marks the position of Lk^0 (*see* **Note 6**, Fig. 2a, lane 2).
5. Divide the remaining volume of the DNA relaxation mix into individual tubes (20 μL or larger volumes) to assay the effects of SMCs.
6. Add increasing amounts of the diluted SMC complex (100 nM) to each divided volume, ensuring that the final molar ratios of SMCs to plasmid DNA (5 nM) are, for example, 0:1, 1:1, 2:1, 5:1, and 10:1.
7. Preincubate the above mixtures at the relaxation temperature (25 °C, 30 °C, or 37 °C) for 5 min before supplementing them with additional cofactors such as ATP (1 mM) or AMPPNP (2 mM). Continue incubation at the relaxation temperature for 25 min (*see* **Note 7**).
8. Quench the reactions with stop buffer and incubate them at the relaxation temperature for 30 min. Keep these samples at room temperature or 4 °C, ready for gel electrophoresis.

3.2 Assay for Dynamic DNA Supercoiling

The setup of this assay is similar to that described above for stoichiometric effects. The main difference is the use of a type 1A instead of a type 1B topoisomerase. An important consideration is that the purified SMC complexes must be free of contaminating topoisomerase activities. This can be verified beforehand by incubating the purified SMC complex with a supercoiled DNA plasmid in DNA reaction buffer. The *Lk* of the supercoiled DNA should remain unaltered. Another consideration is that, in contrast to stoichiometric binding effects, dynamic DNA supercoiling is expected to occur only in the presence of energetic cofactors.

1. Gently dilute the SMC complex (stored at −80 °C) with cold (4 °C) SMC dilution buffer to a final SMC concentration of 100 nM.
2. Prepare an appropriate volume of DNA reaction buffer (depending on the number of reactions planned below) and add supercoiled DNA plasmid to a final concentration of 5 nM (Fig. 2b, lane 1).
3. Add a sufficient amount of type 1A topoisomerase (*E. coli* Topo I) to remove most of the (−) supercoils from the input plasmid (*see* **Note 3**). Most suppliers define 1 U of activity as the amount that relaxes 0.5 μg of plasmid in 1 h at 37 °C. Since the amount of DNA used is small, 1 U of topoisomerase should be sufficient. Importantly, the temperature of this DNA relaxation reaction (25 °C, 30 °C, or 37 °C) should match that used in subsequent SMC reactions (*see* **Note 5**).
4. Take 20 μL of the above relaxation reaction and quench it with 5 μL of stop buffer, followed by incubation at 37 °C for 30 min. Keep this 25 μL at room temperature or 4 °C, ready for gel electrophoresis. This control sample reflects the Lk distribution produced by *E. coli* Topo I in the absence of dynamic twin supercoiled domains (Fig. 2b, lane 2).
5. Divide the remaining volume of the DNA relaxation mix into individual 20 μL or larger volumes to assay the effects of SMCs.
6. Add increasing amounts of the diluted SMC complex (100 nM) to each divided volume, ensuring that the final molar ratios of SMCs to plasmid DNA (5 nM) are, for example, 0:1, 1:1, 2:1, 5:1, and 10:1.
7. Preincubate the above SMC mixtures (all containing plasmid DNA and *E. coli* Topo I) at the relaxation temperature (25 °C, 30 °C, or 37 °C) for 5 min, then supplement them with energetic cofactors (ATP or others). Continue incubation at the relaxation temperature for increasing time points (3, 10, 30, and 90 min).
8. Quench the reactions with stop buffer and further incubate them for 30 min. Keep these samples at room temperature or 4 °C, ready for gel electrophoresis.

3.3 Gel Electrophoresis, Southern Blot, and Lk Analyses

The most accurate way to display distributions of DNA *Lk* topoisomers is through two-dimensional (2D) agarose gel electrophoresis. In 2D gels, ladders of *Lk* topoisomers form an arc, where *Lk* values increase clockwise. Unlike 1D gels, 2D gels separate covalently closed *Lk* topoisomers from nicked and linear molecules and distinguish (+) and (−) supercoiled topoisomers, which otherwise overlap due to similar electrophoretic velocity in 1D gels. The conceptual basis of 2D electrophoresis, the effect of chloroquine on DNA topology, and the detailed procedure have been described in previous *Methods in Molecular Biology* chapters [15, 16].

1. Set up the electrophoresis apparatus: To examine the effects of SMCs on the topology of a 3–8 kb circular DNA, prepare a horizontal electrophoresis apparatus with gel dimensions of at least 15 cm × 15 cm (ideally a gel of 20 cm × 20 cm) (*see* **Note 8**).
2. Cast a 0.7% (w/v) agarose gel (*see* **Note 9**) in 1× TBE buffer containing the chloroquine concentration planned for the first gel dimension (e.g., 0.2 μg/mL chloroquine) (*see* **Note 10**).
3. Run the first gel dimension at 2.5 V/cm for ~20 h in 1× TBE buffer containing 0.2 μg/mL chloroquine (*see* **Note 11**).
4. Soak the gel in TBE buffer containing 1 μg/mL chloroquine with gentle shaking for 1 h at room temperature.
5. Run the second gel dimension (in an orthogonal direction) at 2.5 V/cm for ~4–6 h in 1× TBE buffer containing 1 μg/mL chloroquine.
6. Perform a Southern blot of the entire gel using standard procedures to transfer the electrophoresed DNA to a nylon membrane. Use commercially available kits for chemiluminescent hybridization and DNA detection (*see* **Note 12**).
7. Analyze *Lk* distributions: Follow methods described in previous *Methods in Molecular Biology* chapters [16] and related DNA topology studies [3, 8, 11]. Briefly, determine the midpoint of each *Lk* distribution by quantifying the intensity of individual *Lk* topoisomers. Calculate Lk changes (ΔLk) as the difference (in *Lk* units) between the midpoints of the DNA distributions generated in the absence and presence of SMCs under different reaction conditions.

As an example of the experimental results, Fig. 2a shows the stochiometric effects produced by yeast condensin activity on DNA topology (lanes 4–6). Figure 2b shows that yeast condensin lacks dynamic DNA supercoiling activity (lanes 3–4), which contrasts with that produced by a transcribing RNA polymerase (lanes 5–6).

4 Notes

1. Purification methods should yield mostly negatively supercoiled molecules that are free of proteins and other contaminants. A small fraction of nicked circles (<20%) is acceptable for DNA *Lk* distribution analyses.

2. The specific activity of commercial type 1B topoisomerases is low or decays quickly. An alternative is to use vaccinia virus topoisomerase I (*vvTop1*), a highly active and stable 25 kDa enzyme that can be purified in the laboratory after induction from an expression vector in *E. coli* cells [17].
3. Type 1B topoisomerases fully relax both positive and negative DNA helical tension, producing *Lk* distributions centered at Lk^0. Type-1A topoisomerases (*E. coli* Topo I) do not relax positively supercoiled DNA and only partially relax negatively supercoiled DNA (Fig. 2b, lane 2). This is because type 1A activity requires single-stranded DNA regions, which are only produced in highly or moderately untwisted DNA molecules.
4. The DNA reaction buffer should support SMC complex binding and activity, but its composition may need optimization for specific SMC complexes and effects. Additionally, it must maintain the activity of type 1A and type 1B topoisomerases to ensure proper relaxation of DNA helical stress generated by SMCs.
5. The helical repeat of DNA varies with temperature. To accurately compare Lk^0 and Lk^i, DNA relaxation by topoisomerases should be performed at the same temperature in both the absence and presence of SMCs. Once topoisomerase activity is quenched with a stop buffer, *Lk* values remain unchanged regardless of temperature.
6. DNA should be fully relaxed before adding SMCs to ensure that any deformations induced by SMC interactions are properly observed. If the DNA remains negatively supercoiled, SMCs could stabilize preexisting DNA crossovers or unwound regions, which do not necessarily indicate their ability to induce such deformations.
7. During the 25-min incubation, the previously added type 1B topoisomerase remains active and continues to relax any unconstrained DNA helical tension produced by SMCs. To verify this relaxation activity, a proper control would be to add supercoiled plasmid to one of the incubated mixtures and extend the incubation by a few more minutes.
8. The gel electrophoresis tray, tank, and combs should be thoroughly washed to remove residual intercalators (e.g., chloroquine or ethidium bromide) from previous experiments.
9. The agarose concentration (w/v) should be adjusted to optimize resolution based on plasmid size: 0.8% agarose for 3–4 kb plasmids, 0.7% agarose for 4–6 kb plasmids, and 0.6% agarose for 6–8 kb plasmids.
10. To examine both negative ($Lk^i < Lk^0$) and positive ΔLk values ($Lk^i > Lk^0$), use 0–0.1 μg/mL chloroquine in the first gel

dimension and 1 μg/mL in the second. To resolve highly negative ΔLk values while barely resolving positive ones, use 0.6 μg/mL chloroquine in the first gel dimension and 3 μg/mL in the second.

11. Using 0.2 μg/mL chloroquine in the first gel dimension allows the relaxed DNA (Lk^0) produced by type 1B topoisomerases to migrate faster than nicked DNA circles. In contrast, the partially relaxed DNA distribution produced by *E. coli* Topo I migrates close to the nicked DNA circles.

12. Since the amount of plasmid DNA in each gel sample is small (<10 ng), it is not visible by direct gel staining. Therefore, 2D gels should be blot-transferred to a nylon membrane and hybridized with sequence-specific DNA probes. This will generate light signals that can be recorded using digital imaging devices or radiographic films.

Acknowledgments

This work is supported by the Plan Estatal de Investigación Científica y Técnica of Spain, with grant PID2023-146602NB-I00 to J.R.

References

1. Kimura K, Hirano T (1997) ATP-dependent positive supercoiling of DNA by 13S condensin: a biochemical implication for chromosome condensation. Cell 90(4):625–634
2. Kimura K, Rybenkov VV, Crisono NJ et al (1999) 13S condensin actively reconfigures DNA by introducing global positive writhe: implications for chromosome condensation. Cell 98(2):239–248
3. Martinez-Garcia B, Dyson S, Segura J et al (2023) Condensin pinches a short negatively supercoiled DNA loop during each round of ATP usage. EMBO J 42(3):e111913
4. Janissen R, Barth R, Davidson I et al (2024) All eukaryotic SMC proteins induce a twist of –0.6 at each DNA loop extrusion step. Sci Adv 10(50):eadt1832
5. Bazett-Jones DP, Kimura K, Hirano T (2002) Efficient supercoiling of DNA by a single condensin complex as revealed by electron spectroscopic imaging. Mol Cell 9(6):1183–1190
6. Roca J (2011) The torsional state of DNA within the chromosome. Chromosoma 120(4):323–334
7. Wang JC (1974) The degree of unwinding of the DNA helix by ethidium. I. Titration of twisted PM2 DNA molecules in alkaline cesium chloride density gradients. J Mol Biol 89(4):783–801
8. Segura J, Joshi RS, Díaz-Ingelmo O et al (2018) Intracellular nucleosomes constrain a DNA linking number difference of −1.26 that reconciles the Lk paradox. Nat Commun 9(1):3989
9. Depew DE, Wang JC (1975) Conformational fluctuations of DNA helix. Proc Natl Acad Sci USA 72(11):4275–4279
10. Champoux JJ (2001) DNA topoisomerases: structure, function, and mechanism. Annu Rev Biochem 70:369–413
11. Diaz-Ingelmo O, Martínez-García B, Segura J et al (2015) DNA topology and global architecture of point centromeres. Cell Rep 13(4):667–677
12. Giaever GN, Wang JC (1988) Supercoiling of intracellular DNA can occur in eukaryotic cells. Cell 55(5):849–856
13. Ostrander EA, Benedetti P, Wang JC (1990) Template supercoiling by a chimera of yeast GAL4 protein and phage T7 RNA polymerase. Science 249(4974):1261–1265

14. Lee BG, Merkel F, Allegretti M et al (2020) Cryo-EM structures of holo condensin reveal a subunit flip-flop mechanism. Nat Struct Mol Biol 27(8):743–751
15. Roca J (2009) Two-dimensional agarose gel electrophoresis of DNA topoisomers. Methods Mol Biol 582:27–37
16. Martinez-Garcia B, Diaz-Ingelmo O, Ayats-Fraile A et al (2025) Electrophoretic analysis of DNA supercoiling activities. Methods Mol Biol 2881:259–270
17. Shuman S, Golder M, Moss B (1988) Characterization of vaccinia virus DNA topoisomerase I expressed in Escherichia coli. J Biol Chem 263(31):16401–16407

Chapter 8

The Use of the Homobifunctional Crosslinker BMOE to Investigate Cohesin Function by Measuring Protein–Protein and Protein–DNA Interactions

James Collier and Madhusudhan Srinivasan

Abstract

Cohesin, a eukaryotic SMC complex, plays a crucial role in mediating sister chromatid cohesion and organizing the genome during interphase. To fulfill these functions, cohesin must act as both a stable guardian of genome integrity, capable of holding sister chromatids together for decades in some cell types, and a dynamic molecular machine that rapidly forms and remodels DNA loops in response to chromatin changes. These processes involve a complex series of interactions between cohesin and its many regulatory subunits, and DNA, the details of which are still under investigation. Here, we describe methods used routinely in our laboratory for studying transient protein–protein interactions and protein–DNA interactions, both in vivo and in vitro, using site-specific cysteine–cysteine crosslinking.

Key words Sister chromatid cohesion, Site-specific crosslinking, Protein–Protein interaction, Protein–DNA interaction, SMC complex dynamics

1 Introduction

SMC (Structural Maintenance of Chromosomes) complexes are among the most evolutionarily conserved molecular machines, present across all domains of life [1]. Despite subtle variations in composition, they universally function as architects of genome organization and guardians of genome integrity. SMC proteins are ATP-Binding Cassette (ABC)-type ATPases that use the impetus from binding and hydrolyzing ATP to manipulate DNA topology, orchestrating interactions between different DNA segments, whether within a single molecule (loop formation) or between sister chromatids (cohesion) [2, 3].

Unlike enzymatic modifications that leave chemical marks on DNA, SMC proteins exert their influence through physical restructuring, making their activities challenging to measure. A fundamental question arises: How does one detect the act of physically

Yasuto Murayama and Yasutaka Kakui (eds.), *SMC Proteins: Methods and Protocols*, Methods in Molecular Biology, vol. 2991, https://doi.org/10.1007/978-1-0716-5005-9_8,

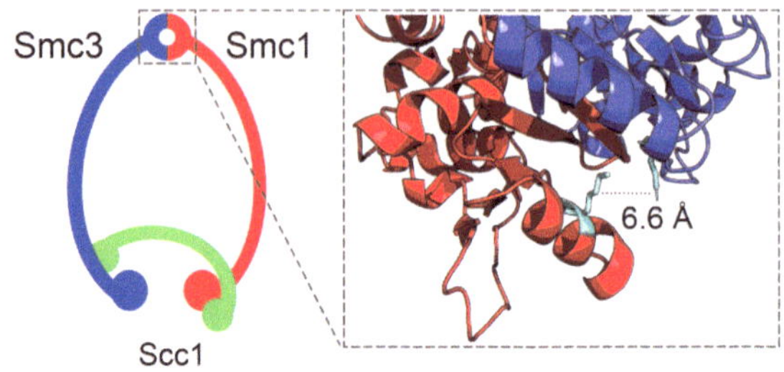

Fig. 1 Schematic representation of the cohesin trimer. The structure of the Smc1/3 hinge domain is shown with the positions of the cysteine substitutions (Smc1 K639C and Smc3 E570C) highlighted

holding different sections of DNA together? In addressing this challenge, homobifunctional crosslinkers have proven to be an essential tool for capturing and analyzing these interactions. Moreover, the integration of crosslinking technology with AI-driven protein structure prediction, such as AlphaFold, has revolutionized our ability to decipher transient protein–protein interactions that regulate SMC function [4].

Here, we describe the in vivo (in *Saccharomyces cerevisiae* cells) and in vitro methods routinely used in our laboratory to probe protein–protein and protein–DNA interactions through site-specific crosslinking with the homobifunctional crosslinker bismaleimidoethane (BMOE). These methods allow us to explore the enzymatic activity of cohesin, a key eukaryotic SMC complex. Cohesin is a ring-shaped protein complex consisting of two SMC proteins (Smc1 and Smc3), which dimerize at the hinge domain. These SMC proteins are bridged by the α-kleisin subunit (Scc1), which interacts with the ATPase domain of Smc1 and a region near the Smc3 ATPase coiled coil [5] (Fig. 1). Cohesin plays a critical role in sister chromatid cohesion by topologically trapping sister DNA molecules, ensuring proper chromosome segregation [6, 7].

2 Materials

2.1 Budding Yeast Strains

1. 26032: *S. cerevisiae smc3-6HA E570C* [6].
2. 26085: *S. cerevisiae smc1 K639C smc3-6HA E570C* [6].
3. 23644 (5C cohesin): *S. cerevisiae smc1 K639C smc3 E570C SCC1-9Pk C56*, carrying 2.3 kb *TRP1-ARS1-CEN4* circular minichromosome [6].
4. 23889 (6C cohesin): *S. cerevisiae smc1 K639C smc3 E570C scc1-9Pk C56 A547C*, carrying 2.3 kb *TRP1-ARS1-CEN4* circular minichromosome [6].

2.2 Stains and Plasmids Used for Protein Purification

1. *Escherichia coli* (*E. coli*) DH10EmBacY.
2. *Spodoptera frugiperda* Sf9 cells.
3. pACEbac1 Smc1 G22C K639C Smc3 E570C S1043C StrepII-Scc1 C56 (5C): The *S. cerevisiae* 5C cohesin trimer expression construct [8] (*see* **Note 1**).
4. pACEbac1 Smc1 G22C K639C Smc3 E570C S1043C Scc1-StrepII C56 *A547C* (6C): The *S. cerevisiae* 6C cohesin trimer expression [8] (*see* **Note 1**).
5. pACEbac1 StrepII-Scc2$_{132-1493}$: The expression construct of the *S. cerevisiae* Scc2 C-terminus (132–1493 amino acids) [8] (*see* **Note 1**).
6. pACEbac1 StrepII-Scc3: The *S. cerevisiae* Scc3 expression construct [8] (*see* **Note 1**).

2.3 Cell Culture and Supplements

1. YPD: 2% (w/v) peptone, 1% (w/v) yeast extract, 1.5% (w/v) glucose.
2. LB: 1% (w/v) tryptone, 0.5% (w/v) yeast extract, 1% (w/v) NaCl, adjust pH to 7.0 with NaOH. Add 1.5% (w/v) ager to prepare solid media.
3. 2xYT: 1.6% (w/v) tryptone, 1% (w/v) yeast extract, 0.5% (w/v) NaCl.
4. 10 mg/mL gentamicin: Prepare in pure water.
5. 10 mg/mL tetracycline: Prepare in ethanol.
6. 50 mg/mL kanamycin: Prepare in ethanol.
7. 100 mg/mL 5-bromo-4-chloro-3-indolyl-β-D-galactoside (X-gal): Prepare in DMSO.
8. 0.1 M isopropyl β-D-1-thiogalactopyranoside (IPTG): Prepare in pure water.
9. Sf900 II SFM media (Thermo Fisher Scientific).
10. Shaking incubator.

2.4 In Vivo Crosslinking Assay

1. 125 mM BMOE stock solution: Prepare 125 mM stock solution in dimethyl sulfoxide (DMSO) (*see* **Note 2**).
2. Yeast lysis buffer: 25 mM HEPES-NaOH (pH 7.5), 150 mM KCl, 0.25% (w/v) Triton X-100, 0.05% (w/v) NP40, cOmplete Protease Inhibitor cocktail (Merck), 1 mM phenylmethylsulfonyl fluoride (PMSF) (*see* **Note 3**).
3. Phosphate buffer saline (PBS): 8.1 mM Na_2HPO_4, 1.47 mM KH_2PO_4, 137 mM NaCl, 2.68 mM KCl, 0.05% (w/v) Tween 20.
4. PBS + 5 mM dithiothreitol (DTT): PBS containing 5 mM DTT. Dissolve DTT just before use.

5. Acid-washed glass beads (425–600 μm).
6. Bradford protein assay kit.
7. 4x LDS sample buffer (Thermo Fisher Scientific).
8. NuPAGE 3–8% or 4–12% gels (Thermo Fisher Scientific).
9. PVDF membrane.
10. Filter paper.
11. Membrane blocking reagent.
12. Anti-HA tag antibody.
13. Beads cell disrupter FastPrep-24 (MP Biomedicals).
14. SDS-PAGE apparatus.
15. Western blot transfer apparatus.

2.5 Protein Purification

1. pLIB/pACEbac1/pFastbac or equivalent baculovirus compatible vectors.
2. P1 buffer: 50 mM Tris-HCl (pH 8), 10 mM EDTA, 100 μg/mL RNase A.
3. P2 buffer: 1% (w/v) SDS, 200 mM NaOH.
4. N3 buffer: 4.2 M guanidine-HCl (pH 4.8), 0.9 M potassium acetate.
5. Ethanol.
6. 70% ethanol.
7. 6-well plate.
8. FuGENE transfection reagent (Promega).
9. Sf9 lysis buffer: 50 mM HEPES-NaOH (pH 7.5), 300 mM NaCl, 1 mM $MgCl_2$, 1 mM Tris(2-carboxyethyl)phosphine (TCEP), 5% (v/v) glycerol, 0.01% (w/v) Triton X-100, 0.01% (w/v) NP40, cOmplete Protease Inhibitor cocktail (Merck), 1 μg/mL RNAse A, 500 U/mL Supernuclease (*see* **Note 4**).
10. Purification buffer: 150 mM NaCl, 50 mM HEPES-NaOH (pH 7.5), 1 mM TCEP, 10% (v/v) glycerol.
11. Elution buffer: Purification buffer plus 100 mM biotin.
12. Purification buffer 50: As purification buffer instead containing 50 mM NaCl.
13. Purification buffer 1000: As purification buffer instead containing 1 M NaCl.
14. Strep-Tactin®XT resin (Cytiva).
15. 5 mL HiTrap Q HP column (Cytiva).
16. Superose 6 Increase 10/300 GL gel filtration column (Cytiva).
17. Liquid nitrogen.

18. 0.8 μm filter paper.
19. 20 mL empty open column.
20. Ultracentrifuge.
21. Dounce homogenizer.
22. FPLC system.

2.6 In Vitro Crosslinking Assay

1. 6.4 mM BMOE stock solution: Prepare 6.4 mM stock solution in dimethyl sulfoxide (DMSO).
2. CB buffer: 50 mM HEPES-NaOH (pH 7.5), 50 mM NaCl, 1 mM $MgCl_2$, 1 mM TCEP, 5% (v/v) glycerol.
3. 10% (w/v) SDS.
4. 4x LDS sample buffer (Thermo Fisher Scientific).
5. NuPAGE 3–8% or 4–12% gradient gel (Thermo Fisher Scientific).
6. Coomassie protein staining solution.
7. SDS-PAGE apparatus.

2.7 In Vivo Cohesin–DNA Interaction Assay

1. 6C lysis buffer: 25 mM HEPES-NaOH (pH 7.5), 100 mM KCl, 1 mM $MgCl_2$, 10 mM trisodium citrate, 25 mM sodium sulfite, 0.25% (w/v) Triton X-100, cOmplete Protease Inhibitor cocktail (Merck), 1 mM PMSF (*see* **Note 3**).
2. Wash buffer: 10 mM Tris-HCl (pH 8.0), 0.25 M LiCl, 0.5% (w/v) NP40, 0.5% sodium deoxycholate, 1 mM EDTA, 1 mM PMSF.
3. Anti-Pk tag antibody (Bio-Rad).
4. Dynabeads Protein G (Veritas).
5. 1x TE: 10 mM Tris-HCl (pH 7.5), 1 mM EDTA.
6. 1% (w/v) SDS DNA loading dye: 1x Purple Gel Loading Dye (NEB), 1% (w/v) SDS.
7. TAE: 40 mM Tris-acetate (pH 8.0), 1 mM EDTA.
8. 0.8% agarose/TAE gel containing 1 μg/mL ethidium bromide.
9. Immobilon-NY+ membrane (Millipore).
10. Magnetic stand.
11. Agarose gel electrophoresis apparatus.
12. Random Primer DNA Labeling Kit Ver. 2 (Takara).
13. G-50 spin columns.
14. ATP, [γ-32P] 6000 Ci/mmol.

2.8 In Vitro Cohesin–DNA Interaction Assay

1. The budding yeast $Scc2_{132\text{-}1493}$ and Scc3, which are purified as previously described [6].
2. 6.4 mM BMOE stock solution: Prepare 6.4 mM stock solution in dimethyl sulfoxide (DMSO).
3. EB buffer: 50 mM HEPES-NaOH (pH 7.5), 20 mM NaCl, 1 mM $MgCl_2$, 1 mM TCEP, 5% (v/v) glycerol.
4. 65 mM ATP.
5. pUC19 plasmid DNA.
6. 0.8% agarose/TAE gel and electrophoresis apparatus.
7. Ethidium bromide solution.

3 Methods

3.1 Measuring Protein–Protein Interactions Using BMOE Crosslinking

Homobifunctional crosslinkers, like BMOE, contain two identical reactive groups that via maleimide chemistry form an irreversible and stable covalent bond between two sulfhydryl (-SH) groups when two cysteines come into proximity within a protein complex or between interacting proteins. The efficiency of crosslinking depends on spatial constraints, as the crosslinker's spacer arm length determines the maximum distance between reactive cysteines. Once formed, the crosslink provides a snapshot of protein interactions, including transient or weak interactions, and preserves the structural relationship between cysteine-containing residues for biochemical analysis.

To achieve effective crosslinking, a cysteine residue should be strategically inserted on each face of the interaction surface. Optimal positioning requires careful consideration to ensure that the cysteines are oriented toward each other and that their separation closely matches the linker length of the chosen crosslinker (e.g., ~8 Å for BMOE). Proper spacing maximizes crosslinking efficiency and preserves the native interaction, providing structurally relevant insights into protein–protein or protein–DNA interactions. An essential consideration when designing cysteine substitutions is the potential impact of the mutation on protein function. It is crucial to ensure that the substitution does not disrupt the protein's activity or negatively affect cell viability. To minimize such risks, we typically introduce the cysteine substitutions in vivo within the endogenous gene loci and then assess the growth characteristics of the modified yeast strains to confirm that the mutations do not compromise cellular fitness or functionality. To crosslink the cohesin hinge domain, the lysine (K639) and glutamic acid (E570) residues of Smc1 and Smc3, respectively, were substituted with cysteines (Fig. 1). These residues are 6.6 Å apart within the Smc1-Smc3 dimerization interface, and importantly, the modification of these residues does not impact the growth or fitness of the mutant yeast cells, ensuring that the substitution does not disrupt cellular viability.

3.1.1 In Vivo Crosslinking Protocol

1. Harvest 40 OD units of *S. cerevisiae* cells (26032 and 26085) and wash them twice with ice-cold PBS.
2. Resuspend the cell pellet in 1 mL ice-cold PBS.
3. Add 42 μL of 125 mM BMOE stock solution to the cell suspension to achieve a final concentration of 5 mM BMOE.
4. Incubate the cells on ice for 6 min to allow for crosslinking.
5. After incubation, wash the cells twice with 2 mL ice-cold PBS containing 5 mM DTT to quench the reaction.
6. Collect the cells by centrifugation and freeze the pellets in liquid N_2 (*see* **Note 5**).
7. Resuspend frozen cell pellets in 800 μL of yeast lysis buffer (*see* **Note 3**).
8. Add equal volume of acid-washed grass beads and lyse the cells using a beads cell disruptor (*see* **Note 6**).

1. Clear the lysate by centrifugation at 13,200 g for 10 min at 4 °C.
2. Quantify protein concentrations from the cleared lysates using a Bradford assay, and normalize the concentrations for further analysis.
3. Add one-third volume of 4x LDS sample buffer and incubate at 95 °C for 5 min.
4. Load 60 μg of total protein onto NuPAGE 3–8% or 4–12% gradient gels, selecting the gel percentage based on the molecular weight of the target proteins for optimal resolution.
5. Transfer the proteins onto PVDF membranes and soak the membranes in blocking reagent.
6. Incubate the membrane with HA tag antibody to probe Smc3-6HA.
7. Detect protein bands using standard Western blotting protocol. A representative result is shown in Fig. 2.

3.1.2 In Vitro Crosslinking Protocol

Expression and Purification of Cohesin Trimer

Our lab expresses cohesin and its regulatory subunits primarily within *Spodoptera frugiperda* Sf9 cells. This involves inserting your expression construct into a bacterial artificial chromosome (bacmid) which are then transfected into *S. frugiperda* cells. Successful transfection leads to the generation and propagation of baculovirus that is used to drive recombinant protein expression [9]. Alternative expression systems are routinely used by others to express cohesin, such as expression within *S. cerevisiae* cells, that can be researched at the readers leisure. Bacterial expression systems,

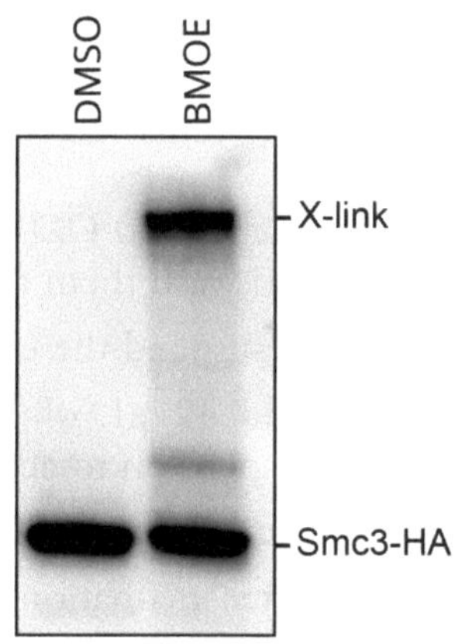

Fig. 2 In vivo crosslinking of Smc1 and Smc3 hinge domains. Strains 26032 (Smc3 E570C) and 26,085 (Smc3 E570C, Smc1 K639C) were subjected to in vivo crosslinking and analyzed by Western blotting using anti-HA antibody

such as expression within *Escherichia coli*, are typically not suitable for cohesin expression, although can be used to express a variety of regulatory subunits. Here, we describe the insect cell expression system and purification protocol used within our laboratory.

Bacmid Preparation

1. Clone the *S. cerevisiae SMC1 and SMC3* complementary DNAs (cDNA), and the *SCC1*, *SCC2*$_{132\text{-}1493}$ and *SCC3* cDNAs, each fused with a 2x StrepII tag at the N-terminus, into pACEbac1 vectors.
2. The cohesin trimer is assembled into a single pACEbac1 plasmid using the bigbac protocol [10].
3. Expression constructs are transformed into *E. coli* DH10EmBacY cells and incubated at 37 °C on LB plates containing 10 μg/mL gentamicin, 10 μg/mL tetracycline, 50 μg/mL kanamycin, 40 μM IPTG, and 100 μg/mL X-gal until blue and white colonies are visible (typically 1–2 days).
4. Inoculate a single white colony in 5 mL 2xTY media with 10 μg/mL gentamicin, 10 μg/mL tetracycline, and 50 μg/mL kanamycin and grow overnight at 37 °C at 200 rpm.
5. Pellet cells by centrifugation at 4000 g for 10 min, discard media.
6. Resuspend cell pellet in 250 μL P1 buffer.
7. Lyse cell pellet by adding 250 μL P2 buffer and incubate for 1–2 min.
8. Neutralize mixture by adding 350 μL N3 buffer and mix by inverting several times.
9. Pellet cell debris by centrifugation at 14,000 g for 10 min and collect lysate.

10. Precipitate the DNA by adding an equal volume of cold isopropanol and incubating at 4 °C for 10 min.
11. Pellet the DNA by centrifugation at 14,000 g for 10 min.
12. Remove the supernatant and wash the DNA pellet with cold 70% ethanol. Repeat centrifugation.
13. Discard supernatant and air-dry the DNA pellet.
14. Resuspend bacmid DNA in 1x TAE and dilute to 1 μg/μL.

Baculovirus Preparation

1. For each expression construct, add one million Sf9 cells to a well of a 6-well plate, in a 2 mL volume of Sf900 II SFM.
2. Add 2 μg bacmid DNA and 6 μL FuGENE transfection reagent to 200 μL Sf900 II SFM media and incubate for 10 min.
3. Add the transfection mix to the cells and top up with an additional 800 μL Sf900 II SFM media.
4. Incubate for 3–5 days at 27 °C to obtain P1 virus (*see* **Notes 7** and **8**).
5. Amplify the virus by infecting a 50 mL culture of Sf9 cells at a density of 1.5–2.0 million cells/mL with 500 μL of P1 virus (1:100 dilution).
6. Incubate the cells for 2–3 days at 27 °C and 120 rpm.
7. Harvest the virus containing fraction by pelleting the cells at 4000 g for 10 min and collecting the supernatant when cell viability drops below 80% (*see* **Notes 9** and **10**).

Protein Expression

1. Infect 1 liter of Sf9 cell culture at a density of 1.5–2.0 million cells/mL with 10 mL P2 virus (1:100 dilution). Cell volume can vary depending on protein yields and individual requirements.
2. Incubate at 27 °C and 120 rpm for 2–3 days.
3. Harvest cells by centrifugation at 1000 g at 4 °C.
4. Freeze cell pellet in liquid N_2 and store at −70 °C.

Protein Purification (Cohesin Trimer)

Typically, our proteins are purified following a three-step purification protocol involving affinity pull-down, ion exchange chromatography, and size exclusion chromatography. Our lab routinely uses the Strep-tag II system for affinity pull-down and the protocol here details its use. Alternative tag and resin combinations can work equally well. Steps can be omitted depending on individual preference and level of purity required. Here, we describe the purification method of the *S. cerevisiae* cohesin trimer.

Cysteines must be in their reduced form in order to be reactive with BMOE and so the presence of reducing agent throughout the purification protocol is necessary. However, BMOE will also react with thiol group containing reducing agents, such as DTT and β-mercaptoethanol, and so proteins must be stored using non-thiol-based reducing agents, such as TCEP.

1. Thaw Sf9 cell pellets in Sf9 lysis buffer (50 mL for each liter of harvested cell culture) at 4 °C (*see* **Note 4**).
2. Lyse cells by dounce homogenization at 4 °C.
3. Clear the lysate by centrifugation at 235,000 g for 1 hour at 4 °C.
4. Filter lysate through 0.8 μm filter paper at 4 °C.
5. Mix the lysate with 2 mL Strep-Tactin®XT resin and incubate for 2 h at 4 °C.
6. Collect the resin into a 20 mL empty open column and wash with 5 column volume of purification buffer.
7. Elute the bound protein by adding 30 mL elution buffer and incubate for 30 min at 4 °C.
8. Collect elution.
9. Load the eluate onto a 5 mL HiTrap Q HP column (*see* **Note 11**).
10. Wash the column with five column volume of purification buffer 50.
11. Elute protein in purification buffer over a gradient of 50 mM to 1 M NaCl.
12. Collect the peak fraction.
13. Load the peak fraction onto a Superose 6 Increase 10/300 GL gel filtration column equilibrated with purification buffer.
14. Collect protein fractions and aliquot. Freeze in liquid N_2 and store at −70 °C.

In Vitro Crosslinking Assay

1. Dilute the purified cohesin in CB buffer to a concentration of 900 nM in a 10 μL reaction volume. When added, ATP and similar ligands are added to a concentration of 10 mM. Any DNA added is typically 2 μg of supercoiled pUC19.
2. Incubate at 24 °C for 10 min.
3. Add 1 μL of 6.4 mM BMOE stock solution.
4. Incubate on ice for 6 min.
5. Add 4x LDS buffer and heat at 70 °C for 10 min.
6. Resolve proteins by 3–8% SDS-PAGE.
7. Stain the gel with Coomassie protein staining solution overnight and then destain thoroughly in pure water. Figure 3 shows a representative result of the in vitro crosslinking assay using the budding yeast cohesin.

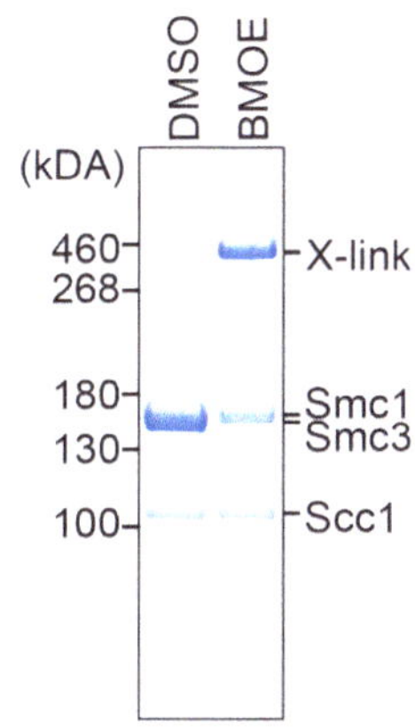

Fig. 3 In vitro crosslinking of Smc1 and Smc3 hinge domains. SDS-PAGE gel showing crosslinking of purified cohesin protein between residues Smc3 E570C and Smc1 K639C

3.2 Measuring Cohesin-DNA Interactions Using BMOE Crosslinking

The elucidation of the molecular structures of the Smc1/Smc3, Smc3/Scc1 and Scc1/Smc1 interfaces has enabled the strategic insertion of cysteine pairs (Smc3 E570C/Smc1 K639C, Smc3 S1043C/Scc1 C56, and Smc1 G22C/Scc1 A547C) allowing for site-specific crosslinking both in vivo and in vitro using bifunctional thiol-specific reagents. Consistent with the notion that cohesin rings entrap DNAs, crosslinking all three interfaces in living cells results in the formation of circular Smc1-Smc3-Scc1 polypeptides, within which circular minichromosome DNAs become topologically entrapped, rendering its retention resistant to SDS denaturation of the SMC ring.

The catenation of individual DNAs by cohesin leads to a modest reduction in electrophoretic mobility, producing catenated monomers (CMs). In contrast, the co-entrapment of sister chromatids results in the formation of catenated dimers (CDs), which migrate more slowly due to their increased molecular size [6] (Fig. 4a).

3.2.1 The 6C Assay in Living *S. cerevisiae* Cells

1. Yeast cells (23644 and 23889) are crosslinked using the in vivo crosslinking protocol described above (*see* Subheading 3.1.1, **steps 1**–7).
2. Resuspend frozen cell pellets in 800 μL of 6C lysis buffer (*see* **Note 3**).
3. Add equal volume of acid-washed grass beads and lyse the cells using a beads cell disruptor (*see* **Note 6**).
4. Clear the lysate by centrifugation at 13,200 g for 10 min at 4 °C.
5. Determine the protein concentration of the lysate using the Bradford assay, and normalize concentrations.

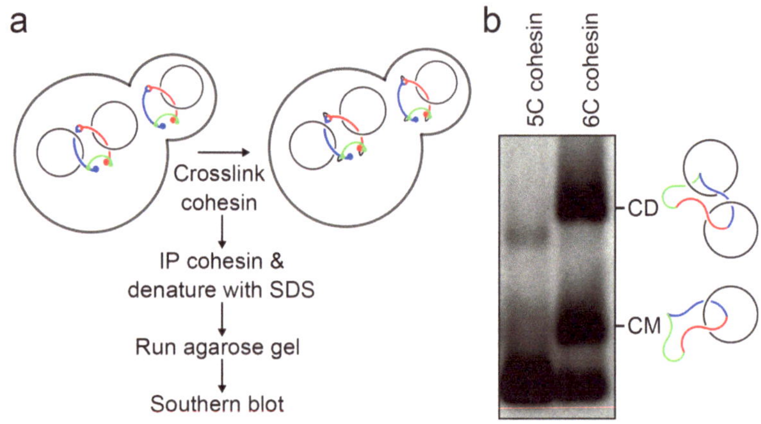

Fig. 4 (**a**) Schematic representation of the in vivo 6C assay. (**b**) The results of the in vivo 6C assay. 6C strain 23,889: With cysteine pairs at all three ring subunit interfaces (2C Smc3: E570C S1043C, 2C Smc1: G22C K639C, and 2C Scc1 C56 A547C). 5C strain: 23644 lacking just one of these cysteines (Scc1 A547C) and carrying a 2.3 kb circular minichromosome were subjected to in vivo cross-linking with BMOE. Scc1-PK6-immunoprecipitated DNA denatured with SDS was detected by Southern blotting. Catenated monomers (CM), catenated dimers (CD)

6. Perform cohesin immunoprecipitation by incubating the lysate with 5 μg of anti-PK antibody for 2 h at 4 °C, followed by incubation with 50 μL slurry of Dynabeads protein G for 2 h at 4 °C with constant mixing.
7. Wash the beads once with 1 mL 6C lysis buffer using a magnetic stand.
8. Wash beads once with 1 mL wash buffer.
9. Wash beads once with 1x TE buffer.
10. Resuspend beads in 30 μL of 1% (w/v) SDS DNA loading dye, and incubate at 65 °C for 4 min.
11. Remove the beads using magnetic stand and take supernatant.
12. Run the supernatant on a 0.8% agarose gel containing ethidium bromide at 1.4 V/cm for 22 hours at 4 °C using TAE buffer.
13. Transfer DNA to Immobilon-NY+ membrane by standard capillary transfer method using alkaline transfer buffer.
14. Detect the DNA bands using standard Southern blotting protocol. Incubate the membrane with ^{32}P-labeled TRP1 DNA fragment to probe 2.3 kb circular minichromosome (*see* **Note 12**). A representative result is shown in Fig. 4b.

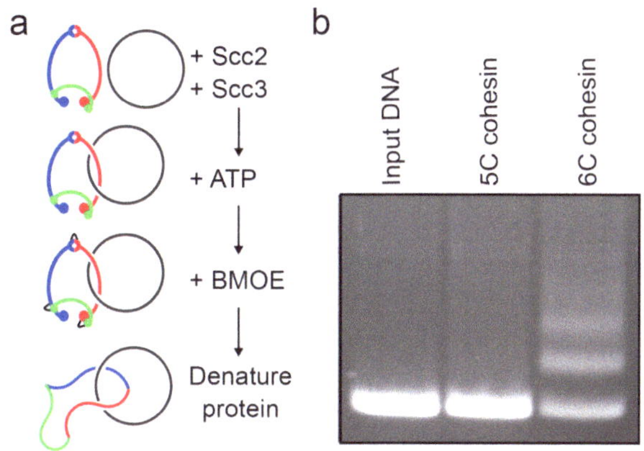

Fig. 5 (**a**) Schematic of the in vitro 6C assay. (**b**) DNA agarose gel showing the entrapment of circular DNA. Comparing cohesin that is covalently circularized (6C) against cohesin lacking one of the cysteines required for covalent circularization (5C)

3.2.2 In Vitro 6C Assay

1. Mix 165 nM 5C or 6C cohesin, 165 nM Scc3, and 165 nM $Scc2_{132\text{-}1493}$ with 9.3 nM pUC19 plasmid DNA in EB buffer, in a 12 μL reaction volume.
2. Add 1 μL of 65 mM ATP to a final concentration of 5 mM in a 13 μL reaction.
3. Incubate the samples at 24 °C for 1 h (*see* **Note 13**).
4. Add 1.5 μL of 6.4 mM BMOE stock solution to each reaction.
5. Incubate samples on ice for 6 min.
6. Add 1.5 μL of 10% (w/v) SDS and heat at 70 °C for 10 min.
7. Separate the reaction products in a 0.8% agarose/TAE gel at 1.5 V/cm at 4 °C for 17 h with 0.1 μg/mL ethidium bromide present.
8. Scan the gel using an UV imager. A representative result is shown in Fig. 5.

4 Notes

1. The SMC1, SMC3, SCC1, $SCC2_{132\text{–}1493}$, and SCC3 cDNAs are codon-optimized for expression in Sf9 cells and synthesized using the Genescript Thermo Fisher service.
2. We typically prepare the stock solutions and freeze aliquots at −20 °C. Avoid multiple freeze/thaw cycles.
3. Add the inhibitor cocktail and PMSF just before use.
4. Add the inhibitor cocktail, RNase A, and Supernuclease just before use.

5. The frozen cell pellets can be stored at −70 °C for further analysis.
6. We typically disrupt yeast cells using a FastPrep-24 (MP Biomedicals) at 6.5 m/s for 3x 1 min intervals, with 800 μL of acid-washed glass beads (425–600 μm, Sigma).
7. Care should be taken to avoid the wells drying out during incubation. This can be achieved by storing the plate within a container and adding wet tissue.
8. Transfection progress can be observed by appearance of yellow fluorescence protein. Virus is harvested by collection of the supernatant after a significant fraction of the cells are observed to fluoresce. This first passage (P1) virus is stored in the dark at 4 °C.
9. We check the cell viably by trypan blue staining.
10. Second passage (P2) virus is stored in the dark at 4 °C and is used to drive protein expression. We have found viruses to remain functional for several years, but this may vary.
11. If proceeding with an ion exchange step, the pH of the buffers should be at least 1 unit above or below the pI of the protein of interest, depending on the resin used. Adjust accordingly.
12. ^{32}P-labeled TRP1 probe is prepared as follows: The TRP1 gene was amplified using oligos Trp1F: 5′--atgtctgttattaatttcacaggtag-3′ and Trp1R: 5′--ctatttcttagcatttttgacgaaatttg-3′ using the genomic DNA of strain 23889 as template. The PCR product was stored at a concentration of 10 ng/μL. The radioactive probe was prepared using the Random Primer DNA Labeling Kit Ver. 2 and ATP, [γ-^{32}P], according to manufacturer's instructions. The labelled probe was purified using G-50 spin columns.
13. Different topological compartments within cohesin have been observed to entrap DNA with different kinetics. Typically, to observe efficient entrapment within the SMC-kleisin ring requires incubation of around 1 h. However, DNA entrapment within compartments created through head engagement, such as the E-S and E-K, are much more rapid and a 2 min incubation is sufficient [7, 8].

Acknowledgments

We acknowledge generous funding by the Wellcome Trust grant 226494/Z/22/Z (M.S.) We thank the editors for their constructive comments and assistance in preparing the manuscript.

References

1. van Hooff JE, Raas MW, Tromer EC et al (2024) Shaping up genomes: prokaryotic roots and eukaryotic diversification of SMC complexes. bioRxiv:2024.01.07.573240
2. Davidson IF, Peters JM (2021) Genome folding through loop extrusion by SMC complexes. Nat Rev Mol Cell Biol 22(7):445–464
3. Murayama Y (2025) Sister chromatid cohesion through the lens of biochemical experiments. Curr Opin Cell Biol 93:102464
4. Nasmyth KA, Lee B, Roig MB et al (2023) What AlphaFold tells us about cohesin's retention on and release from chromosomes. elife 12:RP88656
5. Gruber S, Haering CH, Nasmyth K (2023) Chromosomal cohesin forms a ring. Cell 112: 765–777
6. Srinivasan MS, Scheinost JC, Petela NJ et al (2018) The cohesin ring uses its hinge to organize DNA using non-topological as well as topological mechanisms. Cell 173(6): 1508–1519.e18
7. Collier JE, Nasmyth KA (2022) DNA passes through cohesin's hinge domain as well as its Smc3-kleisin interface. elife 11:e80310
8. Collier JE, Lee BG, Roig MB et al (2020) Transport of DNA within cohesin involves clamping on top of engaged heads by Scc2 and entrapment within the ring by Scc3. elife 9:e59560
9. Luckow VA, Lee SC, Barry GF et al (1993) Efficient generation of infectious recombinant baculoviruses by site-specific transposon-mediated insertion of foreign genes in a baculovirus genome propagated in *Escherichia coli*. J Virol 67(8):4566–4579
10. Weissmann F, Petzold G, Vander Linden R et al (2016) biGBac enables rapid gene assembly for the expression of large multisubunit protein complexes. Proc Natl Acad Sci USA 113(19): E2564–E2569

Chapter 9

Bulk FRET Analysis of the Conformation of the Cohesin Ring

Torahiko L. Higashi

Abstract

Cohesin is a member of the SMC protein family that exerts its chromosome functions by holding district DNA segments together using the ring structure. Topological cohesin loading involves complex, dynamic structural changes; these transitions are tightly linked to the initial cohesin association with DNA and subsequent DNA entrapment, which occurs through the transient opening of the ring at the interfaces between subunits. This chapter describes a method of monitoring the conformational changes of fission yeast cohesin that combines biochemical reconstitution and fluorescence resonance energy transfer (FRET). This method is useful for studying the structural dynamics and protein–protein interactions of not only cohesin but also other SMC proteins.

Key words Cohesin structure, ATP hydrolysis, FRET, In vitro reconstitution, *S. pombe*

1 Introduction

Cohesin, a eukaryotic structural maintenance of chromosomes (SMC) family protein, is known to mediate sister chromatid cohesion, which ensures faithful chromosome segregation during cell division. The ring complex also plays a vital role in other chromosomal structures and functions, including global chromosome organization in interphase cells, transcriptional regulation, and the accurate repair of DNA damage through homologous recombination. Mechanistically, cohesin is thought to function as a topological device that entraps and tethers two distinct DNA segments together [1, 2].

Cohesin is composed of two SMC subunits, Smc1 and Smc3, a kleisin subunit Scc1, and a kleisin-interacting subunit Scc3/SA (in fission yeast, Psm1, Psm3, Rad21, Psc3, respectively) (Fig. 1a). The SMC subunits are both long coiled-coil proteins that form a heterodimer through their interaction at hinge domains. The other ends of the coiled-coil contain ATP binding-cassette (ABC)-type ATPase, known SMC head domain, which engages with each other upon ATP binding. These ATPase heads

Yasuto Murayama and Yasutaka Kakui (eds.), *SMC Proteins: Methods and Protocols*, Methods in Molecular Biology, vol. 2991,
https://doi.org/10.1007/978-1-0716-5005-9_9,

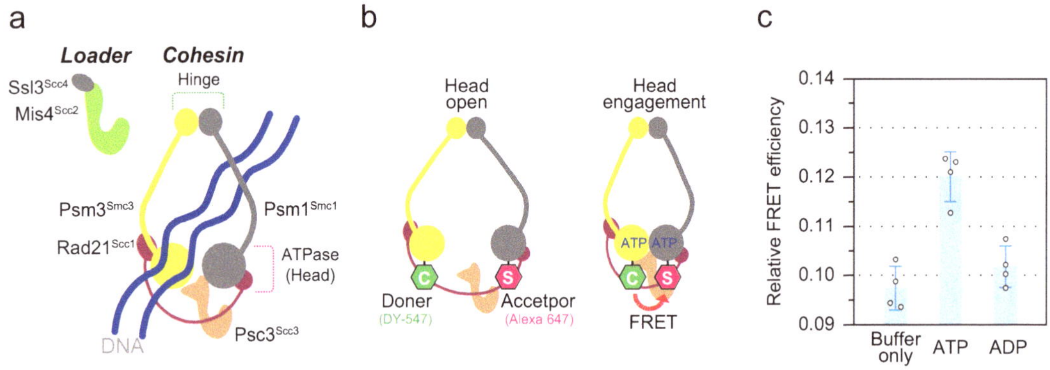

Fig. 1 Bulk FRET measurements of the cohesin ring. (**a**) Schematic of the cohesin complex. (**b**) Schematic of cohesin for head FRET measurement. "C" and "S" denote CLIP and SNAP tag, respectively. (**c**) Relative FRET efficiencies of between Smc1 and Smc3 heads in the condition without or with ATP or ADP. The graph and error bars represent average and standard deviation

are bridged by the asymmetric interactions of the kleisin Scc1 subunit, forming a proteinous ring topology. Scc1 also serves as a docking platform for Scc3. Cohesin loading onto DNA further requires an additional regulatory factor: the Scc2-Scc4 loader complex (in fission yeast, Mis4-Ssl3), which makes numerous physical contacts with cohesin. For DNA entrapment, one of the interfaces between cohesin subunits must open temporarily to allow the DNA strand to pass through. This process couples with ATP hydrolysis by the ATPase heads of cohesin [3–5]. During the DNA loading, the engagement of the SMC ATPase heads is thought to lead to conformational changes throughout cohesin, resulting in the opening and closing of the DNA gate [6–9]. To gain further insights into the DNA loading mechanism, capturing structural changes at each subunit interface is considered highly effective.

Several methods have been employed to capture structural changes in proteins, including atomic force microscopy (AFM), nuclear magnetic resonance (NMR), and cross-linking mass spectrometry (XL-MS). Among these, fluorescence resonance energy transfer (FRET) is a highly sensitive method for detecting changes in distance between two points. By attaching a pair of fluorescent molecules, called a "donor" and an "acceptor" to the two points of interest, the energy transfer efficiency (or FRET efficiency) can be calculated from the fluorescence intensities of the donor and acceptor. This allows changes in the distance between the two points to be detected. Comparing the FRET efficiency at each reaction step enables us to interpret the relative distance and conformational changes of the protein [10].

This chapter describes a method of capturing conformational changes in the cohesin ring using FRET-based distance changes between two points on the cohesin subunits [9]. Utilizing this method in conjunction with functional deficiency mutants or ATP ligand analogs that generate an intermediates state allows the analysis of its structural states. Depending on the choice of fluorescent dyes, this method can be applied to capture structural changes between three points. Furthermore, by applying the same protein constructs used in bulk FRET analysis directly to single-molecule FRET analysis makes it possible to track structural dynamics in real time [11, 12].

2 Materials

2.1 Yeast Strain for Expression of the Fission Yeast Cohesin

1. Cohesin for head FRET measurement: Psm1^{Smc1}-encording complementary DNA (cDNA) fused with SNAP-tag encoding sequence at the C-terminus and Psm3^{Smc3}-encording cDNA fused with CLIP-tag at the C-terminus were cloned into the shuttle vector YIplac211 under the control of *GLA1/GAL10* promotes, respectively. Rad21^{Scc1}-encording cDNA fused with 1x HA-PreScission protease recognition site-2x protein A tag at the C-terminus and Psc3^{Scc3}-encording cDNA fused with 7x histidine tag were cloned into the shuttle vector YIplac128 under the control of *GLA1/GAL10* promotes, respectively. These constructs were sequentially integrated into the budding yeast protease-deficient strain (*pep4Δ*) at *URA3* and *LUE2* genomic loci, receptivity [9].

2.2 Yeast Culture Medium

1. YP-raffinose liquid medium: 2% (w/v) peptone, 1% (w/v) yeast extract, 2% (w/v) raffinose. Prepare in distilled water and sterilized by autoclaving.
2. 20% galactose solution: Prepare in distilled water and sterilized by autoclaving.

2.3 Cohesin Purification and Fluorescent Labelling

1. Cell lysis buffer: 50 mM HEPES-KOH (pH 7.5), 1 mM dithiothreitol (DTT), 300 mM NaCl, 20% (w/v) glycerol, cOmplete-EDTA-free protease inhibitor cocktail (Roche). Add one inhibitor tablet per 50 mL of the buffer immediately before use.
2. IgG buffer: 25 mM HEPES-KOH (pH 7.5), 0.5 mM tris (2-carboxyethyl)phosphine (TCEP), 300 mM NaCl, 10% (w/v) glycerol.
3. R buffer: 20 mM Tris–HCl (pH 7.5), 0.5 mM TCEP, 10% (w/v) glycerol.
4. R buffer containing 1 M NaCl.
5. PreScission protease or HRC3C protease.
6. CLIP-Surface DY-547 (New England Biolabs): Dissolve in dimethyl sulfoxide (DMSO) to a final concentration of 1 mM.

7. SNAP-Surface Alexa Fluor 647 (New England Biolabs): Dissolve in DMSO to a final concentration of 1 mM.
8. 20 mL open column.
9. 50 mL tube.
10. Human IgG-agarose resin (Merck).
11. 1 mL HiTrap heparin HP resin (Cytiva).
12. Superose 6 10/300 GL gel filtration column (Cytiva).
13. Ultrafiltration tube with molecular weight cutoff of 30 kDa.
14. SDS-PAGE apparatus and 7.5% SDS-PAGE gel.
15. Coomassie protein stain.
16. Cryogenic cell grinder.
17. Centrifuges: Large centrifuge for harvest of yeast cell, ultracentrifuge for clarification of cell lysate.
18. FPLC system.
19. V550 UV/VIS spectrophotometer (Jasco).

2.4 Bulk FRET Measurement

1. The $Mis4^{Scc2}$-$Ssl3^{Scc4}$ cohesin loader complex: Mis4-Ssl3 is expressed and purified by following the method described previously [5].
2. pBlusecript KSII (+) plasmid.
3. 20 mM ATP.
4. 20 mM ADP.
5. 10× reaction buffer: 350 mM Tris–HCl (pH 7.5), 5 mM TCEP, 250 mM NaCl, 10 mM $MgCl_2$, and 0.03% (w/v) Tween 20.
6. 50% (w/v) glycerol.
7. 384-well plate.
8. Heat block.
9. CLARIOstar Plus (BMG LABTECH).

3 Methods

3.1 Design of FRET Measurable Cohesin

To measure distance changes between two specific points using FRET, two types of fluorescent molecule must be used: a donor and an acceptor. FRET generally occurs when the linear distance between two fluorophores is within 10 nm (100 Å). FRET can be measured by attaching different combinations of fluorescent proteins, such as CFP and YFP. However, these fluorescent proteins are relatively large, and the retention of fluorophores varies depending on the fused proteins. This can potentially interfere with protein activities and FRET efficiencies. The SNAP and CLIP tags are smaller (15 kDa) and can be covalently conjugated with a variety of fluorescent ligands (*see* **Note 1**). In this chapter, we position the SNAP and CLIP tags at the C-termini of Psm1^{Smc1} and Psm3^{Smc3} respectively, with an 8 amino acid linker separating them (Fig. 1b). This configuration is intended to detect FRET changes between the SMC ATPase head domains of cohesin (*see* **Note 2**). When inserting SNAP and CLIP genes, it is advisable to place a 4–8 amino acid linker between the target protein gene and the SNAP/CLIP gene.

3.2 Cohesin Purification

The FRET-measurable cohesin protein is prepared by incorporating a fluorescent labelling step into the cohesin purification process. Cohesin is initially purified using a 2× protein A tag fused to the Rad21 kleisin subunit, followed by a heparin column affinity chromatography. The purified cohesin then proceeds to the labelling step. First, we label the SNAP tag on Psm1 with an acceptor ligand (Alexa 647) and then conjugate the CLIP tag on Psm3 with a doner ligand (DY-547). SNAP tag exhibits greater specificity and efficiency in ligand conjugation than the CLIP tag. This order prevents nonspecific conjugation of the CLIP ligands with the SNAP tag, enabling specific labelling of the doner and accepter ligands to each tag. Unlabelled, free fluorescent ligands are then removed by gel filtration chromatography (*see* **Note 3**).

The labelling efficiency is calculated from the absorbance of the protein (A_{280}) and each fluorescent molecule. As DY-547 and Alexa 647 have maximum absorption at 555 nm and 655 nm, respectively, the molar concentration of each fluorescent molecule is calculated using their respective molar extinction coefficients ε (DY-547: 150,000 cm^{-1} M^{-1} at 555 nm, Alexa 647: 270,000 cm^{-1} M^{-1} at 665 nm).

In order to determine the molar concentration of the protein, it is necessary to correct for the absorbance of each fluorescent molecule at 280 nm. Therefore, the calibrated protein absorbance ($\text{Absorbance}_{\text{protein}}$) at 280 nm is calculated as follows:

$$(\text{Absorbance}_{280\text{nm}}) - (\text{Absorbance}_{555\text{nm}} \times \text{DY} - 547\ \text{CF}) \\ - (\text{Absorbance}_{655\text{nm}} \times \text{Alexa}\ 647\ \text{CF}).$$

where correction factor (CF) for is DY-547 CF = *0.21* and *Alexa 647* CF = 0.03, respectively. Consequently, the molar concentration of the protein is calculated as follows:

$$\text{Absorbance}_{\text{protein}}/\text{molar extinction coefficient}\ \varepsilon_{\text{protein}}.$$

The labelling efficiency of each fluorescent dye is calculated using the ratio of its molar concentration to that of the protein. In this protocol, the labelling efficiency for both fluorescent dyes is approximately 60–70%.

1. Cultivate the fission yeast cohesin expression strain in 10 liters of YP-raffinose liquid medium at 30 °C until the optical density at 600 nm reaches ~1.0.
2. Add 1 liter of 20% galactose solution (2% final) and cultivate at 30 °C for 2 h.
3. Harvest the cells by centrifugation at 4 °C, wash once with distilled water, then once with cell lysis buffer.
4. Resuspend the cells with 1/5 volume of ice-chilled cell lysis buffer.
5. Drop the cell suspension into liquid nitrogen to create small drops.
6. Disrupt cells by cryogenic grinder (*see* **Note 4**).
7. Suspend the cell powder in twice volume of cell lysis buffer and gently stir at 4 °C for 30 min (*see* **Note 5**).
8. Clarify the cell lysate by centrifugation at 4 °C at 100,000× *g* for 30 min.
9. Transfer the supernatant to 50 mL tube.
10. Add human IgG agarose, which is prewashed with IgG buffer (1 mL bed volume per 50 mL of the clarified cell lysate).
11. Incubate at 4 °C for 2 h with gentle rotation.
12. Transfer the suspension to a 20 mL open column.
13. Wash the resin with 80 bed volume of IgG buffer.
14. Suspend the resin with 5 mL IgG buffer.
15. Add PreScission protease to a final concentration of 5 U/mL, and incubate at 4 °C overnight (~15 h).
16. Collect the eluate and dilute twice with R buffer.
17. Apply the diluted eluate to 1 mL HiTrap heparin HP resin using FPLC system.
18. Wash the column with 10 bed volume of R buffer containing 150 mM NaCl.
19. Elute bound protein with 10 bed volume of a linear NaCl gradient from 150 mM to 1 M. Cohesin is eluate at ~600 mM NaCl.
20. Collect the peak fractions and concentrate them to ~500 μL by ultrafiltration.
21. Add 2 μM SNAP-surface Alexa 647, 1 mM DTT, and 0.003% Tween20 and incubated at 25 °C for 1 h. This step labels Psm1-SNAP with Alexa 647 (*see* **Notes 6** and 7).
22. Add 4 μM CLIP-surface DY-547 and incubate at 4 °C for 16 h to complete the labelling. This step labels Psm3-CLIP with DY-547.

23. Load the labeled fraction to a Superose 6 10/300 GL gel filtration column that was developed in R buffer containing 200 mM NaCl and 0.003% Tween20.
24. Collect the peak fraction and concentrate them to ~500 μL by ultrafiltration.
25. Measurement of the absorbance of the labeled protein to calculate the labelling efficiency using spectrophotometer.
26. Run the purified fraction by 7.5% SDS-PAGE followed by Coomassie staining.

3.3 Bulk FRET Measurement During Cohesin's ATPase Reaction

This section describes a protocol for measuring bulk FRET derived from the ATP-dependent engagement of SMC ATPase heads, as depicted in Subheading 3.2. FRET is measured in solution during the cohesin loading reaction. The reaction mixtures are placed in a fluorescent plate reader and excited using a donor excitation wavelength of 525 nm. Emission spectra from both the donor and acceptor fluorophores are then recorded. The relative FRET efficiencies are then calculated using the following formula:

$$I_A/(I_D + I_A).$$

where I_D and I_A represent the maximum emission intensities of the donor at 565 nm (I_D) and the acceptor at 665 nm (I_A), respectively.

1. Mix 10 nM doubly fluorescent-labelled cohesin, 100 nM Mis4-Ssl3, and 3 nM pBluescript DNA in reaction buffer (35 mM Tris–HCl (pH 7.5), 0.5 mM TCEP, 25 mM NaCl, 1 mM $MgCl_2$, 15% (w/v) glycerol, and 0.003% (w/v) Tween 20) using the stock solutions.
2. Add 1 μL of 20 mM ATP (0.5 mM final) and incubate at 32 °C for 20 min. The final reaction volume is 40 μL.
3. Apply 35 μL of the reaction samples to a 384-well plate.
4. Measure the fluorescence spectra of the reactions using a CLARIOstar high-performance plate reader. Samples are excited at 525 nm and emitted lights are recorded between 560 and 700 nm in 0.5 nm increments.
5. Calculate relative FRET efficiency using the formula of $I_A/(I_D + I_A)$. Figure 1c shows representative results of the ATP-dependent head engagement obtained in the absence or in the presence of ATP or ADP.

4 Notes

1. Although SNAP and CLIP tags are smaller than fluorescent proteins such as GFP, the attachment can sometimes significantly affect protein function or activity. Therefore, thorough

verification is necessary when designing constructs. In this study, DNA binding activity of cohesin constructs was examined prior to bulk FRET measurement, and it was confirmed that DNA binding and ATP hydrolysis activities remained unchanged compared to the wild type [9].

2. The efficiency with which SNAP and CLIP tags are fluorescently labelled varies depending on the construct. If these efficiencies differ significantly, it is difficult to compare FRET efficiencies between different constructs. However, it is meaningful to compare FRET efficiencies between different reaction conditions for the same construct in order to evaluate distance changes between the two points of interest. To compare different constructs, either normalize the labelling efficiencies or use single-molecule FRET instead of bulk FRET to compare FRET efficiencies.
3. Small spin column does not remove free fluorescent ligands sufficiently. These free ligands then interfere with the accurate measurement of labelling efficiency.
4. Frozen yeast cells are disrupted by SPEX CertiPrep 6850 using a setting of Rate 15 cps, 1 min on/1 min off, for 10 cycles.
5. We conduct all purification steps in a cold room at 4 °C.
6. SNAP/CLIP labelling involves the covalent conjugation of a ligand via the serine residue in these tags. Adding a reducing agent, such as DTT, is essential.
7. As fluorescent dyes are susceptible to photobleaching, which reduces labelling efficiency and significantly impacts FRET efficiency, it is essential to handle the proteins in a manner that minimizes photobleaching. During the labelling reaction, the reaction vessel should be kept in the dark. During SEC purification, efforts should be made to minimize exposure to light and UV irradiation of the columns and fractions. Adding photobleaching inhibitors such as oxygen scavengers to the reaction solution can also be effective if they do not inhibit the function of the protein itself.

Acknowledgments

For the establishment of the FRET methods, I would like to thank Simone Kunzelmann, Yasuto Murayama, the Crick Fermentation Science Technology Platform, and Frank Uhlmann, as well as the members of Frank's laboratories for valuable discussions and advice.

References

1. Uhlmann F (2016) SMC complexes: from DNA to chromosomes. Nat Rev Mol Cell Biol 17:399–412
2. Uhlmann F (2025) A unified model for cohesin function in sister chromatid cohesion and chromatin loop formation. Mol Cell 85(6): 1058–1071
3. Arumugam P, Gruber S, Tanaka K et al (2003) ATP hydrolysis is required for cohesin's association with chromosomes. Curr Biol 13:1941–1953
4. Weitzer S, Lehane C, Uhlmann F (2003) A model for ATP hydrolysis-dependent binding of cohesin to DNA. Curr Biol 13:1930–1940
5. Murayama Y, Uhlmann F (2014) Biochemical reconstitution of topological DNA binding by the cohesin ring. Nature 505:367–371
6. Hopfner KP, Karcher A, Shin DS et al (2000) Structural biology of Rad50 ATPase: ATP-driven conformational control in DNA double-strand break repair and the ABC-ATPase superfamily. Cell 101:789–800
7. Murayama Y, Uhlmann F (2015) DNA entry into and exit out of the cohesin ring by an interlocking gate mechanism. Cell 163:1628–1640
8. Gruber S, Arumugam P, Katou Y et al (2006) Evidence that loading of cohesin onto chromosomes involves opening of its SMC hinge. Cell 127:523–537
9. Higashi TL, Eichkhoff P, Sousa JS et al (2020) A structure-based mechanism for DNA entry into the Cohesin ring. Mol Cell 79(6): 917–933
10. Mc Intyre J, Muller EG, Weitzer S et al (2007) In vivo analysis of cohesin architecture using FRET in the budding yeast Saccharomyces cerevisiae. EMBO J 26(16):3783–3793
11. Ticau S, Friedman LJ, Ivica NA et al (2015) Single-molecule studies of origin licensing reveal mechanisms ensuring bidirectional helicase loading. Cell 161(3):513–525
12. Bauer BW, Davidson IF, Canena D et al (2021) Cohesin mediates DNA loop extrusion by a "swing and clamp" mechanism. Cell 84(21): 5448–5464

Part III

Biochemical Assays of SMC Proteins in Chromosomal Processes

Chapter 10

Assessing SMC Complex Function in Replication Fork Progression with DNA Fiber Assays

Neus P. Lorite, Snehlata, Neus Colomina, Jordi Torres-Rosell, Sara Rodríguez-Acebes, and Juan Méndez

Abstract

Chromosome organization undergoes substantial changes during S phase, as parental strands unzip at replication forks to create sister chromatids. Structural Maintenance of Chromosomes (SMC) complexes actively organize chromatin fibers in the wake of the replication fork during a normal S phase, but also in response to replicative stress, thereby influencing replication dynamics. The DNA fiber assay is a relatively simple and cheap technique that allows the study of replication dynamics by sequential incorporation of two nucleotide analogs, followed by DNA spreading and immunodetection of replicated tracks. Here, we present a step-by-step protocol to perform DNA fiber assays in human cell lines, providing a quantitative approach to analyze replication fork progression. As an example, we describe its application in wild-type and SMC5/6 mutant cells, treated or not with hydroxyurea to induce replicative stress. This method can be easily customized to other cell types, alternative strategies for SMC complex inactivation, or treatments that perturb DNA replication. In addition, by modifying culture conditions, this approach can be used to investigate not only fork rate but also fork reversal, fork restart, fork protection, and other aspects of global replication dynamics regulated by SMC activities.

Key words SMC5/6, NSE1, NSMCE1, DNA replication, Replication fork, DNA fiber, CldU, IdU, Replicative stress, Hydroxyurea, DNA repair, Chromosome

1 Introduction

The organization of chromosomes is profoundly altered during passage of the replication fork. Parental DNA strands are separated and used as templates for the synthesis of nascent strands, while sister chromatids are resolved from each other and simultaneously paired in preparation for mitosis. SMC complexes actively participate in the organization of sister chromatids behind forks; for

---

The original version of the chapter has been revised. A correction to this chapter can be found at https://doi.org/10.1007/978-1-0716-5005-9_18

Yasuto Murayama and Yasutaka Kakui (eds.), *SMC Proteins: Methods and Protocols*, Methods in Molecular Biology, vol. 2991, https://doi.org/10.1007/978-1-0716-5005-9_10,

example, sister chromatids are topologically resolved by the prokaryotic SMC complex, while sister chromatid cohesion is established by the eukaryotic cohesin complex [1]. In addition to their roles during a normal S phase, SMC complexes help to reorganize replication structures in response to alterations induced by internal or external factors, such as replicative stress, ultimately affecting global replication dynamics and genome organization.

DNA replication dynamics can be quantitatively studied using DNA fibers. This technique involves the sequential incorporation of two halogenated nucleotide analogues, 5-chloro-20-deoxyuridine (CldU) and 5-Iodo-20-deoxyuridine (IdU), followed by cell lysis and DNA spreading onto a glass slide [2]. The incorporated analogues are then detected by immunofluorescence, using antibodies specific to the first and second labels, to visualize replicated tracks at the single-molecule level. Microscopic analysis of replicated molecules reveals different types of replication structures: Ongoing forks appear as dual-labeled tracks; replication origins show a central first-analogue signal flanked by the second, while new origins contain only the second label; termination events are characterized by a central second-label track flanked by marks from the first label; asymmetric arms indicate impaired fork progression; and fork stalling can be inferred by exclusive first-analogue incorporation, indicating that replication was terminated before the second labeling.

The incidence of replication structures can be drastically altered by replicative stress, a hallmark of cancer cells. In the lab, replicative stress can be induced by drugs interfering with DNA replication such as hydroxyurea, camptothecin, or MMS [3]. Under replicative stress, helicase-polymerase uncoupling leads to excessive ssDNA accumulation, triggering fork remodeling mechanisms that protect against chromosome breakage [4]. These include repriming ahead of blocks, fork reversal, translesion synthesis, or template switching [5]. Fork reversal, regulated by forks remodelers, promotes nascent strand annealing and regressed arm elongation, converting the three-way fork structure into a four-way junction that temporarily halts DNA polymerization and protects the fork from degradation. In the absence of fork reversal, fork rates can be maintained by repriming ahead of the block. Upon recovery, reversed forks return to their normal state and resume replication. In fiber assays, fork reversal can be inferred by the shortening of the second-labeled DNA track upon replicative stress. Fork restart can also be assessed using DNA fibers by intercalating a stress treatment between the two labeling steps, with defects in fork restart detected as a failure to incorporate the second analog. Finally, a variation of this protocol, based on applying replicative stress after both analogs, allows the detection of fork protection mechanisms; in this case, the shortening of the second track is indicative of unprotected fork degradation.

The DNA fiber assay can be used to study the role of SMC complexes in fork dynamics. For example, cohesin deficiency in stem cells reduces replication fork rate [6], suggesting that cohesin is essential for proper fork progression. The active loading and removal of cohesin also seems to be required for fork progression and protection. Mutations that prevent cohesin ring opening at the SMC3-RAD21 interface [7] or defects in SMC3 acetylation, such as those found in Roberts syndrome [8] or STAG2-deficient cells [9, 10], also slow down fork progression and increase fork stalling. Conversely, extending the residence time of cohesin on chromatin by downregulating PDS5A/B or WAPL similarly disrupts fork progression [11], and these factors are critical for fork protection and restart in both mouse and human cells [11–13]. The function of the SMC5/6 complex in DNA replication has also been analyzed using DNA fibers. Early DNA combing experiments, a variation of the DNA fiber assay, demonstrated that SMC5/6 is required to complete replication of the rDNA locus, as well as other genomic loci in response to replicative stress in yeast [14, 15]. More recently, fiber assays have been used to study the effect of human SMC5/6 inactivation during replication. For example, cells mutated in the RING domain of the NSE1 subunit displayed slower replication forks [16]. Similarly, inducible degradation of the SMC5 subunit in mouse neuronal precursors impaired fork progression and increased fork stalling [17]. Patient-derived cells carrying hypomorphic mutations in the SLF2 or SMC5 subunits showed little to moderate reductions in fork speed; however, DNA fiber assays revealed a decrease in the number of ongoing forks, an increase in stalled and asymmetric forks, and overall impaired fork progression in these mutants [18]. Additionally, DNA fiber assays have linked SMC5/6 to fork protection and restart [19, 20].

In this chapter, we provide a step-by-step protocol to perform DNA fiber assays in human cell lines. In addition, we describe an example using HEK293T wild-type and *nse1-ΔR* mutant cells, treated or not with hydroxyurea, to assess the effects of replicative stress on fork progression. This method can be easily adapted to other cell lines, alternative approaches for inactivating SMC complexes, such as RNAi and degrons, or treatments that interfere with DNA replication. Additionally, by adjusting cell culture conditions, this protocol can be used to study not only fork rate but also fork reversal, fork protection, and restart, as global replication dynamics.

2 Materials

2.1 Cell Lines

1. Human embryonic kidney (HEK293T) cells (ATCC, USA, CRL-11268).
2. Mutant *nse1-ΔR* cells: HEK293T cells were gene-edited by CRISPR/Cas9 system to generate mutant cells containing a disruption of the conserved RING domain of NSE1, a subunit of the SMC5/6 complex, resulting in depletion of the NSE1

protein and severe reduction in the expression of all other SMC5/6 subunits [16] (*see* **Note 1**).

2.2 Incorporation of Nucleoside Analogs

1. 6-well plates.
2. Growth media: Dulbecco's modified Eagle medium (DMEM), 10% fetal bovine serum (FBS), 20 U/mL penicillin, 20 μg/mL streptomycin.
3. Phosphate-buffered saline (10× PBS): 0.1 M Na_2HPO_4, 1.37 M NaCl, 27 mM KCl, 18 mM KH_2PO_4 in H_2O. Filtered and diluted 10 times to 1× PBS.
4. 10 mM 5-chloro-20-deoxyuridine (CldU): Dissolve CIdU in H_2O to a final concentration of 10 mM. Store at −20 °C.
5. 10 mM 5-iodo-20-deoxyuridine (IdU): Dissolve IdU in 10 mL of H_2O. Add 50 μL of 10 N NaOH. Store at −20 °C.
6. Hydroxyurea (HU) (*see* **Note 2**).

2.3 Cell Harvesting

1. 10–15 mL tubes to collect cells.
2. 0.25% Trypsin-EDTA (1×).
3. Cold 1× PBS.

2.4 Cell Spreading and Lysis

1. Spreading buffer: 0.5% SDS, 200 mM Tris–HCl (pH 7.4), 50 mM EDTA in H_2O.
2. Microscope slides with frosted margin.
3. 100% ethanol.
4. A humidity chamber.

2.5 Cell Spread Fixation, Denaturation, and Blocking

1. Coplin staining jars.
2. Methanol:acetic acid 3:1 solution.
3. Denaturation solution: 2.5 N hydrochloric acid (HCl) in ultra-pure water.
4. Blocking buffer: 1% bovine serum albumin (BSA), 0.1% Tween 20 in 1× PBS.

2.6 Immunofluorescence

1. Anti-CldU/anti-IdU dilution: Anti-CldU (rat monoclonal anti-BrdU, clone BU1/75 (ICR1)), and anti-IdU (mouse monoclonal anti-BrdU, clone B44) diluted both at 1:100 in blocking buffer. Prepare 80 μL per slide (*see* **Note 3**).
2. Anti-ssDNA: Mouse monoclonal IgG2a, diluted 1:100 to 1: 300 in blocking buffer (*see* **Note 4**).
3. 0.1% Tween20 in 1× PBS.
4. 4% paraformaldehyde (PFA) in 1× PBS.
5. 2ary antibody dilution: Goat anti-rat IgG Alexa Fluor 594 (Texas Red filter) and goat anti-mouse IgG1 Alexa Fluor

488 (FITC filter) diluted both at 1:300 in blocking buffer. Prepare 80 μL per slide.

6. Goat anti-mouse IgG2a Alexa Fluor 647 (Cy5 filter).
7. Micro cover glasses 24 × 60 cm No 1.
8. ProLong Gold.

2.7 Image Acquisition and Analysis

1. Conventional fluorescence microscope equipped with a CCD camera and software for camera control.
2. ImageJ or other image analysis software allowing track length measurement.
3. Standard statistic applications for track length analysis (i.e., Microsoft Office Excel).

3 Methods

3.1 Seeding

1. Seed cells in 6-well plates in complete DMEM supplemented with 10% FBS, 20 U/mL penicillin, and 20 μg/mL streptomycin and grow to exponential phase. Seed two wells with wild-type HEK293T cells and two wells with HEK293T *nse1-ΔR*-modified cells. Seed an additional well for a 1:1 mix of labeled and unlabeled cells to improve fiber spreading.

3.2 Pulse Labeling Nascent DNA with Nucleoside Analogs

1. Dissolve HU in culture media to 10 mM.
2. Prepare 50 μM CldU and 250 μM IdU in culture media (2 mL per well).
3. For replicative stressed cultures, prepare 250 μM IdU and 50 μM HU in culture media (2 mL per well).
4. Pre-warm media containing CldU, IdU or IdU/HU, and 1× PBS at 37 °C.
5. Pulse-label exponentially growing cells with pre-warmed 50 μM CldU for 30 min at 37 °C (*see* **Note 5**).
6. Wash cells 3 times with 2 mL pre-warmed 1× PBS (*see* **Note 6**).
7. Pulse-label cells with pre-warmed 250 μM IdU for 30 min at 37 °C. For replication stress experiments, pulse label cells with media containing both 250 μM IdU and 50 μM HU (*see* **Note 7**).
8. Keep one well without CldU/IdU incorporation for unlabeled cells.

3.3 Cell Harvesting

1. Remove culture media by aspiration and wash cells 3 times with 1× PBS.
2. Add 300 μL of 0.25% Trypsin-EDTA and incubate for 1 min (or until cells detach) at room temperature.

3. Add 1 mL of cold culture media, collect cells, and transfer them to a 10 mL tube previously containing cold culture media to inhibit trypsin. Immediately place them on ice.
4. Centrifuge for 3 min at 1000 rpm.
5. Resuspend cells in 1 mL cold 1× PBS and count the cells (*see* **Note 8**).
6. Centrifuge cells for 3 min at 1000 rpm.
7. Resuspend cells carefully in cold 1× PBS at approximately 1500 cells/μL.
8. Prepare clean tubes to mix labeled and unlabeled cells in a 1:1 ratio and place them on ice (*see* **Note 9**).

3.4 Cell Spreading and Lysis

1. Clean slides with 100% ethanol and dry with towel paper.
2. Label slides on frosted margin with pencil.
3. Place two drops of 2 μL each of the cell suspension on a microscope slide as shown in Fig. 1 (*see* **Note 10**).
4. Lyse cells by adding 10 μL of fresh pre-warmed (30 °C) spreading buffer (*see* **Note 11**).
5. Incubate for 6 min at room temperature in a humidity chamber (*see* **Note 12**).
6. Tilt slides at a 10–15° angle to allow the DNA suspension to spread slowly along the slide (Fig. 1). Let the drop run slowly

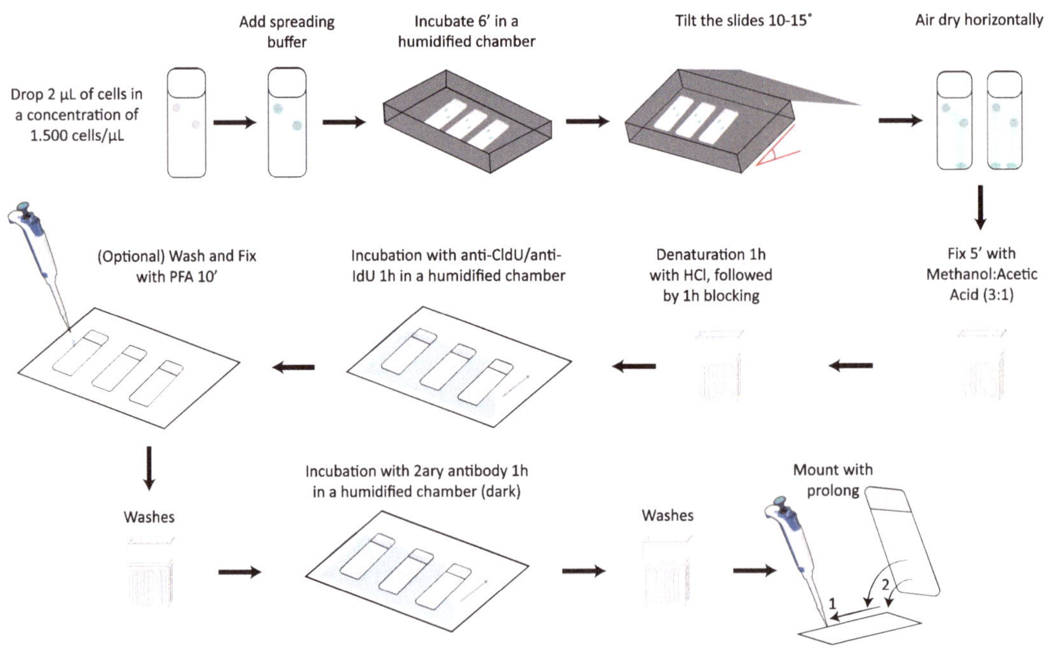

Fig. 1 Workflow of the DNA fiber assay, depicting key steps from labeling to spreading and immunodetection

and at a constant speed. Repeat **steps 3–6** if it stops or does not run straight (*see* **Note 13**).

7. Once the drops have reached the end of the slide, place all slides horizontally and allow them to air-dry for at least 20 min (*see* **Note 14**).

3.5 Cell Spread Fixation

1. Fix cell spreads using cold (−20 °C) methanol:acetic acid 3:1 solution for 5 min at RT in a Coplin staining jar (*see* **Note 15**).
2. Air-dry the slides.
3. At this point, slides can be processed for immunostaining or stored at 4 °C for at least 1 week.

3.6 Denaturation and Blocking

1. Rinse slides 3 times with filtered 1× PBS in a Coplin staining jar.
2. Wash slides once with denaturation solution in a Coplin staining jar and incubate in denaturation solution for 1 h at RT.
3. Rinse slides 4 times with filtered 1× PBS in a Coplin staining jar, incubating slides for 5 min in 1× PBS in the last wash (*see* **Note 16**).
4. Wash slides once with blocking buffer in a Coplin staining jar and incubate them with blocking buffer for 1 h at RT (*see* **Note 17**).

3.7 Immunofluorescence

1. Prepare 80 μL per slide of anti-CldU/anti-IdU dilution.
2. Place the slides horizontally in a humidity chamber, add 76 μL per slide of the anti-CldU/anti-IdU dilution, and cover the slides carefully with parafilm to spread the antibodies uniformly on the slide (Fig. 1).
3. Incubate for 1 h at RT in a humidity chamber.
4. Prepare 80 μL per slide of 2ary antibodies dilution. Keep in the dark at 4 °C.
5. Wash slides once with 0.1% Tween20 in 1× PBS and twice with 1× PBS in a Coplin staining jar.
6. Optional step: Fix primary antibodies with 4% PFA-containing PBS for 10 min at RT in the fume hood (Fig. 1). Place slides horizontally in a box and add 1.5 mL of PFA on each slide. Fix for 10 min and wash 3 times with PBS (*see* **Note 18**).
7. Wash slides once with blocking buffer in a Coplin staining jar.
8. Incubate for 1 h at RT in a humidity chamber with 2ary antibodies dilution in the dark (from this point on, slides should be protected from light).

3.8 Mounting Slides

1. Rinse slides 5 times with 1× PBS in a Coplin staining jar.
2. Air-dry slides completely in the dark.

3. Place the cover glass horizontally, add 40 μL of ProLong on the cover glass (as shown in Fig. 1).
4. Place the immunostained slide over the cover glass. Immediately invert the slide and allow it to dry overnight horizontally at room temperature in the dark (*see* **Note 19**).
5. Store at 4 °C until image acquisition.

3.9 Image Acquisition and Analysis

1. Acquire images with a conventional fluorescence microscope. Use the side of the slide closer to the frosted margin to focus on the image, where more fibers can be observed (*see* **Note 20**).
2. Obtain photos in separated channels, to properly proceed with the image processing, for example, FITC filter to visualize mouse IgG Alexa 488 signal and Texas Red filter to visualize rat IgG Alexa 594 signal.
3. Include a scale in at least the first two images from each slide to convert pixels into μm.
4. Measure a minimum of 200–300 individual tracks in each condition, to estimate the fork rate. Only DNA tracks characterized by continuous CldU-IdU signals should be analyzed to measure replication fork progression ("Ongoing Forks" in Fig. 2).
5. Open the images in ImageJ and merge both channels (Image–Color–Merge Channels–Select photos–Save the image).

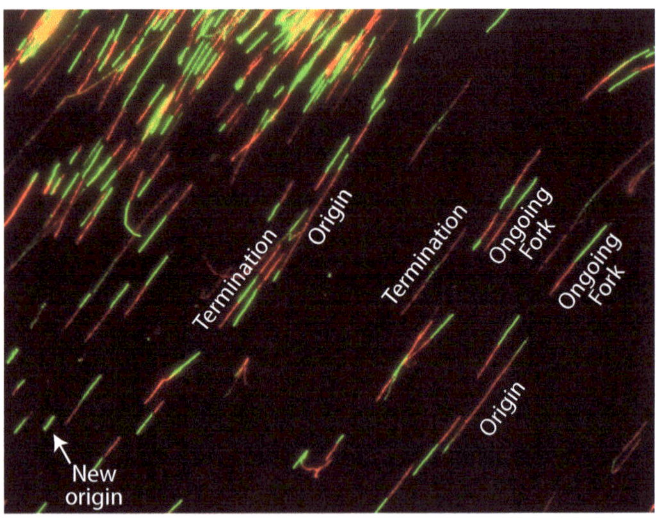

Fig. 2 Representative micrograph of DNA fiber spreading. First label (CldU) is shown in red, and second label (IdU) in green. Most fibers appear as dual-labeled tracks, representing ongoing forks. Origins (red flanked by green tracks), terminations (green flanked by red tracks), and occasional new origins (only green) or stalled forks (only red) may also be observed

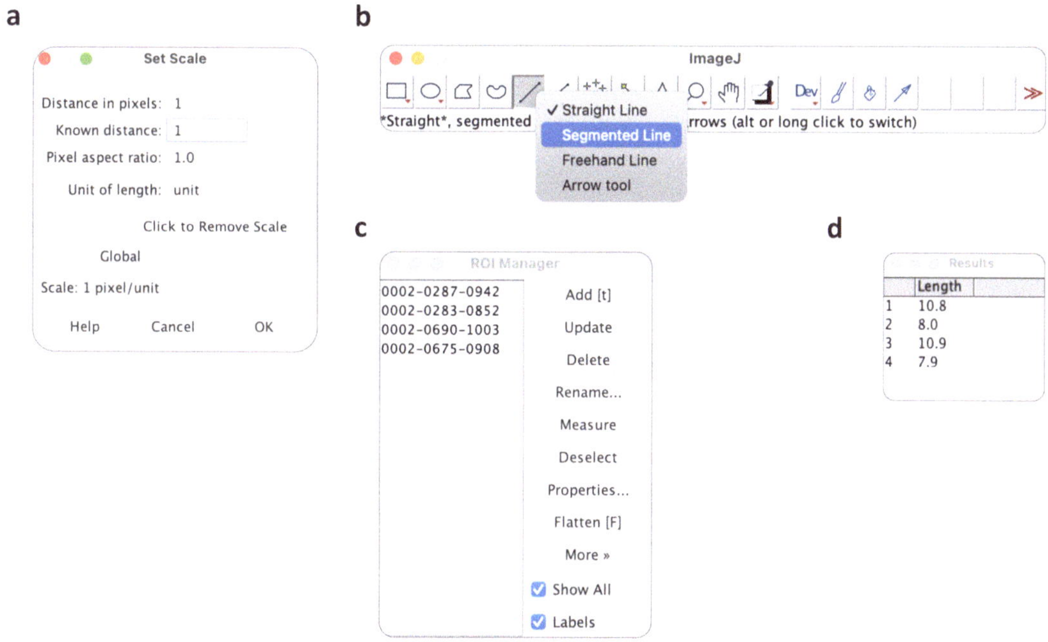

Fig. 3 (**a**) ImageJ setting to set up the scale before analyzing the data. (**b**) Selection of "Segmented Line" option from the ImageJ main menu. (**c**) ROI manager window on ImageJ showing DNA fiber measurements. (**d**) Results window in ImageJ showing alternating CldU and IdU track measurements

6. To properly visualize the fibers, go to Image–Color–Make composite and then to adjust color balance, Image–Color–Color balance.
7. To set up the scale, first, open the merged photo in ImageJ, and click "Analyze" and "set scale" (Fig. 3a). Then change all the parameters to "1," to have "scale: 1 pixel/μm." Select the "segmented line" option in the bar (Fig. 3b) to measure the scale in the picture. Open the ROI manager to know the exact number of pixels in the selected scale (Fig. 3c, d). In "Analyze>Set scale" set the "Distance in pixels" with the obtained measure and the "Known distance" with the scale length. Close the ROI manager (*see* **Note 21**).
8. Use the segmented line (Fig. 3b, c) to trace the fibers, adding them as a new ROI (*see* **Note 22**). Always measure the two labels in each fiber in the same order so that first and second track lengths alternate in the final data file.
9. Click "Measure" in the ROI manager to obtain the length of the measured tracks in μm (Fig. 3d). Transfer this data to an Excel file.
10. Fork progression can be measured by counting the second track of each fiber and comparing the changes in length among different samples. However, in samples under

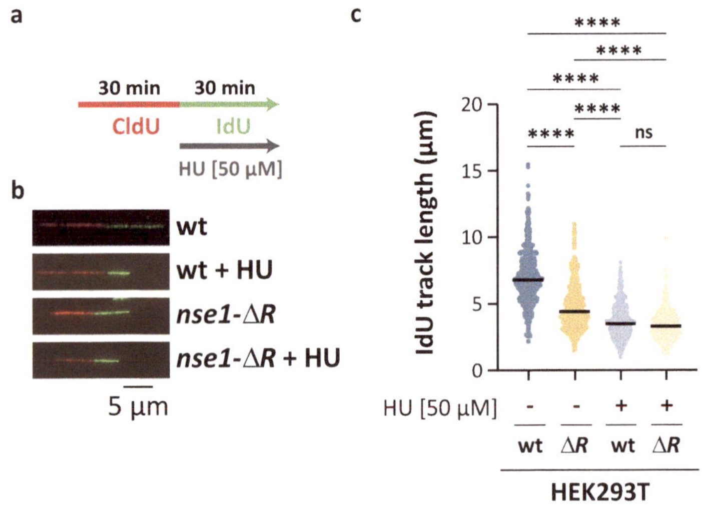

Fig. 4 DNA fiber analysis in HEK293T wild type (wt) and *nse1*-Δ*R* mutant cells under normal conditions and in response to HU to induce replicative stress. (**a**) Scheme of the sequential labeling protocol with CldU (30 min) and IdU (30 min) to evaluate fork progression. Notice that during the second labelling, cells were exposed or not to 50 μM HU. (**b**) Representative fibers from the indicated cultures. (**c**) Quantification and analysis of the IdU track length in micrometers (μm) for wild-type (wt) and *nse1*-Δ*R* (Δ*R*) cells. The median (black line) and the distribution of the IdU tracks length in μm are plotted. At least 300 fibers per condition were measured. (Kruskall-Wallis followed by Dunn's multiple comparisons test, *ns* nonstatistically significant, ****P value <0.0001)

replicative stress conditions (HU treatment during the second analogue incorporation), it is recommended to count both tracks to calculate the ratio of IdU and CldU measurements (IdU/CldU) (*see* **Note 23**).

11. The size distribution of tract lengths, or ratios, may be plotted as a scatter dot plot, with a line representing the median (Fig. 4c).

Figure 4 presents a representative experiment quantifying IdU track length, measured in micrometers (μm), which serves as a proxy for replication fork velocity. Under normal conditions, sequential labeling with CldU and IdU for 30 min resulted in red–green tracks of comparable length (Fig. 4b). However, in *nse1*-Δ*R* mutant cells, red–green track lengths were significantly reduced relative to wild-type HEK293T cells, indicating that SMC5/6 dysfunction impairs replication fork progression (Fig. 4b, c), as previously described [16]. To assess fork velocity under replicative stress, we treated cells with HU, which depletes dNTP pools and slows replication. When HU was applied during IdU incorporation, IdU tracks were notably shorter in wild-type cells, confirming reduced fork velocity (Fig. 4b, c). In contrast, *nse1*-Δ*R* mutant cells showed a more modest effect in response to HU, suggesting that SMC5/6 dysfunction may partially impair fork deceleration under replicative stress (Fig. 4c). These findings highlight the participation of the SMC5/6 complex in regulating replication fork progression, both under normal growth conditions and in response to replicative stress.

4 Notes

1. In addition to the current model, the impact of SMC complexes on replication fork progression can also be investigated using alternative approaches to disrupt their function. In human cell lines, this can be achieved by downregulating specific subunits using siRNAs or shRNAs [6]. Alternatively, degron systems provide a means to analyze the immediate consequences of protein depletion [17, 21].
2. A range of drugs, including aphidicolin, camptothecin, etoposide, and alkylating agents like MMS, offer alternatives to hydroxyurea for studying the effects of blocks on DNA replication.
3. These monoclonal anti-BrdU clones are the most commonly used in fiber labeling experiments. It is important to use antibodies that specifically recognize either CldU or IdU while avoiding cross-reactivity.
4. Labeling the ssDNA is optional. However, it is recommended when counting inter-origin distance because it helps to clearly distinguish replicated DNA from unreplicated regions.
5. Treat wells sequentially to prevent timing discrepancies.
6. This ensures complete removal of the first analog from the media. Perform this step quickly to minimize time gaps between the two labelings.
7. The second nucleoside analog is added at a threefold higher concentration to enhance its incorporation. If using drugs that interfere with DNA replication other than HU during the second labeling, prepare 250 μM IdU in culture media along with the drug.
8. Count both labeled and non-labeled cells to know the exact number of cells before spreading. If different treatments or cell types are used, it is recommended to count all conditions to ensure an adequate cell number for successful spreading.
9. The cell suspension can be stored at 4 °C for up to a week, but fiber quality is optimal when processed immediately. The remaining cells can be used as controls for downregulation or treatment via Western blot analysis.

10. Prepare 2–3 slides per condition as backups in case fiber spreading or quality is not good.
11. Using a 20 μL pipette, carefully dispense 10 μL vertically onto the 2 μL of cell suspension already on the slide. Avoid bubbles to ensure proper fiber spreading.
12. We typically incubate for 6 min for HEK293T and HeLa cells. The incubation timing may vary depending on laboratory conditions and may need to be optimized for other cell types. For example, we use 7 min for HCT116 cells.
13. Always use the same support to tilt the slides for consistency between different experiments.
14. If the cell spreads are not fully dry, they may appear glistening when viewed against the light.
15. Methanol:acetic acid 3:1 solution works better if freshly prepared.
16. It is important to completely remove the HCl, to protect the antibodies used in the next steps.
17. Do not wash with PBS after the blocking step, as the residual blocking buffer helps distribute the antibodies evenly on the slides.
18. It is important to completely remove PFA to protect the antibodies used in the next steps.
19. Avoid creating bubbles when mounting the slides, as they could distort the spread DNA fibers.
20. In the example shown in Fig. 2, images were acquired using Zeiss AxioObserver Z.1 microscope, ZEN3.7 software, with either a 63×/1.4 or 100×/1.4 oil immersion objective.
21. It is important to modify the scale before starting the analysis.
22. It is essential not to count more than 10% of the total number of fibers in any given image. For instance, if the total count is 300 fibers, no more than 30 fibers should be counted in a single photograph.
23. In untreated cells, CldU and IdU tracks should be of similar length, resulting in a ratio close to 1. Ensure at least three independent replicates to support reproducibility and statistical robustness.

Acknowledgments

Work in JT-R lab is supported by grant PID2021-127689NB-I00 from MICIN/AEI/ 10.13039/501100011033/ and European Union Regional Funds (FEDER). The IRBLLEIDA Institute is part of the CERCA. NPL was supported by a "Formación de doctores" fellowship (FPI/ PRE2019-090744) and S by a UdL predoctoral fellowship.

References

1. Hoencamp C, Rowland BD (2023) Genome control by SMC complexes. Nat Rev Mol Cell Biol 24:633–650. https://doi.org/10.1038/s41580-023-00609-8
2. Merrick CJ, Jackson D, Diffley JFX (2004) Visualization of altered replication dynamics after DNA damage in human cells. J Biol Chem 279:20067–20075. https://doi.org/10.1074/JBC.M400022200
3. Zellweger R, Dalcher D, Mutreja K et al (2015) Rad51-mediated replication fork reversal is a global response to genotoxic treatments in human cells. J Cell Biol 208:563–579. https://doi.org/10.1083/jcb.201406099
4. Adolph MB, Cortez D (2024) Mechanisms and regulation of replication fork reversal. DNA Repair (Amst) 141:103731. https://doi.org/10.1016/J.DNAREP.2024.103731
5. Quinet A, Tirman S, Cybulla E et al (2021) To skip or not to skip: choosing repriming to tolerate DNA damage. Mol Cell 81:649–658. https://doi.org/10.1016/J.MOLCEL.2021.01.012
6. Choi E-H, Yoon S, Koh YE et al (2022) Meiosis-specific cohesin complexes display essential and distinct roles in mitotic embryonic stem cell chromosomes. Genome Biol 23:70. https://doi.org/10.1186/s13059-022-02632-y
7. Sakata R, Niwa K, Ugarte La Torre D et al (2021) Opening of cohesin's SMC ring is essential for timely DNA replication and DNA loop formation. Cell Rep 35:108999. https://doi.org/10.1016/J.CELREP.2021.108999
8. Terret ME, Sherwood R, Rahman S et al (2009) Cohesin acetylation speeds the replication fork. Nature 462:231–234. https://doi.org/10.1038/nature08550
9. Mondal G, Stevers M, Goode B et al (2019) A requirement for STAG2 in replication fork progression creates a targetable synthetic lethality in cohesin-mutant cancers. Nat Commun 10:1686. https://doi.org/10.1038/s41467-019-09659-z
10. Tothova Z, Valton A-L, Gorelov RA et al (2021) Cohesin mutations alter DNA damage repair and chromatin structure and create therapeutic vulnerabilities in MDS/AML. JCI Insight 6. https://doi.org/10.1172/JCI.INSIGHT.142149
11. Morales C, Ruiz-Torres M, Rodríguez-Acebes S et al (2020) PDS5 proteins are required for proper cohesin dynamics and participate in replication fork protection. J Biol Chem 295:146–157. https://doi.org/10.1074/JBC.RA119.011099
12. Carvajal-Maldonado D, Byrum AK, Jackson J, Wessel S et al (2019) Perturbing cohesin dynamics drives MRE11 nuclease-dependent replication fork slowing. Nucleic Acids Res 47:1294–1310. https://doi.org/10.1093/nar/gky519
13. Benedict B, van Schie JJM, Oostra AB et al (2020) WAPL-dependent repair of damaged DNA replication forks underlies oncogene-induced loss of sister chromatid cohesion. Dev Cell 52:683–698.e7. https://doi.org/10.1016/J.DEVCEL.2020.01.024
14. Torres-Rosell J, De Piccoli G, Cordon-Preciado V et al (2007) Anaphase onset before complete DNA replication with intact checkpoint responses. Science (1979) 315:1411–1415. https://doi.org/10.1126/science.1134025
15. Bermudez-Lopez M, Ceschia A, de Piccoli G et al (2010) The Smc5/6 complex is required for dissolution of DNA-mediated sister chromatid linkages. Nucleic Acids Res 38:6502–6512. https://doi.org/10.1093/nar/gkq546
16. Lorite NP, Apostolova S, Guasch-Vallés M et al (2024) Crucial role of the NSE1 RING domain in Smc5/6 stability and FANCM-independent fork progression. Cell Mol Life Sci 81:251. https://doi.org/10.1007/s00018-024-05275-3
17. Atkins A, Xu MJ, Li M et al (2020) SMC5/6 is required for replication fork stability and faithful chromosome segregation during neurogenesis. elife 9. https://doi.org/10.7554/eLife.61171
18. Grange LJ, Reynolds JJ, Ullah F et al (2022) Pathogenic variants in SLF2 and SMC5 cause segmented chromosomes and mosaic variegated hyperploidy. Nat Commun 13:6664. https://doi.org/10.1038/s41467-022-34349-8
19. Xu MJ, Jordan PW (2024) SMC5/6 promotes replication fork stability via negative regulation of the COP9 signalosome. Int J Mol Sci 25:952. https://doi.org/10.3390/ijms25020952

20. Fingerman DF, O'Leary DR, Hansen AR et al (2024) The SMC5/6 complex prevents genotoxicity upon APOBEC3A-mediated replication stress. EMBO J 43:3240–3255. https://doi.org/10.1038/s44318-024-00137-x

21. Venegas AB, Natsume T, Kanemaki M et al (2020) Inducible degradation of the human SMC5/6 complex reveals an essential role only during interphase. Cell Rep 31:107533. https://doi.org/10.1016/j.celrep.2020.107533

Chapter 11

Biochemical Reconstitution of Replication-Coupled Cohesin Acetylation

Masashi Minamino, Céline Bouchoux, and Frank Uhlmann

Abstract

Concomitantly with DNA replication, the cohesin complex establishes cohesion between the newly synthesized sister chromatids. Cohesion establishment requires acetylation of the cohesin subunit Smc3 by the Eco1 acetyltransferase. Acetylation is tightly coupled to DNA replication and depends on (i) DNA replication factors, (ii) DNA structures that transiently form at replication forks, as well as (iii) auxiliary cohesin complex components. Here, we describe a protocol to study DNA replication-coupled cohesin acetylation using purified budding yeast proteins.

Key words Chromosome segregation, Sister chromatid cohesion, Cohesin acetylation, DNA replication, In vitro reconstitution, *S. cerevisiae*

1 Introduction

The physical tethering of two replicated chromosomes, known as sister chromatid cohesion, forms the basis for faithful chromosome segregation during cell divisions. Sister chromatid cohesion is mediated by the cohesin complex, a multi-subunit protein complex built on two Structural Maintenance of Chromosomes (SMC) subunits Smc1 and Smc3 [1, 2]. Smc1 and Smc3 form long antiparallel coiled coils which engage each other at both ends. At one end lies the hinge dimerization interface, and at the other end lie the ATPase head domains. Head dimerization depends on ATP, while ATP hydrolysis releases the two heads. An Scc1 subunit additionally bridges these ATPase heads to complete formation of the cohesin ring, within which sister DNAs are entrapped following DNA replication. The Scc1 subunit recruits Scc3, a HEAT repeat containing subunit that is essential for cohesin function. Two additional HEAT repeat containing cohesin regulators also associate with Scc1, Pds5-Wapl, as well as the Scc2-Scc4 cohesin loader complex.

Yasuto Murayama and Yasutaka Kakui (eds.), *SMC Proteins: Methods and Protocols*, Methods in Molecular Biology, vol. 2991, https://doi.org/10.1007/978-1-0716-5005-9_11,

In budding yeast, cohesin loading onto chromosomes requires the Scc2-Scc4 cohesin loader. Loading is counteracted by Pds5-Wapl which unloads cohesin again from chromosomes. To generate stable sister chromatid cohesion, Eco1 acetylates the Smc3 ATPase head. Smc3 acetylation switches off the cohesin ATPase and thereby the Pds5-Wapl-dependent cohesin unloading reaction. However, ATP hydrolysis is likewise required for cohesin loading by Scc2-Scc4, and acetylation also stops the cohesin loading reaction [3]. These considerations explain why cohesin acetylation must be temporally and spatially tightly controlled. Acetylated too early and cohesin cannot load onto DNA, and acetylated too late and sister chromatid cohesion will not last. Indeed, cohesin acetylation is closely coupled to DNA replication and is regulated by several DNA replication factors including PCNA [4–6]. Our recent biochemical analyses revealed that cohesin acetylation is jointly achieved when three requirements meet: (i) DNA replication factors; (ii) DNA structures that transiently arise at replication forks, flaps, or nicks; as well as (iii) auxiliary cohesin complex components, both Pds5 and the Scc2-Scc4 complex [7]. In this chapter, we describe the reconstitution of cohesin acetylation using purified proteins from budding yeast. This assay can serve as a basis to investigate the molecular mechanism by which cohesin and its regulators intersect with DNA replication to control this essential posttranslational modification.

2 Materials

2.1 Culture Media

1. YPD plate: 2% (w/v) peptone, 1% (w/v) yeast extract, and 2% (w/v) glucose, 2% (w/v) agar. Sterile by autoclaving.
2. YP-raffinose: 2% (w/v) peptone, 1% (w/v) yeast extract, and 2% (w/v) raffinose.
3. Galactose solution: 20% (w/v) stock solution, prepared in deionized, purified water.

2.2 Protein Expression Strains and Plasmids

1. **Pds5**: The Pds5 expression construct was integrated into budding yeast (*MAT***a**, *ade2-1, ura3-52, his3-11,15, trp1-1, leu2-3,112, can1-100, pep4Δ::HIS3 wpl1Δ::LEU2 eco1Δ::KANMX6*). The *GAL4* and *PDS5* genes were cloned into the budding yeast expression vector pRSII402 (*ADE2*) in the direction of the *GAL1* and *GAL10* promoters, respectively. 3 × E2a and 2 × Protein A tags were fused to the C terminus of the *PDS5* gene.
2. **Ctf18-RFC**: *RFC2-RFC3, RFC4-RFC5, CTF8-DCC1*, and *CTF18-GAL4* gene expression constructs were integrated into budding yeast (*MAT***a**, *ade2-1, ura3-1, his3-11,15, trp1-1, leu2-3,112, can1-100, bar1::Hyg, pep4::KanMX*). *RFC2-*

RFC3: *RFC2* and *RFC3* genes were cloned into the budding yeast expression vector pRS306 in the direction of the *GAL1* and *GAL10* promoters, respectively. A Calmodulin Binding Protein tag was fused to the N terminus of the *RFC3* gene. *RFC4-RFC5*: *RFC4* and *RFC5* genes were cloned into the budding yeast expression vector pRS304 in the direction or the *GAL1* and *GAL10* promoters, respectively. *CTF8-DCC1*: *CTF8* and *DCC1* genes were cloned into the budding yeast expression vector pRS305. *CTF18-GAL4*: *CTF18* and *GAL4* genes were cloned into the budding yeast expression vector pRS303. An HA and 2 × Protein A tags were fused to the C terminus of the *CTF18* gene.

2.3 Protein Purification

1. Pds5 buffer: 50 mM HEPES-KOH (pH 7.5), 20% glycerol, 0.5 mM TCEP.
2. Pds5 buffer + 0.3 M NaCl + protease inhibitors: As Pds5 buffer containing 0.3 M NaCl, 0.5 mM Pefabloc, cOmplete, EDTA-free Protease Inhibitor Cocktail. Add Pefabloc and the inhibitor cocktail just before use.
3. Pds5 buffer + 0.3 M NaCl: As Pds5 buffer containing 0.3 M NaCl.
4. Pds5 ATP wash buffer: As Pds5 buffer containing 0.3 M NaCl, 1 mM ATP, and 10 mM $MgCl_2$.
5. Pds5 buffer + 10 mM NaCl: As Pds5 buffer containing 10 mM NaCl.
6. Pds5 buffer + 1 M NaCl: As Pds5 buffer containing 1 M NaCl.
7. Pds5 gel filtration buffer: 20 mM Tris–HCl (pH 7.5), 150 mM NaCl, 10% glycerol, 0.5 mM TCEP.
8. Ctf18-RFC buffer: 25 mM HEPES-KOH (pH 7.5), 10% glycerol, 1 mM EDTA, 0.02% (w/v) NP-40, 1 mM DTT.
9. Ctf18-RFC buffer + 0.3 M NaCl + protease inhibitor: As Crf18-RFC buffer containing 0.3 M NaCl, 0.5 mM Pefabloc, cOmplete, EDTA-free Protease Inhibitor Cocktail. Add Pefabloc and the inhibitor cocktail just before use.
10. Ctf18-RFC ATP wash buffer: As Ctf18-RFC buffer containing 0.3 M NaCl, 1 mM DTT, 1 mM ATP, and 10 mM $MgCl_2$.
11. Ctf18-RFC buffer + 1 M NaCl: As Crf18-RFC buffer containing 1 M NaCl.
12. Pefabloc solution: 100 mM stock solution was prepared in distilled water.
13. cOmplete, EDTA-free Protease Inhibitor Cocktail (Merck).
14. Cryogenic freezer mill.
15. 20 mL empty open column.

16. 10 mg/mL RNase A.
17. PreScission protease.
18. Rabbit IgG agarose (Merck).
19. HiTrap Heparin HP 1 mL (Cytiva).
20. HiTrap SP HP 1 mL (Cytiva).
21. Superdex 200 Increase 10/300 GL gel filtration column (Cytiva).
22. Superose 6 Increase 10/300 GL gel filtration column (Cytiva).
23. Bradford protein assay.

2.4 Preparation of Primed-ssDNA

1. *E. coli* NEB 5-alpha (NEB).
2. pBluescript KSII (+) plasmid.
3. Plasmid large-scale purification.
4. Nb.BssSI.
5. Exonuclease III.
6. Phenol–chloroform–isoamyl alcohol (25:24:1).
7. Oligonucleotide (5′-AGTTGGGTAACGCCAGGGTTTTCC CAGTCACGACG-3′).
8. 2× annealing reaction buffer: 100 mM Tris–HCl (pH 7.5), 10 mM $MgCl_2$, 200 mM KCl.
9. Heat block.
10. MicroSpin S-400 HR Columns (Cytiva).

2.5 Primer Extension Assay

1. 2× primer extension buffer: 50 mM HEPES-KOH (pH 7.5), 500 mM potassium-glutamate, 20 mM magnesium acetate, 2 mM DTT, 0.02% (w/v) NP-40.
2. 100 mM ATP solution.
3. 5 mM dNTPs solution.
4. 100 mM acetyl-CoA solution.
5. Budding yeast proteins: Cohesin, Scc2-Scc4, Eco1, Pol ε, Pol δ, RPA, PCNA, Rfc1-RFC. These proteins are purified as described in [8–10].
6. 2× SDS-PAGE sample buffer: 200 mM Tris–HCl (pH 6.8), 100 mM DTT, 4% SDS, 20% glycerol, 0.01% (w/v) bromophenol blue.
7. 8% SDS-PAGE gel.
8. Nitrocellulose membrane.
9. PBST: 8.1 mM Na_2HPO_4, 1.47 mM KH_2PO4, 137 mM NaCl, 2.68 mM KCl, 0.05% (w/v) Tween 20.
10. PBST+ milk: PBS-T containing 5% (w/v) skim milk.

11. Mouse monoclonal anti-acetylated Smc3 antibody (raised in the Shirahige laboratory, The University of Tokyo).
12. HRP conjugated anti-mouse 2ry antibody.
13. Western blot chemiluminescence detection reagent.
14. Deproteinization solution: 10 mM Tris–HCl (pH 7.5), 1 mM EDTA, 50 mM NaCl, 0.75% SDS, 1 mg/mL proteinase K.
15. TAE: 40 mM Tris-acetate (pH 8.2), 1 mM EDTA.
16. DNA loading dye.
17. SYBR Gold (Thermo Scientific).
18. Heat block.
19. SDS-PAGE apparatus.
20. Membrane blotting apparatus.
21. Agarose gel electrophoresis apparatus.
22. Chemiluminescence imager.
23. Fluorescent imager.

3 Methods

We reconstitute replication-coupled cohesin acetylation during elongation of a primer annealed to a circular ssDNA template. The reconstitution assay requires cohesin, Eco1, cohesin loader (Scc2-Scc4), Pds5, PCNA loaders (Rfc1-RFC, Ctf18-RFC), and PCNA, as well as either DNA polymerase ε or DNA polymerase δ. Purification of DNA replication factors followed previously published detailed protocols [8]. Similarly, purification of budding yeast cohesin, Scc2-Scc4, and Eco1 were previously described [9, 10]. Here, we provide a detailed protocol to prepare Pds5 and Ctf18-RFC proteins, as well as the primed-ssDNA template. We then describe the reconstitution of DNA replication-coupled cohesin acetylation.

3.1 Purification of Pds5

1. Cultivate the Pds5 expressing budding yeast strain on a YPD plate at 30 °C for 1 day.
2. Inoculate YP-raffinose medium and cultivate at 30 °C until the optical density at 600 nm reaches 1.0 (we use 3 × 2 liter cultures; *see* **Note 1**).
3. Add 1/10 volume of 20% galactose solution to induce protein expressions and grow the cells further at 30 °C for 4 h.
4. Harvest the cells by centrifugation and wash once with distilled water. Then resuspend cells in an equal cell pellet volume of Pds5 buffer + 0.3 M NaCl + protease inhibitors.

5. Drop the cell suspension into liquid nitrogen to make small drops (usually 3–4 mm diameter). The frozen cell drops can be stored at −70 °C.
6. Break the cells using a cryogenic freezer mill (*see* **Note 2**).
7. Thaw the resulting cell powder in a beaker and resuspend in approximately two volumes of Pds5 buffer + 0.3 M NaCl + protease inhibitors at 4 °C (*see* **Note 3**).
8. Clarify the cell lysate at 4 °C at 200,000× *g* for 1 h.
9. Mix the lysate with Rabbit IgG-agarose (1 mL beads slurry per 50 mL cell lysate), add RNase A to a final concentration of 2.5 μg/mL, and rock at 4 °C for 3 h.
10. Transfer the suspension to an empty open column.
11. Wash the resin with 10 bed volumes of Pds5 buffer + 0.3 M NaCl.
12. Suspend the resin with 20 bed volume of Pds5 ATP wash buffer and then incubate for 15 min at 4 °C. This step removes chaperones that otherwise co-purify with Pds5.
13. Wash the resin with 40 bed volumes of Pds5 buffer + 0.3 M NaCl.
14. Resuspend the resin in 5 bed volumes of Pds5 buffer + 0.3 M NaCl.
15. Add PreScission or HRV-3C protease to a final concentration of 10 μg/mL and then incubate overnight at 4 °C (*see* **Note 4**).
16. Collect the eluate and gently dilute it with Pds5 buffer + 10 mM NaCl to a final NaCl concentration of 150 mM.
17. Load the diluted eluate onto a 1 mL HiTrap Heparin HP column (*see* **Note 5**).
18. Wash the column with 10 bed volume of Pds5 buffer containing 150 mM NaCl.
19. Elute the bound proteins with a linear gradient from 150 mM to 1 M NaCl in Pds5 buffer.
20. Load the peak fractions onto a Superdex 200 Increase 10/300 GL gel filtration column that is developed in Pds5 gel filtration buffer.
21. Collect the peak fractions and determine the Pds5 concentration using Bradford reagents.

3.2 Purification of Ctf18-RFC

1. Cultivate the Ctf18-RFC expressing budding yeast strain on a YPD plate at 30 °C for 1 day.
2. Inoculate YP-raffinose medium and cultivate at 30 °C until the optical density at 600 nm reaches 1.0 (we use 6 × 2 liter cultures).

3. Add 1/10 volume of 20% galactose solution to induce protein expressions and grow the cells further at 30 °C for 2 h.
4. Harvest the cells by centrifugation and wash once with distilled water. Then resuspend cells in an equal cell pellet volume of Ctf18-RFC buffer + 0.3 M NaCl + protease inhibitor.
5. Drop the cell suspension into liquid nitrogen to make small drops (usually 3–4 mm diameter). The frozen cell drops can be stored at −70 °C.
6. Break the cells using a cryogenic freezer mill.
7. Thaw the resulting cell powder in a beaker and resuspend in approximately two volumes of Ctf18-RFC buffer + 0.3 M NaCl + protease inhibitor at 4 °C.
8. Clarify the cell lysate at 4 °C at 200,000× *g* for 1 h.
9. Mix the lysate with IgG-agarose (1 mL beads slurry per 50 mL cell lysate), and rock at 4 °C for 3 h.
10. Transfer the suspension to an empty open column.
11. Wash the resin with 10 bed volumes of Ctf18-RFC buffer + 0.3 M NaCl.
12. Suspend the resin with 20 bed volume of Ctf18-RFC ATP wash buffer and then incubate for 15 min at 4 °C.
13. Wash the resin with 40 bed volumes of Ctf18-RFC buffer + 0.3 M NaCl.
14. Resuspend the resin in 5 bed volumes of Ctf18-RFC buffer + 0.3 M NaCl.
15. Add PreScission protease to a final concentration of 10 μg/mL and then incubate overnight 4 °C.
16. Load the eluate onto a 1 mL HiTrap SP HP column.
17. Wash the column with 10 bed volume of Ctf18-RFC buffer containing 150 mM NaCl.
18. Elute the bound proteins with a linear gradient from 300 mM to 1 M NaCl in Ctf18-RFC buffer.
19. Load the peak fractions onto a Superose 6 Increase 10/300 GL gel filtration column that is developed in Ctf18-RFC buffer containing 400 mM NaCl.
20. Collect the peak fractions and determine the Ctf18-RFC concentration using Bradford reagents.

3.3 Preparation of Primed-ssDNA

1. Purify pBluescript KSII (+) plasmid from 500 to 1000 mL of *E. coli* NEB 5-alpha culture using a plasmid large-scale purification kit.
2. Nick the plasmid DNA using Nb.BssSI by following the manufacturer's protocol (*see* **Note 6**).

3. Treat the nicked DNA solution with phenol–chloroform–isoamyl alcohol (25:24:1) twice.
4. Collect the nicked DNA by ethanol precipitation and dissolve in ultrapure water.
5. Degrade the nicked DNA strand using Exonuclease III by following the manufacturer's protocol (*see* **Note 7**).
6. Treat the ssDNA solution with phenol–chloroform–isoamyl alcohol (25:24:1) twice.
7. Collect the ssDNA by ethanol precipitation and dissolve in ultrapure water.
8. Determine the concentration of the ssDNA by measuring absorbance at 260 nm using 33 μg/mL per A_{260}.
9. Mix 50 nM ssDNA and 150 nM oligonucleotide in 1× annealing reaction buffer using 2× buffer (*see* **Note 8**).
10. Heat the mixture to 90 °C in a heat block for 15 s.
11. Switch off the heat block.
12. Wait until the heat block gradually cools to room temperature.
13. Equilibrate a MicroSpin S-400 HR Column with 600 μL of distilled water.
14. Apply 100 μL of the DNA sample to the MicroSpin S-400 HR Column (*see* **Note 9**).
15. Spin at 700× *g* for 1 min.
16. Collect the eluate and determine the concentration of the primed-ssDNA by measuring absorbance at 260 nm using 33 μg/mL per A_{260}.

3.4 Cohesin Acetylation During Primer Extension

1. Mix 150 nM cohesin, 150 nM Scc2-Scc4, 150 nM Pds5, 35 nM Eco1, 4 nM Pol δ, 400 nM RPA, 70 nM PCNA, 60 nM Rfc1-RFC, 60 nM Ctf18-RFC, and 4 nM primed-ssDNA on ice in reaction buffer (25 mM HEPES-KOH (pH 7.5), 250 mM potassium glutamate, 10 mM magnesium acetate, 1 mM DTT, 0.01% (w/v) NP-40, 5 mM ATP, 80 μM dNTPs, 0.5 mM Acetyl-CoA using 2× primer extension buffer. The final reaction volume is 20 μL (*see* **Note 10**).
2. Incubate at 30 °C for 12 min.
3. Retrieve 6 μL reaction sample to analyze DNA product formation (*see* Subheading 3.5).
4. Add an equal volume of 2× SDS sample buffer to the rest of the primer extension reaction, and heated to 95 °C for 3 min.
5. Separate the denatured proteins by 8% SDS-PAGE and transfer them to a nitrocellulose membrane.
6. Soak the membrane with PBST+ milk for 1 h for blocking.

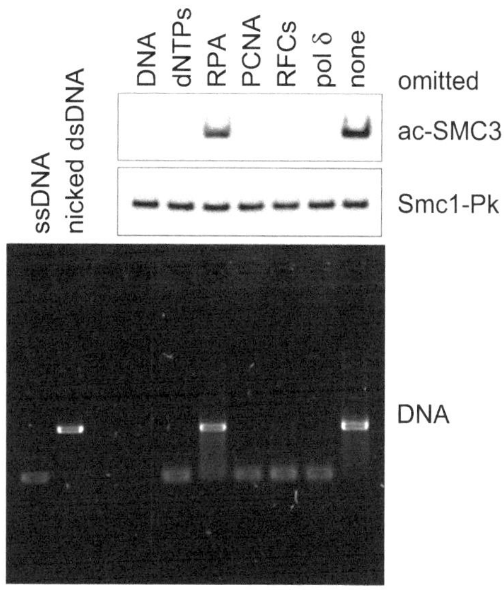

Fig. 1 Biochemical reconstitution of DNA replication-coupled cohesin acetylation. Gel image of primer extension reaction products using DNA polymerase δ, and immunoblot analysis of Smc3 acetylation in the reaction where indicated components are omitted. The Smc1 cohesin subunit, detected via its Pk epitope tag, served as a loading control

7. Incubate the membrane with mouse monoclonal anti-acetylated Smc3 antibody in PBST+ milk overnight at 4 °C. The Smc1 cohesin subunit can be detected via its Pk epitope tag as a control [9].
8. Wash the membrane three times for 5 min with PBST.
9. Incubate the membrane with HRP-conjugated anti-muse 2ry antibody in PBST+milk for 2 h at room temperature.
10. Wash the membrane three times for 5 min with PBST.
11. Detect acetylated Smc3 signal with Western blot chemiluminescence detection reagent and a chemiluminescence imager. A representative result is shown in Fig. 1.

3.5 Confirmation of DNA Conversion

1. Add 14 μL deproteinization solution to 6 μL of the retrieved reaction.
2. Incubate at 50 °C for 20 min.
3. Treat the deproteinized reaction with 20 μL of phenol–chloroform–isoamyl alcohol (25:24:1) once.
4. Add DNA loading dye to the purified DNA and separate the DNA products by 1% agarose/TAE gel electrophoresis.
5. Stain the gel with SYBR™ Gold Nucleic Acid Gel Stain diluted to 1:10,000 in TAE for 1 h.

6. Wash the gel wit TAE for 30 min.
7. Scan the gel using a fluorescent imager. A representative result is shown in Fig. 1.

4 Notes

1. All media and solutions are prepared in distilled water or MilliQ water (Millipore).
2. We use SPEX SamplePrep, Cole-Parmer. Typically, approximately 60 grams of cell drops are disrupted using a setting of rate: 15 cps, 2 min on/2 min off for 6 cycles.
3. All procedures for protein purification including cell lysate preparation and column chromatography are carried out at 4 °C.
4. All purification steps are monitored by resolving fractions by SDS-PAGE, followed by Coomassie brilliant blue staining.
5. We use an ÄKTA Pure protein purification system (Cytiva) for column chromatography.
6. Confirm that DNA is nicked by agarose gel electrophoresis.
7. Confirm that the nicked DNA is converted to ssDNA by agarose gel electrophoresis.
8. We prepare a volume of no more than 100 μL in a 1.5 mL microcentrifuge tube at this step. This restriction is intended to ensure efficient heating of the sample in the next step.
9. We load no more than 100 μL of the reaction onto a MicroSpin S-400 HR Column to ensure efficient cleanup.
10. 20 nM DNA polymerase ε can also be used in place of DNA polymerase δ.

References

1. Uhlmann F (2016) SMC complexes: from DNA to chromosomes. Nat Rev Mol Cell Biol 17:399–412
2. Hoencamp C, Rowland BD (2023) Genome control by SMC complexes. Nat Rev Mol Cell Biol 24:633–650
3. Murayama Y, Uhlmann F (2015) DNA entry into and exit out of the cohesin ring by an interlocking gate mechanism. Cell 163:1628–1640
4. Moldovan GL, Pfander B, Jentsch S (2006) PCNA controls establishment of sister chromatid cohesion during S phase. Mol Cell 23:723–732
5. Rolef Ben-Shahar T, Heeger S, Lehane C et al (2008) Eco1-dependent cohesin acetylation during establishment of sister chromatid cohesion. Science 321:563–566
6. Borges V, Smith DJ, Whitehouse I et al (2013) An Eco1-independent sister chromatid cohesion establishment pathway in *S. cerevisiae*. Chromosoma 122:121–134
7. Minamino M, Bouchoux C, Canal B et al (2023) A replication fork determinant for the

establishment of sister chromatid cohesion. Cell 186:837–849

8. Posse V, Johansson E, Diffley JFX (2021) Eukaryotic DNA replication with purified budding yeast proteins. Methods Enzymol 661:1–33
9. Minamino M, Higashi TL, Bouchoux C et al (2018) Topological *in vitro* loading of the budding yeast cohesin ring onto DNA. Life Sci Alliance 1:e201800143
10. Liu HW, Bouchoux C, Panarotto M et al (2020) Division of labor between PCNA loaders in DNA replication and sister chromatid cohesion establishment. Mol Cell 78:725–738

Chapter 12

A Biochemical Assay to Monitor the Encounter Between Cohesin and DNA Replication

Yasuto Murayama

Abstract

In eukaryotes, newly duplicated chromatids are physically linked together until the onset of chromosome segregation. This physical connection, known as sister chromatid cohesion, is mediated by the cohesin complex, a multisubunit ATPase assembly with a closed-ring conformation that topologically entraps DNA. Cohesin is loaded onto chromatin with the help of regulatory proteins prior to DNA replication and then establishes cohesion when it encounters the replication machinery. This process involves the physical tethering of two replicated sister DNAs and the stabilization of cohesin through acetylation. Here, we describe a biochemical assay to assess the response of cohesin to DNA replication using the purified budding yeast proteins.

Key words Sister chromatid cohesion, DNA replication, Cohesin, Scc2-Scc4, Topological DNA entrapment, Biochemical reconstitution

1 Introduction

The accurate duplication and subsequent transfers of chromosomes to dividing cells is fundamental to cell proliferation. DNA replication is the process that not only duplicates genomic DNA but also coordinates the formation of chromosomal structures, including sister chromatid cohesion, which ensures proper chromosome segregation during both mitosis and meiosis. Cohesin is a member of the structural maintenance of chromosomes (SMC) complexes and mediates sister chromatid cohesion by topologically entrapping DNA [1–5]. The ring-shaped ATPase complex consists of four core subunits (Smc1, Smc3, Scc1, and Scc3), the chromosomal activities of which are regulated by several accessory proteins (Fig. 1a). Topological loading of cohesin is facilitated by the Scc2-Scc4 loader complex, whereas Wapl-Pds5 promotes cohesin dissociation from chromatin [6–9]. Once the cell enters the S phase and initiates DNA replication, cohesin undergoes two changes to

Yasuto Murayama and Yasutaka Kakui (eds.), *SMC Proteins: Methods and Protocols*, Methods in Molecular Biology, vol. 2991, https://doi.org/10.1007/978-1-0716-5005-9_12,

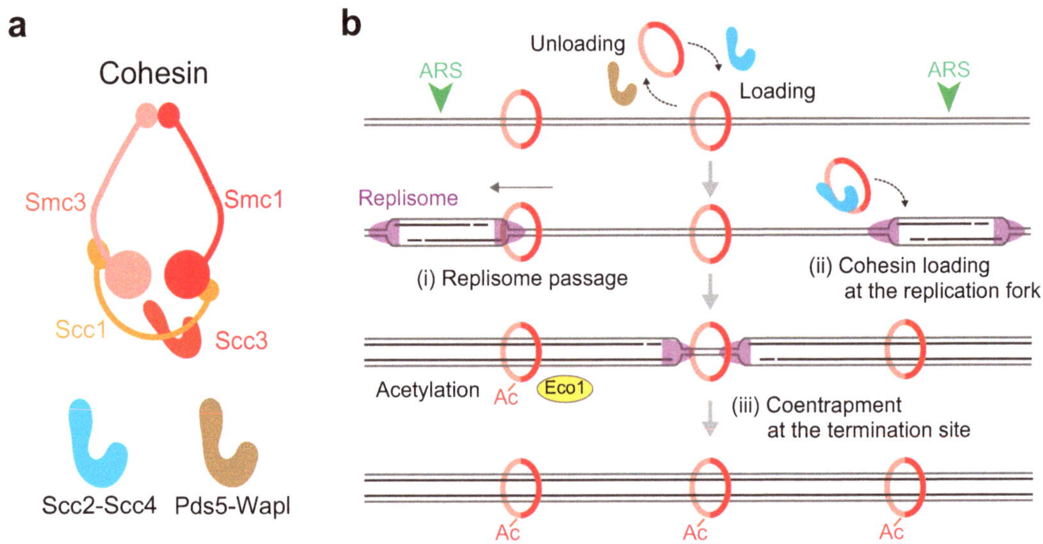

Fig. 1 Cohesin and the establishment of sister chromatid cohesion. (**a**) The cohesin complex and its regulatory factors, Scc2-Scc4 and Pds5-Wapl. (**b**) Molecular models for cohesion establishment. Mechanically distinct, but not mutually exclusive, models are proposed for the co-entrapment of sister DNAs by cohesin during DNA replication

establish sister chromatid cohesion. First, cohesin physically links two duplicated sister DNAs by exploiting topological entrapment (Fig. 1b) [10–13]. This can be achieved when cohesin, in the state of DNA entrapment, accommodates the completion of DNA replication. Alternatively, cohesin is newly recruited to the ongoing replisome and sequentially entraps two sister DNAs at the replication fork. After the sister DNAs have been co-entrapped, chromatin-bound cohesin is acetylated by Eco1 and establishes the stable cohesion that persists until the onset of anaphase [14–17]. Several replisome-associated proteins are also involved in cohesion establishment, including Chl1, Ctf4, Csm3-Tof1, Mrc1, and Ctf18-RFC [12, 18].

Eukaryotic DNA replication takes place at multiple sites along the genome marked by the origin–recognition complex (ORC), which loads two minichoromosome maintenance helicase core complexes (Mcm2-7) in a head-to-head orientation [19–22]. In the budding yeast, ORC recognizes the specific DNA segments defined as autonomously replication sequences (*ARS*). The Mcm2-7 double hexamer is activated via phosphorylation by the Dbf4-dependent Cdc7 kinase complex (DDK) to form the replicative Cdc45-Mcm2-7-GINS (CMG) helicase complex, with the aids of firing factors and the S-phase cyclin-dependent kinase (S-CDK). Once activated, two CMG helicases proceed bidirectionally, accompanying both continuous leading strand and discontinuous lagging strand synthesis mediated by DNA polymerases. The advancing replisome converges with another replisome orienting from the next origin, completing DNA replication.

We have previously investigated how cohesin responds to DNA replication by reconstituting both biochemical activities using the purified budding yeast proteins [23]. This biochemical reconstitution showed that cohesin is intrinsically tolerant to DNA replication, allowing sister DNA duplication to be completed. This chapter describes a bulk biochemical assay to monitor a cohesin-replisome encounter by reconstituting a functional cohesin and replisome using the purified 23 budding yeast proteins/protein complexes, which can serve as a basis for investigating the molecular mechanism underlying the establishment of sister chromatid cohesion.

2 Materials

2.1 Protein Expression Strains

1. Cohesin (Smc1, Smc3, Scc1, Scc3): YST344 (*SMC1* cDNA fused with 3× Pk tag at the C-terminus and *SMC3* cDNA were cloned into the budding yeast expression vector YIplac211 (*URA3*) in the direction of the *GAL1* and *GAL10* promoters, respectively. *SCC1* cDNA fused with 1× E2-PreScission protease recognition site-2× protein A tag at the C-terminus and *SCC3* cDNA fused with 1× HA-7× histidine tag were cloned into the budding yeast expression vector YIplac128 (*LEU2*) under the *GAL1* and *GAL10* promoters, respectively. These DNA constructs were integrated into the budding yeast protease-deficient strain (*pep4Δ*, *prb1Δ*, *prc1Δ*) at *URA3* and *LEU2* genomic loci, respectively) [23]. (*See* **Notes 1** and **2**).
2. Scc2-Scc4: YST320 (*SCC2* cDNA fused with 1× E2-PreScission protease recognition site-2× protein A tag at the C-terminus and *SCC4* cDNA were cloned into the budding yeast expression vector YIplac204 (*TRP1*) in the direction of the *GAL1* and *GAL10* promoters, respectively. This DNA construct is integrated into the budding yeast protease-deficient strain (*pep4Δ*, *prb1Δ*, *prc1Δ*) at *TRP1* genomic locus) [23].

2.2 Culture Media and Supplements

1. SD glucose medium plate: 0.67% (w/v) yeast nitrogen base without amino acid (BD), 2% (w/v) D-glucose, 20 μg/mL of adenine, arginine, histidine, methionine, tryptophan, uracil, 30 μg/mL of tyrosine, uracil, leucine, lysine, isoleucine, 60 μg/mL of phenylalanine, 100 μg/mL glutamic acid, aspartic acid, 150 μg/mL valine and 400 μg/mL serine, dissolved in distilled water and adjusted to pH 5.8. Add 2% (w/v) ager and sterile by autoclaving.

2. SD glucose –URA–LEU plate: As SD glucose medium plate but lacking uracil and leucine.
3. SD glucose –TRP plate: As SD glucose medium plate but lacking tryptophan.
4. YP raffinose medium: 2% (w/v) peptone, 1% (w/v) yeast extract, 1.5% (w/v) raffinose, and 20 μg/mL adenine, dissolved in distilled water. Prepare 1.5 liters of YP raffinose in a 5-liter flask and sterile by autoclaving.
5. 30% (w/v) galactose: 30% solution is prepared in distilled water and sterilized by autoclaving.
6. LB medium: 1% (w/v) tryptone, 0.5% (w/v) yeast extract, 1% (w/v) NaCl, dissolved in distilled water and adjusted to pH 7.0 with NaOH. 1.5% (w/v) ager is added to make the solid medium. Sterile by autoclaving.
7. 100× ampicillin: 100 mg/mL stock solution is prepared in distilled water and sterilized by filtration.
8. Shaking incubator.

2.3 Protein Purification

1. 0.2 M phenylmethylsulfonyl fluoride (PMSF) stock solution: 0.2 M stock solution is prepared in 2-propanol.
2. 0.5 M Tris(2-carboxyethyl)phosphine (TCEP) stock solution: 0.5 M stock solution is prepared in ultrapure water and adjusted to pH 7.0 with hydrochloride (HCl).
3. G buffer: 40 mM HEPES-KOH (pH 7.5), 0.3 M NaCl, 0.5 mM PMSF. Add PMSF from the stock solution just before use (*see* **Note 3**).
4. L1 + I buffer: 40 mM HEPES-KOH (pH 7.5), 10% (v/v) glycerol, 0.3 M NaCl, 0.5 mM TCEP, 0.5 mM PMSF, and protease inhibitor cocktail. Add TCEP, PMSF (from the stock solutions), and protease inhibitor cocktail just before use (*see* **Note 4**).
5. L2 buffer: 20 mM HEPES-KOH (pH 7.5), 10% (v/v) glycerol, 0.3 M NaCl, 0.01% (w/v) Tween 20, 0.5 mM TCEP. Add TCEP from the stock solution just before use.
6. Cleavage Elution solution: L2 buffer containing 1 mM TCEP, 10 μg/mL RNase A, and 5 U/mL PreScission protease (or HRV 3C protease). Mix TCEP, RNase A, and PreScission protease with L2 buffer just before use.
7. CH buffer: 20 mM HEPES-KOH (pH 7.5), 10% (v/v) glycerol, 0.5 mM TCEP. Add TCEP from the stock solution just before use.
8. CH buffer + 1 M NaCl: Same as CH buffer but containing 1 M NaCl.
9. Liquid nitrogen.

10. Open columns (*see* **Note 5**).
11. Human IgG conjugated resin (*see* **Note 6**).
12. Glutathione conjugated resin (*see* **Note 7**).
13. 1 mL heparin column (*see* **Note 8**).
14. 10 × 300 mm (24 mL) gel filtration column with a retention range of 5–5000 kDa (*see* **Note 9**).
15. 10 × 300 mm (24 mL) gel filtration column with a retention range of 10–600 kDa (*see* **Note 10**).
16. A centrifugal ultrafiltration tube with a molecular weight cutoff (MWCO) of 30 kDa (*see* **Note 11**).
17. Cryogenic cell grinder.
18. Centrifuge: High-speed centrifuge for cell harvest, ultracentrifuge for preparation of cell lysate, low-speed centrifuge for centrifugal ultrafiltration.
19. Peristaltic pump.
20. FPLC system.

2.4 DNA Substrate Preparation

1. *Escherichia coli* (*E. coli*) DH5α chemical competent cell.
2. pARS1–4.1: The 4.1 kbp plasmid DNA harboring the *ARS1*-containing region [23].
3. Plasmid preparation kit (*see* **Note 12**).
4. TE: 10 mM Tris–HCl (pH 7.5), 1 mM ethylenediaminetetraacetic acid (EDTA).
5. CsCl: Ultracentrifugation analysis grade.
6. 5 mg/mL ethidium bromide (EtBr).
7. Phenol/chloroform/isoamyl alcohol (25:24:1).
8. NaCl-saturated isopropanol.
9. Ethanol.
10. Heat-sealable ultracentrifugal tubes (*see* **Note 13**).
11. 1 mL syringe with needle.
12. A dialysis cassette with a molecular weight cutoff (MWCO) of 10 kDa (*see* **Note 14**).
13. Ultracentrifuge.
14. Photospectrometer.

2.5 DNA Marker Preparation

1. Primer 1819: 5′-CAGCGCAACACCCTTATCTGG-3′.
2. Primer 1823: 5′-CAGCGCAACACCCTTATCTGG-3′.
3. Primer 1824: 5′-TGAGCCATCATGACGCCGATG-3′.
4. Primer 1825: 5′-CTGAATG CAGCCATAGGTGCG-3′.
5. Primer 1826: 5′-GTGGTCCCGGCACCTTTCAC-3′.

6. Lambda phage DNA.
7. High-fidelity PCR polymerase.
8. PCR product purification kit.
9. Thermal cycler.

2.6 Biochemical Assays

1. 10× Loading buffer: 250 mM HEPES-KOH (pH 7.5), 10 mM DTT, 0.1% (w/v) NP-40.
2. 10× Replication buffer: 250 mM HEPES-KOH (pH 7.5), 10 mM DTT, 0.1% (w/v) NP-40, 75 mM magnesium acetate ($Mg(OAc)_2$).
3. 50% (v/v) glycerol.
4. 10 mM and 140 mM $Mg(OAc)_2$.
5. 1 M potassium acetate (KOAc).
6. 100 mM ATP: 100 mM stock solution is prepared with pure water and adjusted pH to 7.5.
7. 20 mM CTP/GTP/TTP mixture: 100 mM nucleotide stock solutions are mixed and dissolved in pure water to a final concentration of 20 mM each.
8. 8 mM dATP/dCTP/dGTP mixture: 100 mM deoxynucleotide stock solutions are mixed and dissolved in pure water to a final concentration of 8 mM each.
9. 1 mM dTTP: 100 mM dTTP stock solution is dissolved in pure water to a final concentration of 1 mM.
10. 1 mM biotin-16-dUTP (*see* **Note 15**).
11. 10 mg/mL bovine serum albumin (BSA).
12. Cohesin loading buffer: 25 mM HEPES-KOH (pH 7.5), 1 mM DTT, 0.01% (w/v) NP-40, 5% (v/v) glycerol, and 1 mM $Mg(OAc)_2$. Prepare using 10× Loading buffer, the stock solutions (listed in Subheading 2.6), and ultrapure water.
13. MCM loading buffer: 25 mM HEPES-KOH (pH 7.5), 1 mM DTT, 0.01% (w/v) NP-40, 5% (v/v) glycerol, 100 mM KOAc, 7.5 mM $Mg(OAc)_2$, 0.1 mg/mL BSA, 5 mM ATP. Prepare using 10× Loading buffer, the stock solutions (listed in Subheading 2.6), and ultrapure water.
14. Replication buffer: 25 mM HEPES-KOH (pH 7.5), 1 mM DTT, 0.01% (w/v) NP-40, 5% (v/v) glycerol, 100 mM KOAc, 7.5 mM $Mg(OAc)_2$, 0.1 mg/mL BSA, 5 mM ATP, 0.1 mM CTP, 0.1 mM GTP, 0.1 mM TTP, 40 μM dATP, 40 μM dCTP, 40 μM dGTP, 25 μM dTTP, and 15 μM biotin-dUTP. Prepare using 10× Replication buffer, the stock solutions (listed in Subheading 2.6), and ultrapure water.
15. pARS1-4.1: The 4.1 kbp plasmid DNA harboring the *ARS1*-containing region [23].

16. Budding yeast replication proteins: ORC, Cdt1-Mcm2-7, Cdc6, DDK, S-CDK, Sld2, Sld3-Sld7, Dpb11, Cdc45, GINS, Mcm10, Csm3-Tof1, Mrc1, Ctf4, RFC, PCNA, RPA, polymerase a (Pol a), Polymerase e (Pol e), Top1, Pif1. These proteins are purified as previously described [19, 20, 23–26].
17. Anti-Pk (V5) tag antibody (*see* **Note 16**).
18. Protein A-conjugated magnetic beads (*see* **Note 17**).
19. *Pst*I.
20. *Bgl*II.
21. IPH buffe: 20 mM HEPES-KOH (pH 7.5), 0.25% (w/v) Triton X-100, 750 mM NaCl, 1 mM ETDA.
22. IPL buffer: 20 mM HEPES-KOH (pH 7.5), 0.1% (w/v) Triton X-100, 100 mM NaCl, 1 mM ETDA.
23. TE/SDS: 10 mM Tris–HCl (pH 7.5), 1 mM EDTA, 0.5% (w/v) sodium dodecyl sulfate (SDS).
24. Stop solution: 10 mM HEPES-KOH (pH 7.5), 0.3% (w/v) SDS, 60 mM EDTA, 0.25 mg/mL Proteinase K. Mix proteinase K just before use.
25. 2× SDS-PAGE sample buffer: 80 mM Tris–HCl (pH 6.8), 2% (w/v) SDS, 10% (v/v) glycerol, 0.01% (w/v) bromophenol blue (BPB), 200 mM DTT.
26. 10% (w/v) SDS.
27. 250 mM EDTA.
28. RE buffer: 50 mM Tris–HCl (pH 7.5), 100 mM NaCl, 10 mM $MgCl_2$.
29. G-50 rein in RE buffer: 5 g of G-50 resin powder is suspended in 75 mL RE buffer and satirized by autoclaving.
30. 0.5 mL empty spin columns.
31. DNA dye: 10 mM HEPES-KOH (pH 7.5), 15% (w/v) ficol, 0.05% (w/v) orange G.
32. Agarose, gel electrophoresis grade.
33. TAE: 40 mM Tris-acetate (pH 8.2), 1 mM EDTA.
34. 1/2× TBE: 44.5 mM Tris-borate, 1 mM EDTA.
35. 0.2 M HCl.
36. 0.2 M NaOH/0.5 M NaCl.
37. Filter paper.
38. DNA blotting membrane (*see* **Note 18**).
39. Tris-buffered saline (TBS): 20 mM Tris–HCl (pH 7.4), 150 mM NaCl.
40. TBS + Tween: TBS containing 0.1% (w/v) Tween 20.

41. TBS + Tween+SDS: TBS containing 0.1% (w/v) Tween 20 and 0.01% (w/v) SDS.
42. Blocking solution: Blocking reagent is dissolved in TBS + Tween (*see* **Note 19**).
43. DNA gel staining solution (*see* **Note 20**).
44. Protein gel staining solution (*see* **Note 21**).
45. IRDye 680-conjugated streptavidin (*see* **Note 22**).
46. Magnetic stand.
47. Tube incubator.
48. Agarose gel electrophoresis system.
49. Ultraviolet crosslinker.
50. Polyacrylamide gel electrophoresis system.
51. Wet transfer membrane blotting system.
52. Fluorescent imager.

3 Methods

3.1 Protein Purification

The budding yeast cohesin and DNA replication proteins are expressed using the budding yeast or *E. coli* overproduction expression system and are purified as previously described [19, 20, 23–26]. Here, we describe the purification procedures for the budding yeast cohesin and the Scc2-Scc4 loader complex. Cohesin is purified as a protein complex from the budding yeast cells expressing all cohesin subunits (Smc1, Smc3, Scc1, Scc3) using a galactose-inducible system. Scc2-Scc4 is expressed and purified using the same protein expression system (*see* **Notes 23** and **24**).

3.1.1 Cohesin Purification

1. Cultivate the budding yeast strain YST344 on three SD glucose –URA–LEU plates at 30 °C for 1 day.
2. Inoculate the colonies into YP raffinose (1.5 liters × 5 cultures) and cultivate at 30 °C until the cell density reaches 1–2 × 10^7 per mL.
3. Add 1/15 volume of 30% (w/v) galactose (100 mL per 1.5-liter culture) to induce protein expression and further grow the cells at 30 °C for 2 h.
4. Harvest the cells by centrifugation at 4 °C, and wash once with ice-chilled G buffer. Then resuspend the cells in a volume of the same buffer equal to one quarter of the cell pellet.
5. Drop the cell suspension directly into liquid nitrogen to form small drops (usually 2–3 mm in diameter).
6. Disrupt the cells using a cryogenic grinder under liquid nitrogen (*see* **Note 25**).

7. Thaw the cell powder in a 500 mL beaker and dissolve them in twice the cell powder volume of L1 + I buffer with gentle stirring for 1 hour.
8. Centrifuge the cell lysate at 4 °C at 45,000× *g* for 1 h and transfer the supernatant to a new tube (*see* **Note 26**).
9. Add 1.2 mL bed volume of human IgG agarose resin to an open column and wash the resin with 10 mL of L2 buffer.
10. Apply the clarified cell lysate to the IgG agarose using a peristaltic pump at a flow rate of 0.5 mL/min.
11. Wash the resin with 80 bed volumes of L2 buffer.
12. Resuspend the resin in 1 mL of Cleavage Elution solution (containing 10 μg/mL RNase A and 5 U/mL PreScission protease) and incubate overnight (12–15 h) at 4 °C.
13. Apply the eluate to the glutathione-conjugated resin (0.2 mL bed volume) to remove the PreScission protease.
14. Collect the pass-through and dilute gently with CH buffer to a final NaCl concentration of 0.15 M.
15. Apply the diluted eluate to a 1 mL heparin column (*see* **Note 27**).
16. Wash the column with 10 bed volumes of CH buffer containing 0.15 M NaCl.
17. Elute the bound proteins with steps of 0.25, 0.6, and 1 M NaCl in CH buffer (5 bed volume each). Cohesin is eluted at 0.6 M NaCl.
18. Collect the peak fractions and concentrate to ~600 μL using a centrifugal ultrafiltration tube (30 kDa MWCO).
19. Load the concentrated peak fraction onto a gel filtration column with a retention range of 5–5000 kDa (24 mL bed volume) that is developed in CH buffer containing 0.2 M NaCl. Cohesin is eluted at ~12 mL (*see* **Note 9**).
20. Collect the peak fractions and concentrate to ~500 μL using a centrifugal ultrafiltration tube (30 kDa MWCO).
21. Determine the concentration of cohesin by measuring the absorbance at 280 nm with an extinction coefficient of 2.69×10^5/cm.
22. Freeze the aliquots with liquid nitrogen and store at −80 °C.

3.1.2 Scc2-Scc4 Purification

1. Cultivate the budding yeast strain YST320 on three SD glucose –TRP plates at 30 °C for 1 day.
2. Induce Scc2-Scc4 expression and prepare the cell powder in the same way as for cohesin expression (described in Subheading 3.1.1, **steps 2–6**).

3. Purify Scc2-Scc4 using human IgG agarose following the same procedure as for cohesin (Subheading 3.1.1, **steps 7–14**).
4. Apply the diluted IgG agarose eluate to a 1 mL heparin column.
5. Wash the column with 10 bed volumes of CH buffer containing 0.15 M NaCl.
6. Elute the bound proteins with 10 bed volume of a linear gradient of 0.15–1 M NaCl in CH buffer (10 bed volume). Scc2-Scc4 is eluted at ~0.6 M NaCl.
7. Collect the peak fractions and concentrate to ~600 μL using a centrifugal ultrafiltration tube (30 kDa MWCO).
8. Load the concentrated peak fraction onto a gel filtration column with a retention range of 10–600 kDa (24 mL bed volume) that is developed in CH buffer containing 0.2 M NaCl. Scc2-Scc4 is eluted at ~10.5 mL (*see* **Note 10**).
9. Collect the peak fractions and concentrate to ~500 μL using a centrifugal ultrafiltration tube (30 kDa MWCO).
10. Determine the concentration of Scc2-Scc4 by measuring the absorbance at 280 nm with an extinction coefficient of 2.13×10^5/cm.
11. Freeze the aliquots with liquid nitrogen and store at −80 °C.

3.2 Preparation of Covalently Closed Circular DNA (cccDNA) of pARS1-4.1

1. Transform the pARS1-4.1 plasmid DNA into *E. coli* DH5α strain and cultivate the cells in 100 mL of LB medium containing 100 μg/mL ampicillin at 37 °C overnight (~15 h).
2. Purify pARS1–4.1 using a commercially available plasmid purification kit (*see* **Note 12**).
3. Dissolve the purified plasmid in 8 mL TE.
4. Add 8.64 g of CsCl (1.59 g/mL final) to the DNA solution and dissolve the CsCl completely.
5. Add 640 μL of 5 mg/mL EtBr and transfer the DNA solution to two heat-sealable ultracentrifugal tubes.
6. Centrifuge the tubes at 20 °C at 230,000× *g* for 15 h (*see* **Note 28**).
7. Visualize DNA by blue light irradiation.
8. Aspirate the cccDNA fraction using 1 mL syringe with needle.
9. Mix with an equal volume of NaCl-saturated isopropanol to remove EtBr.
10. Centrifuge at room temperature at 2500× *g* for 1 min and discard the isopropanol layer (upper layer). Repeat **steps 9** and **10** until the upper layer becomes transparent.

11. Dialyze the DNA solution against 2 liters of TE at room temperature for 2 h.
12. Treat the DNA solution twice with phenol/chloroform/isoamyl alcohol (25:24:1).
13. Collect the DNA by ethanol precipitation and dissolve in 200 μL of TE.
14. Dialyze the DNA solution against 500 mL of TE at 4 °C for >5 h.
15. Determine the DNA concentration by measuring A_{260} using 50 μg/mL per A_{260}.

3.3 Preparation of Molecular Weight Standard Marker of Biotinylated DNA

1. Mix 5 ng of lambda phage DNA with a pair of the following primers (0.5 μM each) and high-fidelity PCR DNA polymerase in PCR reaction buffer containing 0.2 mM dATP, 0.2 mM dCTP, 0.2 mM dGTP, dTTP, and 10 μM biotin-dUTP in 50 μL; 1819 + 1823 for 2 kbp DNA, 1819 + 1824 for 4 kbp DNA, 1819 + 1825 for 6 kbp DNA, and 1819 + 1826 for 8 kbp DNA.
2. Amplify the biotinylated DNA by PCR using a thermal cycler (*see* **Note 29**).
3. Purify the amplified DNA fragment using a PCR product purification kit.
4. Determine the DNA concentration of each DNA fragment (2, 4, 6, 8 kbp) by measuring A_{260} using 50 μg/mL per A_{260}.
5. Mix equal amounts of the four biotinylated DNA fragments (2, 4, 6, 8 kbp DNA) in TE to a final concentration of 1.6 ng/μ L each.
6. Add one-fifth volume of DNA dye.

3.4 Cohesin Loading Assay

The cohesin loading assay monitors the topological entrapment of a circular DNA substrate by purified cohesin (Fig. 2a) [23, 27]. Budding yeast cohesin is incubated with a circular DNA and the cohesin–DNA complex is collected by immunoprecipitation through the 3× Pk tag on the Smc1 subunit. The cohesin-bound DNA is analyzed by agarose gel electrophoresis. Efficient topological DNA loading is seen when cohesin is incubated with ATP and the Scc2-Scc4 loader complex in the low salt buffer (Fig. 2b, c). The concentrations of all components are given as final concentrations in the reaction mixture.

1. Mix 30 nM cohesin, 30 nM Scc2-Scc4, and 1.5 nM cccDNA of pARS1-4.1 on ice in the cohesin loading buffer using the stock solutions (10× Loading buffer, 50% (v/v) glycerol, 10 mM Mg $(OAc)_2$ and ultrapure water). The volume of the mixture is 19 μL.

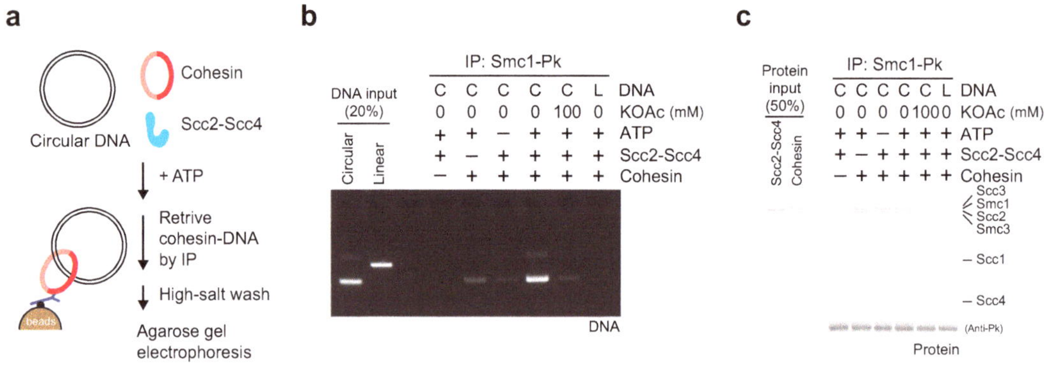

Fig. 2 Cohesin loading assay using the purified budding yeast proteins. (**a**) Schematic of the topological DNA loading assay by cohesin. (**b**) Gel image showing a representative result of the cohesin loading assay in the presence or absence of the indicated components. C and L denote circular and liberalized DNA, respectively. Efficient cohesin loading was seen in the presence of ATP and Scc2-Scc4. Note that cohesin loading was inhibited in the presence of salt (100 mM KOAc). (**c**) Confirmation of cohesin-IP by SDS-PAGE analysis of the same samples

2. Add 1 μL of 100 mM ATP (5 mM final) and incubate for 30 min at 30 °C (the total reaction volume of the cohesin loading mixture is 20 μL).
3. Add 500 μL of ice-chilled IPH buffer (750 mM NaCl) to the reaction.
4. Add 10 μL of anti-Pk pre-bound protein A conjugated magnetic beads (*see* **Note 30**).
5. Rotate the reaction tube at 5 rpm at 4 °C for 2 h.
6. Wash the beads three times with 750 μL of IPH buffer (750 mM NaCl) using a magnetic stand.
7. Wash the beads once with 500 μL of IPL buffer (100 mM NaCl). Completely remove the wash buffer using a magnetic stand.
8. Add 20 μL of TE/SDS, and incubate it for 10 min at 37 °C.
9. Remove the beads using a magnetic stand and transfer the eluate to a new tube.
10. For DNA analysis, mix 10 μL of the eluate with 0.5 μL of 10 mg/mL Proteinase K, and incubate for 15 min at 37 °C.
11. Add 2.5 μL of DNA dye and run the sample on 1% (w/v) agarose gel electrophoresis with TAE buffer at 3.6 V/cm at room temperature for 1 h.
12. Soak the gel in DNA gel staining solution and leave it for 1 h at room temperate in the dark (without shaking) (*see* **Note 31**).
13. Wash the gel twice with TAE for 30 min each time.

14. Scan the gel with an imager and quantify the intensity of the DNA bands using an imaging software (*see* **Note 32**).
15. For protein analysis, mix 8 μL of the eluate (**step 9**) with an equal volume of 2× SDS-PAGE sample buffer, and incubate it for 3 min at 98 °C.
16. Analyze the sample by 7.5% SDS-PAGE.
17. Soak the gel in protein gel staining solution for 1 hour at room temperature in the dark (*see* **Note 21**).
18. Wash the gel three times with pure water for 30 min each time.
19. Scan the gel with an imager and quantify (*see* **Note 32**).

3.5 Cohesin-Replisome Encounter Assay

Cohesin establishes sister chromatid cohesion by mediating the physical tethering of two nascent sister DNAs during DNA replication. The replication-coupled biochemical assay monitors the topological entrapment of replicated DNA by cohesin using purified replication proteins and a circular DNA substrate containing the replication initiation site (*ARS1*) (Fig. 3) [23]. Following topological DNA loading of cohesin, the Mcm2-7 helicase core complex is loaded onto the cohesin-bound DNA by incubation with the O complex and Cdc6. Following phosphorylation of the Mcm2-7 double hexamer by DDK, DNA replication is initiated by the addition of replication proteins, including firing factors, S-CDK, and DNA polymerases. Nascent DNA strands of both leading and lagging strands are labelled by incorporation of biotin-16-dUTP. Two replisomes converge at the opposite end of *ARS1* to complete DNA replication. Reconstituted replication tends to halt before the termination, accumulating replication intermediates derived from

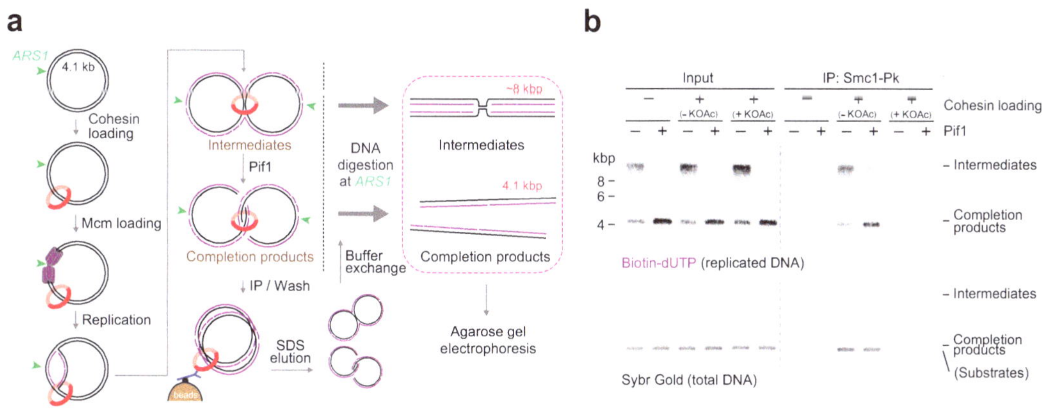

Fig. 3 Cohesin-replisome encounter assay using the purified budding yeast proteins. (**a**) Schematic of the biochemical reconstitution assay to monitor cohesin-replisome encounter during DNA replication. (**b**) Gel image showing a representative result of the cohesin-replisome encounter assay in the presence or absence of the indicated components. Nascent DNA strands were spiked by incorporation of biotin-dUTP and were detected by IRdye 680 conjugated streptavidin

two replicated DNA molecules that are base-paired at a non-replicated region (Fig. 3). The termination step is facilitated by the Pif1 helicase, generating two replicated circular DNAs that are connected by DNA catenation (Fig. 3) [21]. After recovering the cohesin-replicated DNA complex by immunoprecipitation, the cohesin-bound DNA is analyzed by agarose gel electrophoresis. To separate the intermediate and completion products of DNA replication, the DNA products are digested at the initiation site (*ARS1*) by a restriction enzyme. This digestion converts the intermediate into an X-shaped structure with a molecular size of ~8 kbp, whereas the catenated termination products are released into linear DNAs of 4.1 kbp (Fig. 3). The cohesin-bound DNA includes the replication completion products (4.1 kbp), indicating that cohesin remains associated with DNA until sister DNA duplication is complete. The concentrations of all components are given as final concentrations in the reaction mixture.

3.5.1 DNA Replication Reaction Following Cohesin Loading

1. Mix 56 nM cohesin, 56 nM Scc2-Scc4, and 8 nM cccDNA of pARS1-4.1 on ice in the cohesin loading buffer using the stock solutions (10× Loading buffer, 50% (v/v) glycerol, 10 mM Mg $(OAc)_2$, and ultrapure water). The volume of the mixture is 19 μL.
2. Add 1 μL of 100 mM ATP (5 mM final) and incubate for 30 min at 30 °C (the total reaction volume of the cohesin loading mixture is 20 μL).
3. Dilute 18 μL of the cohesin loading reaction mixture twice on ice in MCM loading buffer containing 30 nM ORC, 60 nM Cdt1-Mcm2-7, and 30 nM Cdc6 (*see* **Note 33**).
4. Incubate the reaction mixture for 15 min at 30 °C.
5. Add 0.9 μL of 1.6 μM DDK (40 nM final) to the reaction mixture, and incubate further for 15 min at 30 °C (the total reaction volume of the MCM loading mixture is 36 μL).
6. Dilute 15 μL of the cohesin/MCM loading reaction mixture fourfold at 30 °C in DNA replication buffer containing replication proteins in final concentrations of 7 nM cohesin, 7 nM Scc2-Scc4, 3.75 nM ORC, 15 nM Cdt1-MCM2–7, 3.75 nM Cdc6, 10 nM DDK, 4 nM S-CDK, 30 nM Sld2, 20 nM Sld3-Sld7, 30 nM Dpb11, 40 nM Cdc45, 12.5 nM GINS, 20 nM Csm3-Tof1, 10 nM Mrc1, 20 nM Ctf4, 20 nM RFC, 20 nM PCNA, 100 nM RPA, 5 nM Polα, 20 nM Polε, 10 nM Top1, and 5 nM Pif1 (*see* **Note 34**).
7. Incubate the reaction mixture for 2 min at 30 °C.
8. Add 1 μL of 300 nM Mcm10 (5 nM final) to the reaction mixture, and incubate for 20 min at 30 °C (the total reaction volume of the DNA replication mixture is 60 μL).

3.5.2 Reaction Termination of the Input Sample

1. Transfer 8.5 μL of the reaction sample to a new tube.
2. Add an equal volume of stop solution, and incubate for 30 min at 37 °C.

3.5.3 Purification of Cohesin-Bound Replicated DNA by Immunoprecipitation

1. Transfer 50 μL of the reaction sample to a new tube and mix it with 500 μL of ice-chilled IPH buffer (750 mM NaCl).
2. Add 20 μL of anti-Pk pre-bound protein A conjugated magnetic beads (*see* **Note 30**).
3. Rotate the reaction tube at 5 rpm at 4 °C for overnight (~15 h).
4. Wash the beads three times with 750 μL of IPH buffer (750 mM NaCl) using a magnetic stand.
5. Wash the beads once with 500 μL of IPL buffer (100 mM NaCl). Completely remove the wash buffer using a magnetic stand.
6. Add 17 μL of stop solution, and incubate it for 15 min at 37 °C.
7. Remove the beads using a magnetic stand and transfer the eluate to a new tube.

3.5.4 Restriction Digestion of Replication Products

1. Add 500 μL slurry of G-50 resin equilibrated with RE buffer to a 0.5 mL empty spin column, and centrifuge it for 1 min at 700× *g* to remove the excess amount of the buffer. Place the G-50 spin column onto a new tube.
2. Mix the input (Subheading 3.5.2, **step 2**) and IP samples (Subheading 3.5.3, **step 7**) with 10 μL of phenol/chloroform/isoamyl alcohol (25:24:1), and centrifuge them at 20,000× *g* for 5 min.
3. Take 15 μL of the aqueous layer and pass through the G-50 spin column for 1 min at 700× *g*.
4. Mix 14 μL of the pass-through fraction with 1 μL of *Pst*I/*Bgl*II mixture (20 U each) and incubate it for 30 min at 37 °C.
5. Add 1 μL of 10% (w/v) SDS and 1 μL of 250 mM EDTA to the reaction mixture to stop restriction digestion.
6. Add 3.2 μL of DNA dye to the reaction mixture.

3.5.5 Detection of DNA Replication Products

1. Separate the digested DNA samples by 0.8% (w/v) agarose gel electrophoresis with TAE buffer at 3.6 V/cm at room temperature for 2 h.
2. Soak the gel in 0.25 M HCl for 30 min and then wash twice with pure water.
3. Soak the gel in 0.2 M NaOH/0.5 M NaCl solution for 30 min and then dip twice in 1/2× TBE buffer for 10 min each time.

4. Transfer the DNA to a DNA blotting membrane using a wet transfer blotter with 1 liter of 1/2× TBE at 80 V for 50 min at 4 °C (*see* **Note 35**).
5. Cross-link the DNA to the membrane using an ultraviolet crosslinker (120 mJ/cm^2).
6. Soak the membrane in blocking solution (*see* **Note 36**).
7. Wash the membrane three times with TBS + Tween for 5 min each time.
8. Incubate the membrane with IRDye 680-conjugated streptavidin diluted at 1/10,000 in TBS + Tween + SDS at room temperature for 30 min with gentle agitation.
9. Wash the membrane three times with TBS + Tween + SDS for 5 min each and then twice with TBS for 1 min each.
10. Scan the gel image with an imager and quantify the band intensities of the DNA molecules using an imaging software (*see* **Note 37**).

4 Notes

1. The genotype of the budding yeast strain used for protein expression is *MATa*, *ade2-1*, *trp1-1*, *lue2-3,12*, *his3-11,15*, *ura3-52*, *pep4Δ::HIS3MX*, *prb1Δ::hisG*, *prc1Δ::hisG*, *can1-100*.
2. YIplac plasmid series are the shuttle vectors used for targeted integration into the budding yeast genome.
3. All buffers and solutions are prepared in ultrapure water (MilliQ, Millipore), respectively, unless otherwise stated.
4. We use cOmplete EDTA-free protease inhibitor cocktail (Merck).
5. We use Econopac Chromatography Columns (Bio-Rad).
6. We use Human IgG-agarose resin (Merck).
7. We use Glutathione Sepharose 4B (Cytiva).
8. W use HiTrap Heparin HP column (1 mL) (Cytiva).
9. We use Superose 6 Increase gel filtration column (10 × 300, 24 mL) (Cytiva).
10. We use Superdex 200 Increase gel filtration column (10 × 300, 24 mL) (Cytiva).
11. We use Amicon Ultra-4 Centrifugal filter Units, 30 K MWCO, 15 mL (Millipore).
12. We use QIAGEN Plasmid Maxi kit (Qiagen).
13. We use 13 × 51 mm Quick-Seal Centrifuge Tubes (Beckman Coulter).

14. We use a dialysis cassette or a microdialyzer with a molecular weight cutoff of 10,000 (Thermo Fisher Scientific).
15. We use 1 mM biotin-16-dUTP (Jena Bioscience).
16. We use Anti-Pk (V5) tag monoclonal antibody (Bio-Rad).
17. We use Dynabeads protein A (Veritas).
18. We use Zeta-Probe Blotting Membrane (Bio-Rad).
19. We use ECL Prime Blocking Reagent (Cytiva).
20. We use SYBR Gold Nucleic Acid Gel stain (Thermo Fisher Scientific).
21. We use Oriole Fluorescent Gel Stain (Bio-Rad).
22. We used IRDye 680-conjugated streptavidin (LI-COR biotech).
23. All protein purification procedures, including cell lysate preparation (except centrifugation) and column chromatography, are performed in a 4 °C cold room.
24. All purification fractions are checked by SDS-PAGE and Coomassie Brilliant Blue staining.
25. We use SPEX CertiPrep 6850. Typically, we disrupt up to 30 grams of cell drops using a setting of [rate: 15 cps, 1 min on/1 min off] × 24 cycles.
26. Ultracentrifugation is performed with an Optima XE-90 using a type 45 Ti rotor and 38 × 102 mm polycarbonate centrifuge bottles (Beckman Coulter).
27. We use an ÄKTA pure 25 chromatography system (Cytiva).
28. Ultracentrifugation is performed using a NVT100 rotor (Beckman Coulter).
29. We use KOD polymerase kit (TOYOBO). To amplify the marker DNAs, the PCR reaction is performed using the setting [98 °C for 5 s–59 °C for 10 s–68 °C for 90 s] × 25 cycles.
30. To prepare anti-Pk prebound protein A conjugated magnetic beads, we first wash 50 μL slurry of Dynabeads protein A (Veritas) three times with 1 mL of TBS buffer containing 0.01% (w/v) Tween 20. The beads are resuspended in 400 μL of TBS buffer containing 0.01% (w/v) Tween 20 and mixed with 10 μL of 1 mg/mL anti-Pk solution. After 2 h rotation at room temperature, the beads are washed once with 500 μL of IPH buffer and then resuspended in 50 μL of IPH buffer. Occasionally, the volume of the preparation is increased with the same antibody:beads ratio.
31. SYBR Gold Nucleic Acid Gel stain stock solution is diluted at 1/33,000 in TAE.

32. A LAS-4000 imager and MultiGauge software (FujiFilm) were used in our experiments.
33. For the Mcm2-7 loading reaction, we first prepare 17.1 μL of a buffer mixture containing ORC, Cdt1-Mcm2–7, and Cdc6 in 25 mM HEPES-KOH (pH 7.5), 200 mM KOAc, 14 mM Mg $(OAc)_2$, 0.01% (w/v) NP-40, 5% (v/v) glycerol, 0.2 mg/mL acetylated BSA, 5 mM ATP in a separate tube on ice using the stock solutions (10× Loading buffer, 50% glycerol, 140 mM $Mg(OAc)_2$, 1 M KOAc, 100 mM ATP, 10 mg/mL BSA and ultrapure water). 18 μL of cohesin loading reaction (Subheading 3.5.1, **step 2**) is added to the protein mixture and incubates it at 30 °C. After 15 min incubation, DDK is added.
34. For the DNA replication reaction, we first prepare 44 μL of a mixture containing replication proteins and chemical reagents in a DNA replication buffer in a separate tube on ice using the stock solutions (10× Replication buffer, 50% (v/v) glycerol, 1 M KOAc, 100 mM ATP, 20 mM CTP/GTP/TTP mixture, 8 mM dTAP/dCTP/dGTP mixture, 1 mM dTTP, 1 mM biotin-dUTP, 10 mg/mL BSA and ultrapure water). After 2 min incubation at 30 °C, 15 μL of cohesin/MCM loading reaction (Subheading 3.5.1, **step 4**) is added to the replication protein mixture. DNA replication is initiated by adding 1 μL of the Mcm10 stock solution.
35. We use two pieces of Whatman 17 CHR filter paper (Cytiva) to support the gel and the membrane during electrophoresis.
36. ECL Prime Blocking Reagent (Cytiva) is dissolved in TBS containing 0.1% (w/v) Tween 20 to a final concentration of 2% (w/v). The membrane is incubated at room temperature for 20 min with gentle agitation.
37. A ChemiDoc Touch imager (Bio-Rad) was used in our experiments.

Acknowledgments

This study was supported by JSPS KAKENHI grants (23 K23814, 24H02282, and 25H00980 to Y.M.) and Takeda Science Foundation (to Y.M.).

References

1. Haering CH, Farcas A-M, Arumugam P et al (2008) The cohesin ring concatenates sister DNA molecules. Nature 454:297–301
2. Uhlmann F (2016) SMC complexes: from DNA to chromosomes. Nat Rev Mol Cell Biol 17:399–412
3. Srinivasan M, Scheinost JC, Petela NJ et al (2018) The Cohesin ring uses its hinge to organize DNA using non-topological as well as topological mechanisms. Cell 173:1508–1519.e18

4. Yatskevich S, Rhodes J, Nasmyth K (2019) Organization of chromosomal DNA by SMC complexes. Annu Rev Genet 53:445–482
5. Hoencamp C, Rowland BD (2023) Genome control by SMC complexes. Nat Rev Mol Cell Biol 24:633–650
6. Ciosk R, Shirayama M, Shevchenko A et al (2000) Cohesin's binding to chromosomes depends on a separate complex consisting of Scc2 and Scc4 proteins. Mol Cell 5:243–254
7. Kueng S, Hegemann B, Peters BH et al (2006) Wapl controls the dynamic association of Cohesin with chromatin. Cell 127:955–967
8. Murayama Y, Uhlmann F (2014) Biochemical reconstitution of topological DNA binding by the cohesin ring. Nature 505:367–371
9. Murayama Y, Uhlmann F (2015) DNA entry into and exit out of the Cohesin ring by an interlocking gate mechanism. Cell 163:1628–1640
10. Lengronne A, Katou Y, Mori S et al (2004) Cohesin relocation from sites of chromosomal loading to places of convergent transcription. Nature 430:573–578
11. Murayama Y, Samora CP, Kurokawa Y et al (2018) Establishment of DNA-DNA interactions by the Cohesin ring. Cell 72:465–477.e15
12. Srinivasan M, Fumasoni M, Petela NJ et al (2020) Cohesion is established during DNA replication utilising chromosome associated cohesin rings as well as those loaded de novo onto nascent DNAs. elife 9:56611
13. Cameron G, Gruszka DT, Gruar R et al (2024) Sister chromatid cohesion establishment during DNA replication termination. Science 384: 119–124
14. Unal E, Heidinger-Pauli JM, Kim W et al (2008) A molecular determinant for the establishment of sister chromatid cohesion. Science 321:566–569
15. Rolef Ben-Shahar T, Heeger S, Lehane C et al (2008) Eco1-dependent cohesin acetylation during establishment of sister chromatid cohesion. Science 321:563–566
16. Rowland BD, Roig MB, Nishino T et al (2009) Building sister chromatid cohesion: smc3 acetylation counteracts an antiestablishment activity. Mol Cell 33:763–774
17. Minamino M, Bouchoux C, Canal B et al (2023) A replication fork determinant for the establishment of sister chromatid cohesion. Cell 186:837–849.e11
18. Xu H, Boone C, Brown GW (2007) Genetic dissection of parallel sister-chromatid cohesion pathways. Genetics 176:1417–1429
19. Yeeles JTP, Deegan TD, Janska A et al (2015) Regulated eukaryotic DNA replication origin firing with purified proteins. Nature 519:431–435
20. Hizume K, Endo S, Muramatsu S et al (2018) DNA polymerase ε-dependent modulation of the pausing property of the CMG helicase at the barrier. Genes Dev 32:1315–1320
21. Deegan TD, Baxter J, Bazán MÁO et al (2019) Pif1-family helicases support fork convergence during DNA replication termination in eukaryotes. Mol Cell 74:231–244.e9
22. Costa A, Diffley JFX (2022) The initiation of eukaryotic DNA replication. Annu Rev Biochem 91:107–131
23. Murayama Y, Endo S, Kurokawa Y et al (2024) Coordination of cohesin and DNA replication observed with purified proteins. Nature 626: 653–660
24. Ayyagari R, Impellizzeri KJ, Yoder BL et al (1995) A mutational analysis of the yeast proliferating cell nuclear antigen indicates distinct roles in DNA replication and DNA repair. Mol Cell Biol 15:4420–4429
25. Bastin-Shanower SA, Brill SJ (2001) Functional analysis of the four DNA binding domains of replication protein A. The role of RPA2 in ssDNA binding. J Biol Chem 276: 36446–36453
26. Hizume K, Kominami H, Kobayashi K et al (2017) Flexible DNA path in the MCM double hexamer loaded on DNA. Biochemistry 56: 2435–2445
27. Minamino M, Higashi TL, Bouchoux C et al (2018) Topological in vitro loading of the budding yeast cohesin ring onto DNA. Life Sci Alliance 1:e201800143

Chapter 13

An In Vitro Assay to Study RSC Remodeling Activity Stimulation by the Cohesin Loader

Sofía Muñoz

Abstract

The cohesin loader complex, consisting in Scc2 and Scc4 subunits, allows cohesin access to chromosomes, and therefore, it is required for cohesin's essential roles in many structural and functional aspects of chromosomes such as sister chromatid cohesion, genome organization, gene transcription, and DNA repair. The cohesin loader interacts with the RSC chromatin remodeling complex directing cohesin loading to nucleosome-free regions. In addition to stimulate cohesin loading and define the cohesin loading sites along the genome, the cohesin loader plays a role in nucleosome positioning at gene promoters, suggesting an interplay between cohesin loading and chromatin remodeling. Here, we describe an in vitro assay to assess the ability of the cohesin loader to stimulate RSC nucleosome sliding activity.

Key words RSC, Scc2-Scc4, Chromatin remodeling, Nucleosome, Histones, Cohesin loader, Nucleosome sliding

1 Introduction

Cohesin is the SMC complex that holds sister chromatids together from DNA replication until metaphase in a process known as sister chromatid cohesion, which is essential for correct chromosome segregation during mitosis [1, 2]. Besides sister chromatid cohesion, cohesin has critical roles in other structural and functional aspects of chromosomes, including genome organization during interphase, gene transcription, and DNA repair [3, 4]. Cohesin loading onto chromosomes requires an additional factor, the cohesion loader, encoded by the *SCC2* and *SCC4* genes in *S. cerevisiae* [5]. The cohesin loader is composed of two distinct functional modules: a globular head that corresponds to Scc4 bound to the N-terminus of Scc2 (Scc4-Scc2N) and a HEAT repeat-rich hook-like tail corresponding to the remaining C-terminal portion of Scc2 (Scc2C) that holds the catalytic activity [6]. Scc2C stimulates ATPase hydrolysis by cohesin-enabling DNA entrapment in a

Yasuto Murayama and Yasutaka Kakui (eds.), *SMC Proteins: Methods and Protocols*, Methods in Molecular Biology, vol. 2991, https://doi.org/10.1007/978-1-0716-5005-9_13,

sequence unspecific manner [7, 8], whereas Scc4-Scc2N directs the cohesin loader to chromatin receptors such as the kinetochore, the RSC (Remodels the Structure of Chromatin) chromatin remodeling complex or replication forks [9–11]. In addition to catalyze cohesin loading onto DNA and define the cohesin loading sites along the genome, the cohesin loader has been described to play a role in the regulation of the nucleosome landscape at gene promoters [12], suggesting a two-way functional crosstalk between cohesin loading and chromatin remodeling, mediated by the cohesin loader and the RSC chromatin remodeler.

Chromatin remodelers are molecular machines that utilize ATP hydrolysis to translocate DNA. All remodelers include a SNF2-related ATPase domain that is highly conserved throughout all remodeler families and share a similar mechanism of DNA translocation that underlies the different larger-scale remodeling outcomes characteristic of the remodeler category (sliding, ejection, spacing, or histone exchange) [13]. RSC is the yeast ortholog of human BAF and PBAF complexes, which are members of the conserved SWI/SNF family of ATP-dependent chromatin remodelers. They are large complexes with multiple subunits that move DNA along the histone octamer, leading to nucleosome sliding or eviction, rendering DNA accessible to factors involved in various aspects of DNA metabolism, including transcription and DNA repair [13, 14].

We have recently described that the cohesin loader binds to the RSC ATPase module [15] and adapted an in vitro RSC translocation assay [16] to demonstrate that the cohesin loader stimulates RSC remodeling activity in vitro. Mutations in NIPBL, the human cohesin loader, cause the neurodevelopmental disorder Cornelia de Lange (CdLS). This rare genetic disease clinical features are thought to be the consequence of changes in the expression of multiple genes during embryonic development [17, 18]. This assay can be performed to evaluate the capacity that different cohesin loader versions bearing CdLS mutations have in activating RSC remodeling activity to ascertain whether the ability of the cohesin loader in stimulating RSC is impaired in CdLS. Moreover, additional factors such as cohesin can be included in the assay to gain insight about their impact in the nucleosome landscape.

2 Materials

2.1 Cell Culture Media and Supplements

1. LB medium: 1% (w/v) tryptone, 0.5% (w/v) yeast extract, and 1% (w/v) NaCl. 2% (w/v) agar is added to make the solid medium (*see* **Note 1**).

2. YPD medium: 2% (w/v) peptone, 1% (w/v) yeast extract, and 2% (w/v) dextrose. 2% (w/v) agar is added to make the solid medium.
3. YP-raffinose medium: 2% (w/v) peptone, 1% (w/v) yeast extract, and 2% (w/v) raffinose.
4. Galactose solution: 20% (w/v) stock solution is prepared in water and dissolved and sterilized in the autoclave.
5. Ampicillin: 1000x stock solution (100 mg/mL) is prepared in water and sterilized by filtration.
6. Chloramphenicol: 1000x stock solution (34 mg/mL) is prepared in ethanol.
7. Spectinomycin: 1000x stock solution (100 mg/mL) in DMSO/H_2O, 1:1.
8. 1 M isopropyl β-D-1-thiogalactopyranoside (IPTG): 1 M stock solution is prepared in water and sterilized by filtration.

2.2 Strains and Plasmids for Protein Expression

1. Yeast histones: *E. coli* BL21(DE3) RIL co-expressing pCDFduet.H2A-H2B and pETduet.H3-H4 [19] was a gift from M.R. Singleton (Fig. 1a).
2. Cohesin loader: pMM9: *SCC2* gene followed by two tandem protein A tags at the C-terminus was cloned under the control of the *GAL1* promoter into pRSII402. The plasmid also contained the *GAL4* gene under the control of the *GAL10* promoter. p1012: Scc4-encoding DNA, fused to a triple HA epitope tag at the C-terminus, was cloned into YIplac204. The linearized pMM9 and p1012 were sequentially integrated into a budding yeast strain lacking the Pep4 protease at *ADE2* and *TRP1* loci, respectively, generating strain Y5278 [8] (*see* **Note 2**, Fig. 1b).
3. RSC: Y5542, endogenously tagged tandem affinity purification (TAP) Rsc2 strain [16], was a gift from J.F.X. Diffley (Fig. 1c).
4. There are two distinct RSC complexes in *S. cerevisiae* defined by the presence of one of the two paralogous Rsc1 or Rsc2 subunits. Rsc1-TAP (Y6235) was generated by gene targeting using a PCR product.

2.3 Protein Purification

1. Histones Lysis Buffer: 20 mM Tris-HCl (pH 8), 500 mM NaCl, 5 mM EDTA, 10 mM β-mercaptoethanol and protease inhibitors (2.3. materials 18 and 19).
2. Histones Buffer A: 20 mM Tris-HCl (pH 8), 5 mM EDTA, 10 mM β-mercaptoethanol.
3. Histones Buffer B: 20 mM Tris-HCl (pH 8), 2 M NaCl, 5 mM EDTA, 10 mM β-mercaptoethanol.

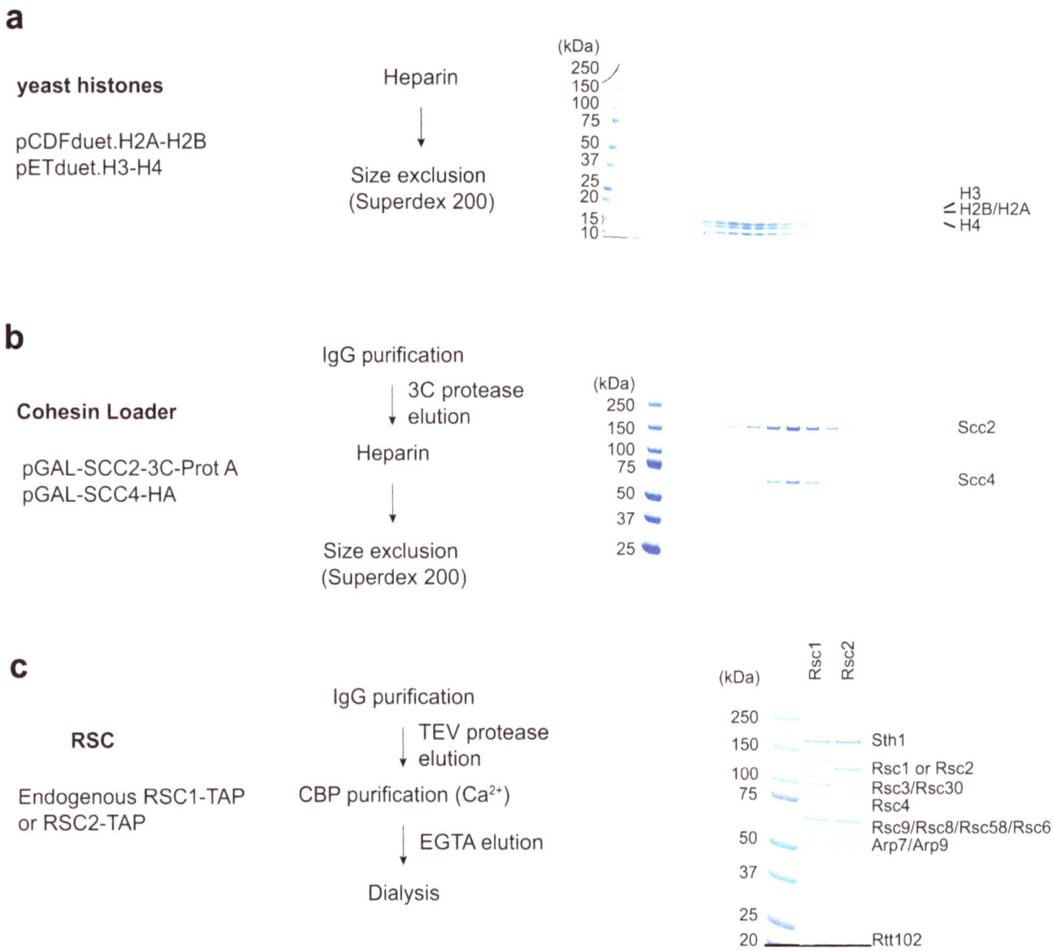

Fig. 1 Protein purification strategies. (**a**) (Left) Yeast histones were expressed in *E. coli* BL21(DE3) RIL cells co-expressing pCDFduet.H2A-H2B and pETduet.H3-H4. (Middle) Cell lysates were loaded to a heparin column followed by gel filtration. (Right) Fractions from the gel filtration purification step were loaded to an SDS-PAGE gel and stained with Coomassie. (**b**) (Left) The Scc2-Scc4 complex was purified following overexpression under control of galactose-inducible promoters in budding yeast. (Middle) The cohesin loader was purified by sequential column chromatography: protein A affinity adsorption on IgG-agarose followed by 3C Protease elution, heparin, and Superdex 200 gel filtration. (Right) Fractions from the gel filtration purification step were loaded to an SDS-PAGE gel and stained with Coomassie. (**c**) (Left) RSC was purified from budding yeast cells expressing an endogenously either TAP-tagged Rsc1 or Rsc2 subunits. (Middle) RSC was purified by binding to IgG-agarose, followed by TEV protease elution, then bound to calmodulin beads in the presence of calcium, eluted in the presence of EGTA and dialyzed. (Right) Purified RSC1 and RSC2 complexes, visualized by SDS-PAGE and Coomassie Blue staining

4. Histones Gel Filtration Buffer: 20 mM Tris-HCl (pH 8), 2 M NaCl, 0.1 mM EDTA, 10 mM β-mercaptoethanol.
5. Loader Lysis Buffer: 50 mM Tris-HCl (pH 8), 300 mM NaCl, 10% glycerol, 2.5 mM $MgCl_2$, 0.5 mM TCEP, 25 U/mL Benzonase, 10 μg/mL RNase A and protease inhibitors.

6. Loader Buffer A: 50 mM Tris-HCl (pH 8), 10% glycerol, 2.5 mM $MgCl_2$, 0.5 mM TCEP.
7. Loader Buffer B: 50 mM Tris-HCl (pH 8), 1 M NaCl, 10% glycerol, 2.5 mM $MgCl_2$, 0.5 mM TCEP.
8. Loader Gel Filtration Buffer: 20 mM Tris-HCl (pH 7.5), 150 mM NaCl, 10% glycerol, 0.5 mM TCEP.
9. RSC Lysis Buffer: 50 mM HEPES-KOH (pH 7.5), 400 mM KOAc, 10% glycerol, 10 mM EDTA, 0.5 mM DTT and protease inhibitors.
10. RSC Ig Wash Buffer: 10 mM Tris-HCl (pH 7.5), 400 mM KOAc, 10% glycerol, 0.1% NP-40, 0.5 mM EDTA, 0.5 mM DTT.
11. RSC TEV cleavage Buffer: 10 mM Tris-HCl (pH 7.5), 200 mM KOAc, 10% glycerol, 0.1% NP-40, 0.5 mM EDTA, 0.5 mM DTT.
12. RSC CBP Binding Buffer: 25 mM Tris-acetate (pH 7.5), 250 mM KOAc, 1 mM MgOAc, 1 mM imidazole, 2 mM $CaCl_2$, 10% glycerol, 0.1% NP-40, 0.5 mM DTT.
13. RSC CBP Wash Buffer 1: 25 mM Tris-acetate (pH 7.5), 400 mM KOAc, 1 mM MgOAc, 1 mM imidazole, 2 mM $CaCl_2$, 10% glycerol, 0.1% NP-40, 0.5 mM DTT.
14. RSC CBP Wash Buffer 2: 25 mM Tris-HCl (pH 7.5), 150 mM NaCl, 2 mM $CaCl_2$, 10% glycerol, 0.1% NP-40, 0.5 mM DTT.
15. RSC CBP Elution Buffer: 25 mM Tris-HCl (pH 7.5), 150 mM NaCl, 2 mM EGTA, 10% glycerol, 0.1% NP-40, 0.5 mM DTT.
16. RSC Dialysis Buffer: 25 mM Tris-HCl (pH 7.5), 200 mM KOAc, 10% glycerol, 0.5 mM DTT.
17. 3 M KOAc
18. 10% polyethylenimine: Prepare 10% from 50% and adjusted to pH 7.2 with HCl.
19. cOmplete EDTA-Free Protease Inhibitor Cocktail (Merck).
20. Pefabloc SC (Roche).
21. PreScission protease.
22. AcTEV Protease (Thermo Fisher).
23. Rabbit immunoglobulin G (IgG) resin.
24. Calmodulin Affinity resin.
25. Glutathione Sepharose 4B (Cytiva).
26. HiTrap Heparin HP 1 mL column (Cytiva).
27. HiTrap Heparin HP 5 mL column (Cytiva).
28. Superdex 200 Increase 10/300 GL gel filtration column (Cytiva).

29. HiLoad 16/600 Superdex 200 gel filtration column (Cytiva).
30. Slide-A-Lyzer Dialysis Cassettes, 3.5 K MWCO (molecular weight cutoff), 12 mL (Thermo Fisher).
31. Amicon Ultra-4 centrifuge filter unit, 15 ml 100 K MWCO (Millipore).
32. Amicon Ultra-4 centrifuge filter unit, 15 ml 30 K MWCO (Millipore).
33. Vivaspin 20 centrifugal concentrator, 100 K MWCO (Sartorius).
34. Coomassie-based ready to use protein stain for polyacrylamide gels (*see* **Note 3**).
35. Precast 4–12% or 4–15% polyacrylamide gradient gels.

2.4 Preparation of the Mononucleosome Substrate

1. Plasmid p1380, containing the Widom 601 sequence, was a gift from P. Cherepanov.
2. Oligo primers: SM365 (5′-CCCCGGGCTGCAGGA-3′) and SM366 (5′- CCTCGAGGTCGACGGTATCG-3′).
3. High-fidelity PCR DNA polymerase.
4. 2 mM dNTP mixture: 2 mM each of dATP, dCTP, dGTP, and dTTP dissolved in pure water.
5. A PCR purification kit.
6. Slide-A-Lyzer MINI Dialysis Devices, 3.5 K MWCO (Thermo Fisher).
7. Salt Gradient Dialysis Buffers: 10 mM Tris HCl (pH 7.5), 0.5 mM TCEP containing 2 M, 1 M, 0.8 M, 0.6 M, 0.4 M, 0.2 M, 0.1 M, and 50 mM NaCl, respectively.
8. Qubit 4 fluorometer (Thermo Fisher).

2.5 RSC Sliding Assay

1. Sliding Assay Buffer: 10 mM HEPES-NaOH (pH 7.5), 20 mM KOAc, 45 mM NaCl, 1 mM $MgCl_2$, 0.1 mg/ml BSA.
2. 100 mM ATP.
3. pBluescript plasmid.
4. 50% glycerol.
5. Precast 10% native polyacrylamide gels (*see* **Note 4**).
6. SYBR Gold Nucleic Acid Gel Stain (Thermo Fisher) (*see* **Note 5**).
7. A fluorescent imager (*see* **Note 6**).

3 Methods

3.1 Histone Purification

1. Inoculate *E. coli* BL21(DE3) RIL co-expressing pCDFduet. H2A-H2B and pETduet.H3-H4 in 50 ml of LB supplemented with 100 μg/mL ampicillin, 34 μg/mL chloramphenicol, and 100 μg/mL spectinomycin and grow over night at 37 °C.
2. Dilute the 50 ml of culture in 2 liters of LB supplemented with the same concentration of ampicillin, chloramphenicol, and spectinomycin. The OD600 after diluting will be around 0.1.
3. After 3 h incubation at 37 °C at 200 rpm (the culture OD600 will be 0.5–0.7), induce histone expression by adding freshly prepared 1 M IPTG to a final concentration of 0.5 mM and further culture for 4 h at 37 °C.
4. Harvest the cells by centrifugation.
5. Discard well supernatant and resuspend the cell pellet in 40 mL Histone Lysis Buffer. Split the cell suspension in two 50 mL tubes and freeze at −80 °C.
6. Thaw the cell suspensions and disrupt the cells by sonication at 40% output 3 cycles of 30″ on/30″ off. Check lysis under the microscope, and if it is not complete, add two sonication cycles (*see* **Note 7**).
7. Spin the cell lysate for 1 h at 260,000 g.
8. Take the supernatant and keep it on ice.
9. Load the supernatant in a HiTrap Heparin HP 5 ml column equilibrated with 25% Histones Buffer containing 500 mM NaCl (25% Histone buffer B) (*see* **Note 8**).
10. Wash with 5 column volumes (CV) of Histones Buffer A (0 mM NaCl).
11. Elute the bound proteins with a gradient of 0–2 M NaCl in Histones Buffer (0–100% of Histone Buffer B) in a total of 15 CV, collecting the eluate in 1 ml fractions. Histones elute at 45–65 mL.
12. Analyze 8 μl of each fraction by SDS-PAGE using a 4–15% gradient gel and Coomassie staining (*see* **Notes 3** and **9**).
13. Collect the peak fractions (typically 4 fractions) to low binding microcentrifuge tubes and spin 10′ at maximum speed to remove debris.
14. Load the resulting eluate to a HiLoad 16/600 Superdex 200 column equilibrated with Histones Gel Filtration Buffer, collecting the eluate in 1 mL fractions. Histones elute in a peak at 60–70 mL.
15. Analyze 8 μl of each fraction by SDS-PAGE using a 4–15% gradient gel and Coomassie staining (Fig. 1a).

16. Collect the peak fractions and concentrate them using a centrifugal filtration tube (Amicon Ultra-4 centrifuge filter unit of 30 K MWCO).
17. Measure the concentration by a Bradford protein assay, freeze aliquots in liquid nitrogen, and store at −80 °C (*see* **Note 10**).

3.2 Cohesin loader Purification

1. Inoculate Y5278 in YP-raffinose medium and cultivate at 30 °C overnight.
2. Dilute the preculture to 8 liters of YP-raffinose medium at OD600 = ~0.5 and cultivate for 2 h.
3. Induce the loader expression by adding 800 mL galactose solution (2% final) and culture for 1.5 h.
4. Harvest the cells, wash them with water once, and resuspend them in Loader Lysis Buffer.
5. Drop the cell suspension into liquid nitrogen forming small drops and store them at −80 °C.
6. Break the cells using a cryogenic grinder under liquid nitrogen (*see* **Note 11**).
7. Thaw the resulting cell powder in a 500 mL beaker and resuspend them in two volumes of Loader Lysis Buffer at 4 °C with gentle stirring for 1 h.
8. Spin the cell lysate for 30′ at 30,000 g.
9. Transfer the supernatant to ultracentrifuge tubes and centrifuge it 1 h at 140,000 g.
10. Equilibrate 1 mL bed volume of rabbit IgG-agarose resin in Loader Lysis Buffer.
11. Add the rabbit IgG resin (1 mL resin slurry per 80 ml lysate) to the clarified cell lysate and incubate for 2 h.
12. Transfer the suspension to an empty open column (*see* **Note 12**) and wash the resin with 10 bed volumes of Loader Lysis Buffer.
13. Resuspend the resin in 5 bed volumes of Loader Lysis Buffer, add 5 U/ml PreScission protease, and incubate it at 4 °C overnight.
14. Collect the eluate. Apply 5 bed volumes of Loader Lysis Buffer twice to the IgG resin to collect residual proteins.
15. Load the eluate in a HiTrap Heparin HP 1 ml column equilibrated with 30% Loader Buffer B (300 mM NaCl).
16. Wash the column with 10 column volumes of Loader Buffer containing 300 mM NaCl (30% of Loader Buffer B) and elute with a linear gradient of 300 mM to 1 M NaCl in Loader Buffer (30–100% Loader Buffer B) collecting 1 mL fractions.

17. Check the elution fractions by SDS-PAGE using a 4–12% gradient gel followed by Coomassie staining.
18. Collect the cohesin loader containing fractions and concentrate them to 500 μL using a centrifugal filtration tube (Amicon Ultra-4 centrifuge filter unit of 100 K MWCO).
19. Load the concentrated fractions onto a Superdex 200 Increase 10/300 GL column equilibrated with Loader Gel Filtration Buffer, collecting the eluate in 0.5 mL fractions.
20. Check the fractions by SDS-PAGE using a 4–12% gradient gel followed by Coomassie staining (Fig. 1b).
21. Collect the cohesin loader containing fractions and concentrate them using a centrifugal filtration tube (Amicon Ultra-4 centrifuge filter unit of 100 K MWCO).
22. Measure the protein concentration by a Bradford protein assay, freeze aliquots in liquid nitrogen, and store at −80 °C.

3.3 RSC Remodeler Purification

1. Inoculate Y5542 (for RSC2) or Y6235 (for RSC1) in YPD medium at 30 °C overnight.
2. Grow the cells in 8 liters of YPD at 30 °C until the culture reaches to the stationary phase (OD600 = 2.5).
3. Harvest the cells, wash them with water once, and resuspend them in RSC Lysis Buffer.
4. Drop the cell suspension into liquid nitrogen forming small drops and store them at −80 °C.
5. Break the cells using a cryogenic grinder under liquid nitrogen (*see* **Note 11**).
6. Thaw the resulting cell powder in a 500 mL beaker and resuspend them in two volumes of RSC Lysis Buffer at 4 °C with gentle stirring for 1 h.
7. Centrifuge the lysate for 30′ at 30,000 g
8. Transfer the supernatant to a beaker at 4 °C with gentle stirring, and adjust salt concentration to 400 mM KOAc using a 3 M stock (23 ml of 3 M KOAc for 400 mL supernatant).
9. While stirring, precipitate the nucleic acids by dropwise addition of 10% polyethylenimine to a final concentration of 0.1%.
10. Spin the lysate for 1 h at 140,000 g.
11. Equilibrate 1.2 mL bed volume of rabbit IgG-agarose resin in RSC Lysis Buffer.
12. Add the rabbit IgG resin (1 ml resin slurry per 80 ml lysate) to the clarified cell lysate and incubate for 2 h.
13. Transfer the suspension to an empty open column and wash the resin with 10 bed volumes of RSC Ig Wash Buffer.

14. Wash the resin further with 10 bed volumes of RSC TEV cleavage Buffer and resuspend the resin in 3 mL of RSC TEV cleavage Buffer, add 50 μL AcTEV protease (10 U/μL final), and incubate it at 4 °C overnight.
15. Collect the eluate and then apply 5 bed volumes of RSC TEV cleavage Buffer twice to the IgG resin to collect residual proteins. Add 1 M $CaCl_2$ to achieve 3 mM final. Then dilute it with 3 volumes of RSC CBP Binding Buffer.
16. Equilibrate 1.2 mL bed volume of Calmodulin Affinity Resin in RSC CBP Binding Buffer.
17. Add the Calmodulin Affinity Resin (1 mL resin slurry per 25 mL lysate) to the diluted eluate and incubate for 3 h.
18. Transfer the suspension to an empty open column and wash the beads four times with 10 mL RSC CBP Binding Buffer, twice with 10 mL RSC CBP Wash Buffer 1, and once with 10 mL RSC CBP Wash Buffer 2.
19. Resuspend in 5 mL RSC CBP Elution Buffer and collect the eluate after 5′ incubation.
20. Add 5 mL RSC CBP Elution Buffer to the resin and collect the eluate in 1 mL fractions after 10′ incubation.
21. Check the elution fractions by SDS-PAGE followed by Coomassie staining (Fig. 1c).
22. Collect the RSC containing fractions and dialyze against 4 liters of RSC Dialysis Buffer for 2 h using a dialysis cassette (12 mL Slide-A-Lyzer Dialysis Cassettes of 3.5 K MWCO).
23. Concentrate them using a centrifugal filtration tube (Vivaspin 20 centrifugal concentrator of 100 K MWCO).
24. Measure the concentration by a Bradford protein assay, freeze aliquots in liquid nitrogen, and store at −80 °C.

3.4 Preparation of a Mononucleosome Substrate

1. Prepare 100 μL of PCR reaction mixture containing 5 ng of the plasmid p1380, 0.2 μM each of SM365 and SM366 oligo primers, 0.2 mM dNTP and high-fidelity PCR DNA polymerase.
2. Amplify the 200-bp DNA fragments containing the centrally located 601 strong positioning sequence by PCR using 45 cycles of 15" denaturation at 94°C, 20" annealing at 55 °C, and 20″ synthesis at 72 °C.
3. Purify the 200-bp DNA products with a PCR purification kit. From 32 tubes of 100 μL reaction, after PCR cleanup, we usually obtain 186 ng/μL of the 200-bp DNA in total 320 μL.
4. Mix 15.6 μg of the 200-bp DNA (120 pmol) with 15.8 μg of histone octamers (144 pmol, 1:1.2 DNA: octamer ratio) in a

150 μL reaction volume in 10 mM Tris-HCl (pH 7.5), 2 M NaCl, 0.5 mM TCEP.

5. Transfer the mixture to a micro dialyzer (Slide-A-Lyzer MINI Dialysis Devices of 3.5 K MWCO).
6. Dialyze it at 4 °C against 2 liters of Salt Gradient Dialysis buffers sequentially with the following NaCl concentrations and times: 2 M NaCl for 2 h, 1 M NaCl overnight, 0.8 M NaCl for 2 h, 0.6 M NaCl for 2 h, 0.4 M NaCl for 2 h, 0.2 M NaCl for 2 h, 0.1 M NaCl overnight, and finally 50 mM NaCl for 2 h.
7. Transfer the sample to a 1.5 mL tube and incubate it for 30′ at 37 °C to position the histone octamer on the Widow 601 sequence.
8. Quantify DNA concentration after dialysis using a Qubit 4 fluorometer.
9. Run a 1 μL aliquot in a 10% native polyacrylamide gel, stain it with SYBR Gold, and scan on a fluorescent imager to make sure that all DNA molecules are packed into nucleosomes and there is not a "Free DNA" population (*see* **Notes 4–6**).

3.5 RSC Sliding Assay

1. Mix 20 nM mononucleosome substrate and 10 nM RSC, with or without 15 nM Cohesin loader in of Sliding Assay Buffer in a reaction volume of 50 μL.
2. Add 100 mM ATP to a final concentration of 1 mM and incubate at 30 °C with shaking at 500 rpm.
3. Take 10 μL aliquots at the interest time points and transfer them to fresh tubes containing 200 ng pBluescript plasmid (2 μL of a 100 ng/μL prep) that stops the reaction by competing for RSC binding with the mononucleosome substrate.
4. Add 3 μL of 50% glycerol (10% final concentration) and separate the reaction samples by 10% native polyacrylamide gel electrophoresis in TBE.
5. Stain the gel with SYBR Gold diluted to 1:10,000 in TBE for 30′.
6. Wash the gel with TBE for 5′ and scan it on a fluorescent imager (*see* **Note 6**, Fig. 2).

4 Notes

1. All media and solutions are prepared in MilliQ water (Millipore), unless denoted otherwise.
2. pRSII402 and YIplac204 are integrative plasmids for *S. cerevisiae*.

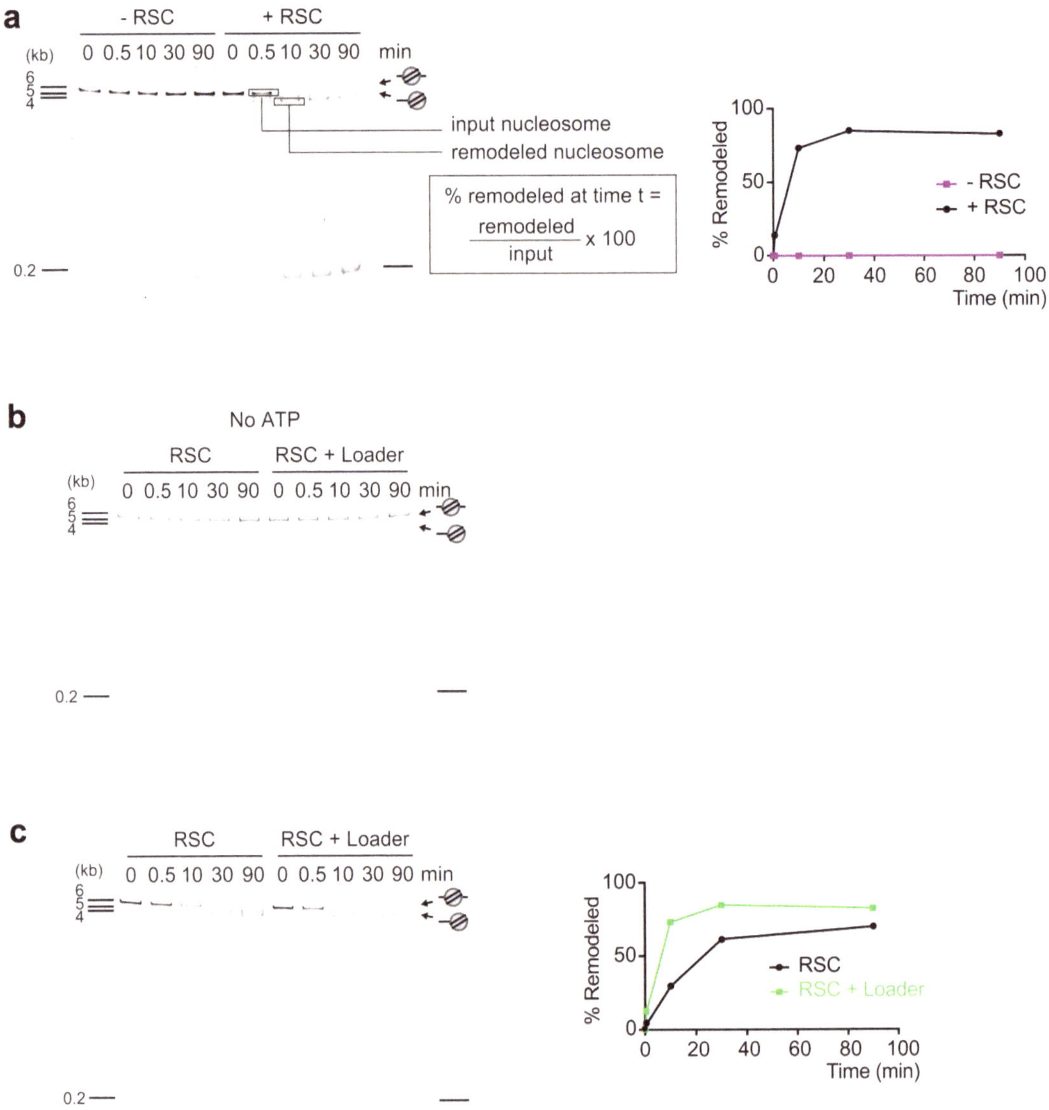

Fig. 2 In vitro RSC sliding assay. (**a**) Native polyacrylamide gel analysis of a mononucleosome substrate incubated with or without RSC. The appearance of a faster migrating band, indicative of the remodeled product following nucleosome sliding, depended on the presence of RSC. The intensity of the indicated bands was quantified over time and the % of the remodeled product was calculated as indicated. (**b**) RSC remodeling reaction depends on ATP. The above assay was repeated, with or without added cohesin loader, but in the absence of ATP. (**c**) RSC remodeling occurs faster in the presence of the cohesin loader

3. We use InstantBlue Coomassie Protein Stain from Expedeon.
4. We use Novex TBE, 10% native polyacrylamide gels from Invitrogen.
5. SYBR Gold is a fluorescent compound with an excitation peak at 496 nm and an emission peak at 539 nm.

6. We use a Typhoon FLA 9500 imager (GE Healthcare).
7. All procedures for protein purification including cell lysate preparation and open and packed column chromatography are carried out at 4 °C. We perform all chromatography steps in a 4 °C cold room.
8. We use an ÄKTA purifier system (GE Healthcare) for column chromatography.
9. All purification fractions are checked by SDS-PAGE, followed by Coomassie staining.
10. From 2 liters culture, we typically obtain around 30 mg of histones at a concentration of 5 mg/ml.
11. We use SPEX CertiPrep 6850 freezer mill.
12. We use Econopac Chromatography Columns (Bio-Rad).

Acknowledgments

I would like to thank M.R. Singleton, A. Ballandras-Colas and P. Cherepanov, C. Casas-Delucchi and J. Diffley, S. Mouilleron, and M. Minamino for reagents and the Crick Fermentation Science Technology Platform, F. Uhlmann, C. Bouchoux, as well as all the Crick Chromosome Segregation Laboratory members for discussions and technical advice. SM is funded by the Spanish Ministry of Science Innovation and Universities grants RYC2022-037334-I (funded by MICIU/AEI and FSE+) and PID2023-149921NA-I00 (funded by MICIU/AEI and FEDER, UE).

References

1. Peters JM, Nishiyama T (2012) Sister chromatid cohesion. Cold Spring Harb Perspect Biol 4(11). https://doi.org/10.1101/cshperspect.a011130
2. Nasmyth K, Haering CH (2009) Cohesin: its roles and mechanisms. Annu Rev Genet 43:525–558. https://doi.org/10.1146/annurev-genet-102108-134233
3. van Ruiten MS, Rowland BD (2021) On the choreography of genome folding: a grand pas de deux of cohesin and CTCF. Curr Opin Cell Biol 70:84–90. https://doi.org/10.1016/j.ceb.2020.12.001
4. Dorsett D, Strom L (2012) The ancient and evolving roles of cohesin in gene expression and DNA repair. Curr Biol 22(7):240. https://doi.org/10.1016/j.cub.2012.02.046
5. Ciosk R, Shirayama M, Shevchenko A et al (2000) Cohesin's binding to chromosomes depends on a separate complex consisting of Scc2 and Scc4 proteins. Mol Cell 5(2):243–254
6. Chao WC, Murayama Y, Munoz S et al (2015) Structural studies reveal the functional modularity of the Scc2-Scc4 cohesin loader. Cell Rep 12(5):719–725. https://doi.org/10.1016/j.celrep.2015.06.071
7. Murayama Y, Uhlmann F (2014) Biochemical reconstitution of topological DNA binding by the cohesin ring. Nature 505(7483):367–371. https://doi.org/10.1038/nature12867
8. Minamino M, Higashi TL, Bouchoux C et al (2018) Topological in vitro loading of the budding yeast cohesin ring onto DNA. Life Sci Alliance 1(5). https://doi.org/10.26508/lsa.201800143
9. Hinshaw SM, Makrantoni V, Harrison SC et al (2017) The kinetochore receptor for the cohesin loading complex. Cell 171(1):72–84 e13. https://doi.org/10.1016/j.cell.2017.08.017

10. Muñoz S, Minamino M, Casas-Delucchi C et al (2019) A role for chromatin remodeling in cohesin loading onto chromosomes. Mol Cell 74(4):664–673 e5. https://doi.org/10.1016/j.molcel.2019.02.027
11. Minamino M, Bouchoux C, Canal B et al (2023) A replication fork determinant for the establishment of sister chromatid cohesion. Cell 186(4):837–849.e11. https://doi.org/10.1016/j.cell.2022.12.044
12. Lopez-Serra L, Kelly G, Patel H et al (2014) The Scc2-Scc4 complex acts in sister chromatid cohesion and transcriptional regulation by maintaining nucleosome-free regions. Nat Genet 46(10):1147–1151. https://doi.org/10.1038/ng.3080
13. Eustermann S, Patel AB, Hopfner KP et al (2024) Energy-driven genome regulation by ATP-dependent chromatin remodellers. Nat Rev Mol Cell Biol 25(4):309–332. https://doi.org/10.1038/s41580-023-00683-y
14. Clapier CR, Iwasa J, Cairns BR et al (2017) Mechanisms of action and regulation of ATP-dependent chromatin-remodelling complexes. Nat Rev Mol Cell Biol 18(7):407–422. https://doi.org/10.1038/nrm.2017.26
15. Muñoz S, Jones A, Bouchoux C et al (2022) Functional crosstalk between the cohesin loader and chromatin remodelers. Nat Commun 13(1):7698. https://doi.org/10.1038/s41467-022-35444-6
16. Wittmeyer J, Saha A, Cairns B (2004) DNA translocation and nucleosome remodeling assays by the RSC chromatin remodeling complex. Meth Enzymol 377:322–343. https://doi.org/10.1016/s0076-6879(03)77020-7
17. Fryns JP (1986) On the nosology of the Cornelia de Lange and coffin-Siris syndromes. Clin Genet 29(3):263–264
18. Krantz ID, McCallum J, DeScipio C et al (2004) Cornelia de Lange syndrome is caused by mutations in NIPBL, the human homolog of Drosophila melanogaster nipped-B. Nat Genet 36(6):631–635. https://doi.org/10.1038/ng1364
19. Kingston IJ, Yung JS, Singleton MR (2011) Biophysical characterization of the centromere-specific nucleosome from budding yeast. J Biol Chem 286(5):4021–4026. https://doi.org/10.1074/jbc.M110.189340

Chapter 14

Purification and In Vitro Analysis of *Escherichia coli* RecN

Shunsuke Noda and Takashi Hishida

Abstract

The RecN protein, a member of the SMC (structural maintenance of chromosomes) family, is highly conserved among bacteria and plays a critical role in the repair of DNA double-strand breaks (DSBs). Like other SMC proteins, RecN forms a ringlike structure and exhibits topological DNA-binding activity, enabling it to encircle DNA molecules. This activity is thought to be essential for maintaining nucleoid integrity at DSB sites and facilitating RecA-mediated recombination reactions. Here, we describe the purification process of the *E. coli* RecN protein and the assays used to evaluate its ability to mediate intermolecular DNA tethering through topological entrapment, as well as its role in facilitating strand exchange reactions. These biochemical characterizations provide insights into the function of RecN during RecA-mediated DSB repair.

Key words RecN, RecA, SMC, DSBs, Topological DNA binding

1 Introduction

DNA double-strand breaks (DSBs) are a particularly deleterious form of DNA damage that threaten genome integrity. If left unrepaired or repaired incorrectly, DSBs can result in mutagenesis, chromosomal rearrangements, and cell death. Homologous recombination is an evolutionarily conserved mechanism that plays a pivotal role in the repair of DSBs [1, 2]. This process employs intact homologous DNA sequences as templates to carry out faithful repair of DSBs, involving the coordinated action of multiple proteins [3, 4].

In *Escherichia coli*, DSBs are initially processed by the RecBCD helicase-nuclease complex to generate 3′ single-stranded DNA (ssDNA) tails, onto which RecA assembles as a nucleoprotein filament [5, 6]. The RecA filament then searches for a homologous sequence in the intact sister chromosome and mediates homologous pairing, strand invasion, and strand exchange to form a recombination intermediate [7–10]. The invading 3′-ssDNA then serves as a primer for DNA repair synthesis, leading to the

Yasuto Murayama and Yasutaka Kakui (eds.), *SMC Proteins: Methods and Protocols*, Methods in Molecular Biology, vol. 2991, https://doi.org/10.1007/978-1-0716-5005-9_14,

formation of a Holliday junction. Finally, the structure-specific helicase RuvAB promotes branch migration of the Holliday junction, while the RuvC endonuclease resolves it, ensuring proper segregation of the repaired chromosome [11–13].

Structural maintenance of chromosome (SMC) family proteins play diverse roles in the integrity and dynamics of chromosomes, including sister chromatid cohesion, chromosome condensation, and DNA repair [14–20]. *E. coli recN*, a typical SOS-regulated gene, encodes a protein that belongs to the SMC family of proteins [21–23]. Mutants lacking RecN are highly sensitive to ionizing radiation and DNA cross-linking agent mitomycin C (MMC), as well as to exhibit an abnormal nucleoid morphology in the presence of MMC, suggesting a specific role in DSB repair [24–26]. In vivo studies using fluorescent microscopy demonstrate that when DSBs are induced by MMC, GFP-RecN assembles into nucleoid-associated foci in RecA-dependent manner and colocalizes with RecA [24, 26, 27]. Additionally, RecN preserves sister-chromatid interactions during MMC exposure and modulates the dynamics of RecA nucleoprotein filaments during homology search [28–30]. Together, these results suggest that RecN plays both structural and functional roles in coordinating chromosome organization and facilitating homologous recombination during DSB repair.

To better understand the function of RecN in these processes, further studies are needed, with a particular focus on its biochemical properties. Here, we describe procedures for purifying *E. coli* RecN and analyzing its biochemical properties. Specifically, we will focus on assays to assess RecN-catalyzed topological DNA entrapment and RecN-enhanced DNA recombination. These assays will help elucidate how the two distinct roles of RecN—maintaining sister chromatid interactions and facilitating RecA-mediated reactions—are linked during DSB repair.

2 Materials

2.1 Purification

2.1.1 Strain, Plasmid, Media, and Other Reagents

1. BL21(DE3).
2. RecN expression plasmid contains N-terminally hexahistidine-fused RecN, in which *recN* gene is cloned into pET15b [31].
3. LB medium: 1% (w/v) tryptone, 0.5% (w/v) yeast extract, 1% (w/v) NaCl, adjusted to pH 7.0 with NaOH. 1.5% (w/v) agar is added to make the solid medium.
4. 500 mM isopropyl-β-D-thiogalactopyranoside (IPTG).
5. Ampicillin: 1000x stock solution (100 mg/mL) is prepared in distilled water and sterilized by filtration.

2.1.2 Protein Purification

1. Resuspension buffer: 20 mM Tris-HCl (pH 7.5), 150 mM NaCl, and 20% (w/v) sucrose.
2. Lysis buffer: 20 mM Tris-HCl (pH 7.5), 1 M NaCl, 1 mM EDTA, 10% glycerol, 7 mM 2-mercaptoethanol, and 1 mM phenylmethylsulfonyl fluoride (PMSF) (*see* **Note 1**).
3. N buffer: 20 mM Tris-HCl (pH 7.5), 1 M NaCl, 5% glycerol, and 2 mM 2-mercaptoethanol.
4. Storage buffer: 20 mM Tris-HCl at pH 7.5, 200 mM NaCl, 1 mM EDTA, 50% glycerol, and 1 mM dithiothreitol (DTT).
5. Ammonium sulfate (*see* **Note 2**).
6. 2 M imidazole (*see* **Note 3**).
7. Ni^{2+}-affinity column.
8. Co^{2+}-affinity column.

2.2 Biochemical Characterization

2.2.1 Proteins, DNA, and Other Reagents

1. RecA protein (*see* **Note 4**).
2. SSB protein (*see* **Note 4**).
3. ΦX174 Replication form I DNA.
4. ΦX174 Virion single-stranded circular DNA.
5. 10 units/μL *Pst*I.
6. MicroSpin S-400 HR columns (Cytiva).
7. 100 mM ATP.
8. Creatine phosphate.
9. Creatine kinase.
10. Proteinase K, recombinant PCR grade.
11. Dynabeads M-280 Streptavidin (Thermo Fisher).
12. SYBR Gold nucleic acid gel stain (Thermo Fisher).
13. CBB protein stain solution.
14. PCR DNA purification kit.

2.2.2 Buffers

1. Annealing mixture: 50 mM Tris-HCl (pH 7.5), 5 mM $MgCl_2$, 100 mM KCl, 0.05 μM ΦX174 circular ssDNA (CssDNA), 0.15 μM biotinylated 72 mer oligonucleotide (5′_biotin-TG GCTTGAACGCGTCATGGAAGCGATAAAACTCTG CAGGTTGGATACGCCAATCATTTTTATCGAAGCGCGC _3′). Nucleic acid concentrations are given per DNA molecule.
2. TET buffer: TE containing 100 mM KCl and 0.05% (w/v) Tween20.
3. EM buffer: 20 mM Tris-HCl (pH 7.5), 0.01% (w/v) NP-40, and 1 mM DTT, 50 mM KCl, 2 mM $MgCl_2$, 5% glycerol.

4. T50 buffer: 20 mM Tris-HCl (pH 7.5), 0.01% (w/v) NP-40, 7 mM 2-mercaptoethanol, 50 mM KCl, 2 mM $MgCl_2$, 2 mM ATP, 5% glycerol.
5. T100 buffer: 20 mM Tris-HCl (pH 7.5), 0.01% (w/v) NP-40, 7 mM 2-mercaptoethanol, 100 mM KCl, 2 mM $MgCl_2$, 2 mM ATP, 5% glycerol.
6. SE buffer: 33 mM Tris-HCl (pH 7.5), 40 mM NaCl, 10 mM $MgCl_2$, 1 mM DTT, 2% glycerol.
7. ATP regeneration system: 3 mM ATP, 8 mM creatine phosphate, and 8 U/mL creatine kinase in SE buffer.
8. Elution solution: 10 mM Tris-HCl (pH 7.5), 1 mM EDTA, 1% (w/v) SDS.
9. Stop buffer: 10% (w/v) SDS, 0.1 M EDTA, 10 mg/mL Proteinase K.
10. 6x DNA loading buffer: 20 mM Tris-HCl (pH 7.5), 15% (w/v) Ficoll, 0.05% (w/v) bromophenol blue, 0.05% (w/v) Xylene Cyanol FF.
11. 5x SDS loading buffer: 40 mM Tris-HCl (pH 6.8), 2% (w/v) SDS, 10% glycerol, 270 mM 2-mercaptoethanol, 0.02% (w/v) bromophenol blue.
12. 1x TAE: 40 mM Tris-acetate, 1 mM EDTA.
13. 0.5x TBE: 22.5 mM Tris-borate, 0.5 mM EDTA.

2.3 Equipment

1. Incubators and shakers appropriate for growing bacteria culture and maintaining enzymes at the optimal temperature for their activity.
2. High-speed centrifuges.
3. Ultrasonic disruptor.
4. Spectrophotometer for protein quantification.
5. Column chromatography system (*see* **Note 5**).
6. Heat block.
7. Electrophoresis apparatus for running SDS-PAGE and native PAGE.
8. Magnetic stand for efficient separation of all types of magnetic beads.
9. A fluorescent gel imager (*see* **Note 6**).

3 Methods

3.1 Purification of RecN

1. Introduce RecN expression plasmid into BL21(DE3). Cultivate on LB plates containing 100 μg/mL ampicillin overnight at 37 °C.

2. Inoculate a single colony in 5 mL of LB liquid medium containing 100 μg/mL ampicillin and grow overnight at 37 °C.
3. Inoculate 4 mL of the overnight culture into 400 mL of LB liquid medium containing 100 μg/mL ampicillin in a 1 L baffled flask.
4. Incubate at 37 °C with 200 rpm shaking until OD_{600} reaches 0.4.
5. Induce protein expression by adding IPTG to a final concentration of 100 μM and continue to grow for 2 hours at 37 °C (*see* **Note 7**).
6. Centrifuge the cells at 4,000 g for 15 min at 4 °C (*see* **Note 8**).
7. Resuspend the cell pellets in 20 mL of Resuspension buffer and centrifuge at 4,000 g for 15 min at 4 °C (*see* **Note 9**).
8. Resuspend the cell pellets in 20 mL of lysis buffer.
9. Disrupt the cells by sonication (*see* **Notes 10** and **11**).
10. Centrifuge the cell crudes at 12,000 g for 30 min at 4 °C (*see* **Note 12**).
11. Transfer supernatant to a glass beaker and add ammonium sulfate to achieve 30% salt saturation.
12. Incubate with gentle agitation for 1 hour at 4 °C, and centrifuge the cell crudes at 12,000 g for 30 min.
13. Discard the supernatant and resuspended the pellet in 40 mL of N buffer (*see* **Note 13**).
14. Load the suspension onto a 5 mL of Co^{2+}-affinity column equilibrated in N buffer, wash with 10 column volume (CV) of N buffer containing 10 mM imidazole, and elute protein with a linear gradient of 10–500 mM imidazole over 5 CV (*see* **Note 14**).
15. Collect the peak fractions and dialyze overnight with N buffer at 4 °C to remove imidazole (*see* **Note 15**).
16. Load sample onto a 1 mL Ni^{2+}-affinity column equilibrated in N buffer, wash with 10 CV of N buffer containing 10 mM imidazole followed by 5 CV of N buffer containing 70 mM imidazole, and elute protein with N buffer containing 500 mM imidazole (*see* **Note 16**).
17. Pool the peak fractions and concentrate it to 500 μL by centrifugal ultrafiltration.
18. Dialyze overnight in 1 L of storage buffer at 4 °C (*see* **Note 15**).
19. Freeze aliquots in liquid nitrogen and store sample at −80 °C.
20. Determine proteins concentration using Bradford assay (*see* **Note 17**).

3.2 Electrophoretic Mobility Shift Assay for Topological DNA Binding of RecN

Like other SMC proteins, RecN is thought to form a ringlike structure around DNA, leading to topological DNA entrapment [31]. In this mode of DNA binding, RecN is capable of sliding along the DNA while remaining bound to circular DNA without dissociation. In contrast, in linear DNA, DNA binding of RecN is rarely observed because it is displaced at the DNA ends. This section describes a method for assessing the circular dsDNA (Cds)-RecN complexes through the topological entrapment. The high-molecular-weight Cds-RecN complexes observed in the wells of agarose gel electrophoresis reflect the formation of large protein–DNA complexes facilitated by RecN-RecN interactions, in addition to RecN-mediated topological DNA entrapment, as they are eliminated by restriction enzyme-mediated linearization (Fig. 1).

1. Prepare 10 μL reaction mixture on ice in EM buffer containing 1.5 nM ΦX174 RF I DNA (as a DNA molecule), 2 mM ATP, and 1 μM RecN.
2. Incubate reaction mixture for 10 min at 37 °C.
3. Add 2 μL of 10 units/μL *Pst*I restriction enzyme (*Pst*I +) or 2μL of storage buffer (*Pst*I −), and further incubate for 15 min at 37 °C.
4. Stop the reactions on ice and add 2.5 μL of 6x DNA loading buffer.
5. Run 10 μL of each reaction on 0.8% agarose gel electrophoresis with 0.5x TBE buffer.
6. Soak the gel in 0.5x TBE buffer containing SYBR Gold stain, diluted 1/10,000, and stain for 1 h with gentle shaking.
7. Visualize the gel using a fluorescent gel imager (Fig. 1).

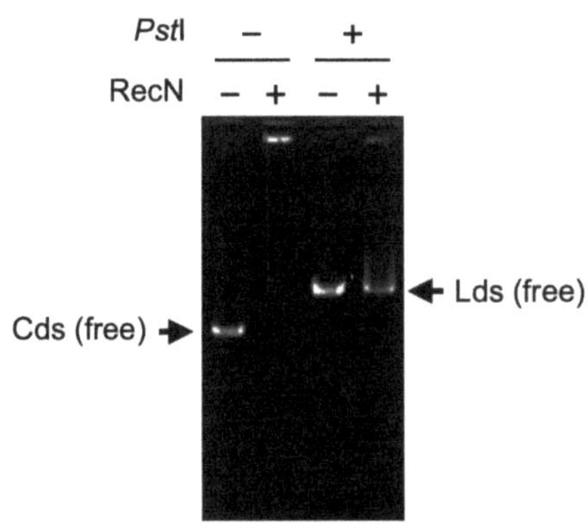

Fig. 1 RecN topologically binds to circular double-stranded DNA (CdsDNA). Assays are as described in Subheading 3.2.1. Electrophoretic mobility shift assays in which RecN is incubated with circular dsDNA (Cds) and then incubates with or without the single-cut restriction enzyme *Pst*I, which generates linear dsDNA (Lds)

3.3 Pull-Down Assay to Assess Intermolecular DNA Tethering Activity of RecN

RecN binds to ssDNA and dsDNA through topological entrapment with a clear preference for ssDNA [31]. Furthermore, when RecN is loaded onto ssDNA, it tethers a second dsDNA molecule, with this binding activity being enhanced by the presence of ATP [31]. This binding activity may play a crucial role in RecA-mediated homologous pairing during synapsis by facilitating proximity between ssDNA and homologous dsDNA during DSB repair. This section describes a pull-down assay utilizing bead-bound circular ssDNA. In this assay, following the formation of a complex between bead-bound circular ssDNA and RecN, the complex is mixed with circular dsDNA. The fraction bound to circular ssDNA is collected using magnetic beads (Fig. 2a).

3.3.1 Immobilize Circular ssDNA (CssDNA)

1. Prepare the following annealing mixture on ice (total volume: 100 μL): 50 mM Tris-HCl (pH 7.5), 5 mM $MgCl_2$, 100 mM KCl, 0.05 μM ΦX174 circular ssDNA (CssDNA), 0.15 μM biotinylated 72 mer oligonucleotide (5′_biotin-TGGCTTGA ACGCGTCATGGAAGCGATAAAACTCTGCAGGTTGGA TACGCCAATCATTTTTATCGAAGCGCGC _3′). Nucleic acid concentrations are given per DNA molecule.
2. Heat the annealing mixture at 90 °C for 15 s and then gradually cool to 25 °C in a thermal cycler.
3. Remove excess oligonucleotides from the mixture by a size exclusion spin column (MicroSpin S-400 HR) (*see* **Note 18**).
4. Add 100 μL of annealed DNA mixture to the washed streptavidin beads (*see* **Note 19**).

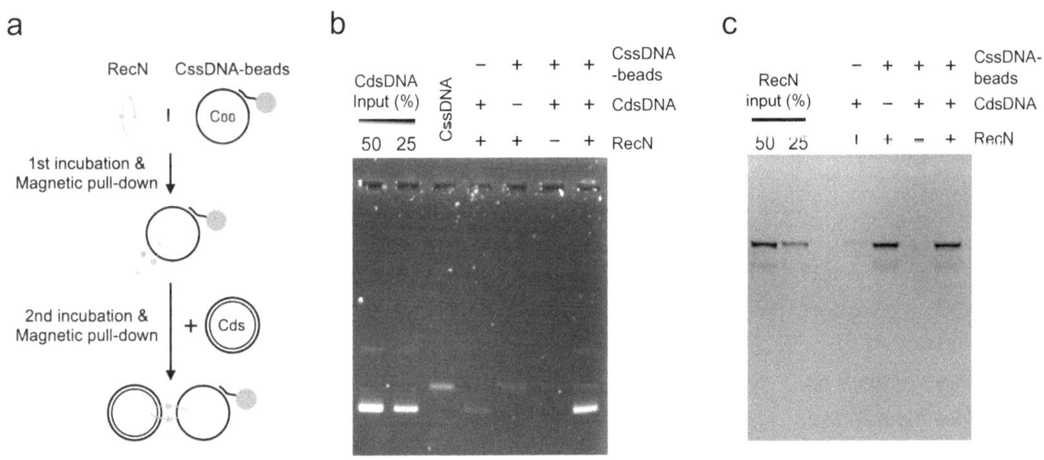

Fig. 2 CssDNA-bound RecN captures second CdsDNA. (**a**) Schematic illustration of two-step pull-down assay using CssDNA-conjugated magnetic beads. (**b**, **c**) Assays are described in Subheading 3.2.2. RecN is incubated with DNA-free beads or CssDNA-beads. The beads were then washed and incubated with CdsDNA as a secondary dsDNA substrate. After splitting the beads-bound materials into two, they are analyzed by agarose gel electrophoresis (**b**) or SDS-PAGE (**c**)

5. Incubate for 1 h at room temperature with gentle rotation of the tube.
6. Place the tube in a magnet for 2–3 min and discard the supernatant.
7. Wash the immobilized CssDNA beads three times in TET buffer.
8. Resuspend immobilized CssDNA beads with 100 μL of TET buffer, and store at 4 °C (*see* **Note 20**).

3.3.2 Intermolecular DNA Tethering Assay (Two-Step Pull-Down Assay)

1. Mix 1 μM RecN and immobilized CssDNA beads (0.75 nM as a DNA molecule) in T100 buffer on ice (total volume: 10 μL).
2. Mix the sample by tapping, incubate for 10 min at 37 °C, and stand on ice for 2 min.
3. Place the tube in a magnetic stand for 2 min and discard the supernatant carefully (*see* **Note 21**).
4. Wash the CssDNA beads with 100 μL of T100 buffer.
5. Add 10 μL of T50 buffer containing 0.75 nM ΦX174 circular dsDNA (Cds DNA) (0.75 nM as a DNA molecule).
6. Incubate for 10 min at 37 °C with occasional tapping.
7. Place the tube in a magnetic stand for 2 min and discard the supernatant.
8. Wash the CssDNA beads in 100 μL of T50 buffer.
9. Add 20 μL of Elution solution and incubate for 10 minutes at 37 °C with occasional tapping.
10. Split 20 μL samples into 2 for DNA and protein analysis.
11. To analyze DNA, add 2 μL of 6x DNA loading buffer, and run on the 0.8% agarose gel electrophoresis with 1x TAE buffer.
12. Stain and visualize the gel with SYBR Gold as described above (Fig. 2b).
13. To analyze protein, add 2.5 μL of 5x SDS loading buffer and heat samples at 95 °C for 3 min.
14. Run the sample on a 10% SDS-PAGE until the dye reaches to the edge of the gel, and detect the protein bound to the CssDNA beads by CBB staining (Fig. 2c).

3.4 Effect of RecN on RecA-Mediated Strand Exchange

Strand exchange reactions with RecA are carried out using circular ssDNA (Css) and linear dsDNA (Lds) as shown previously [32, 33]. In this experiment, circular ssDNA is pre-incubated with RecA to form nucleoprotein filaments, and subsequently, SSB is added. The reaction is initiated with an addition of Lds to the mixture. The stimulation of the recombination reaction is quantified by measuring the amounts of recombination intermediate (RI) and nicked circular DNA (NC) (Fig. 3a).

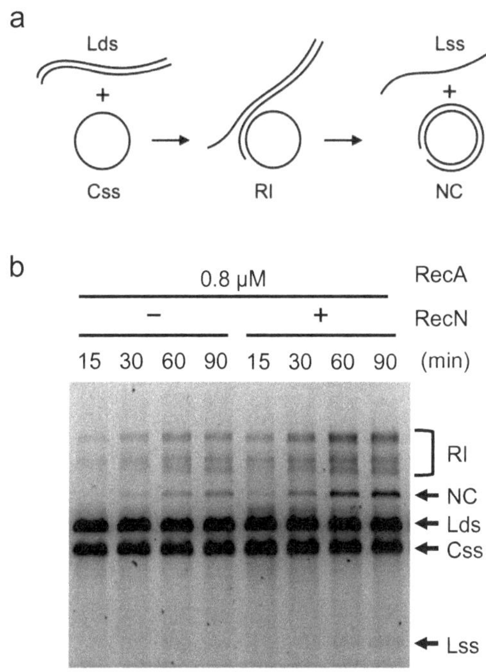

Fig. 3 (**a**) Schematic illustration of the strand exchange assay. A three-strand exchange reaction between homologous circular ssDNA (Css) and a linear dsDNA (Lds) produces a recombination intermediate (RI), a nicked circular dsDNA (NC), and a displaced linear ssDNA (Lss). (**b**) RecN promotes RecA-mediated DNA strand exchange. Assays are as described in Subheading 3.3. The reactions were initiated by adding Lds and incubated for 90 min. 10 ul aliquots of the sample were taken at indicated time points (15, 30, 60, and 90 minutes). Each aliquots were analyzed by agarose gel electrophoresis and SYBR Gold staining. The gel image was inverted in black and white tones

To investigate the impact of RecN on the RecA-mediated strand exchange reaction, it is necessary to perform experiments at a RecA concentration (0.8 μM) that is below the optimal conditions. In this section, we describe a time-course experiment to evaluate the effect of RecN on RecA-mediated strand exchange reaction. An alternative approach involves performing a RecN titration experiment, where the RecA reaction (0.6 μM) is monitored at different concentrations (0.5–1.5 μM RecN) [31].

3.4.1 Preparation of PstI-Linearized ΦX174 Double-Stranded DNA

1. Digest 100 μg of ΦX174 Replication form I DNA with *Pst*I.
2. Purify the digest DNA using commercially available PCR purification kit.
3. Adjust DNA concentration to 100 μM (as nucleotide) with H_2O.

3.4.2 DNA Strand Exchange Assay

1. The final reaction volume is 50 μL and the final concentration is expressed as a value per 50 μL. Prepare 30 μL of reaction mixtures containing 10 μM ΦX174 circular ssDNA (as a nucleotide) and ATP regeneration system (3 mM ATP, 8 mM creatine phosphate, and 8 U/mL creatine kinase) in SE buffer (*see* **Note 22**).
2. Add 5 μL of 8 μM RecA (0.8 μM final).
3. Add 5 μL of 10 μM RecN (1 μM final) or 5 μL of storage buffer as a control.
4. Incubate for 15 min at 37 °C.
5. Add 5 μL of 20 μM SSB (2 μM final).
6. Incubate for 10 min at 37 °C.
7. Add 5 μL of 100 μM *Pst*I-linearized ΦX174 dsDNA (10 μM final as a nucleotide) to initiate the reaction.
8. Incubate at 37 °C and then 10 μL aliquots of the reaction mixture are taken at indicated time point (15, 30, 60, and 90 min), and immediately mix well with 1 μL of Stop buffer.
9. Incubate each aliquot for 30 min at 30 °C.
10. Add 2.2 μL of 6x DNA loading buffer.
11. Run 10 μL of each reaction on 0.8% (w/v) agarose gel electrophoresis with 1x TAE buffer at 4.6 V/cm for 100 minutes at room temperature.
12. Stain the gel using SYBR Gold stain and visualize the gel described as above (Fig. 3b).
13. The intensity of DNA bands is quantified with Image J software.

4 Notes

1. PMSF is freshly added at 0.5 mM final concentration from a 100 mM stock solution in 2-propanol.
2. Grind the dry ammonium sulfate into a fine powder using a mortar and pestle.
3. Dissolve powdered imidazole in distilled water and adjust to pH 7.5 using HCl.
4. We use commercially available RecA and SSB (bioacademia).
5. We use ÄKTA FPLC (Cytiva).
6. We use Gel Doc EZ imager (Bio-Rad).
7. Prolonged induction time reduces protein stability.
8. All procedures thereafter are carried out at 4 °C.
9. Cells can be frozen in liquid nitrogen and store in −80 °C until ready to use.

10. Sonication of *E. coli* cells using Ultrasonic disruptor UD-201 (TOMY SEIKO) as follows: output: 5, duty: 60% with 2 min resets in between 1 min sonication. The cells are disrupted in total 7 min sonication. This sonication is carried out on ice.
11. The ionic strength of the lysis buffer affects the solubility of E. coli RecN so that sonication should be performed at high salt concentrations of 1 M or higher.
12. Keep a small aliquot for checking solubility of proteins.
13. Pellets can be stored at −80 °C for 1 month.
14. We use an ÄKTA pure FPLC system (Cytiva) for column chromatography.
15. We use a Slide-A-Lyzer Dialysis cassettes for dialysis (Thermo Fisher).
16. The use of two chelating columns, one for nickel and one for cobalt, effectively eliminates the potential for contamination by nucleases.
17. The typical yield from 400 mL of *E. coli* cells is ~0.5–1 mg of ~10–30 μM protein.
18. The sample is applied to the MicroSpin S-400 column and eluted by centrifugation at 735 g for 2 min at 4 °C.
19. Resuspend 150 μL of Dynabeads M-280 streptavidin in the tube, wash twice with 100 μL of TE containing 100 mM KCl and 0.05% (v/v) Tween20, and discard the supernatant. The beads are ready to be coated with the biotinylated CssDNA molecule.
20. To measure the concentration of CssDNA, take 10 μL of CssDNA beads solution, discard the supernatant, and add 10 μL of 0.1 M NaOH to the solution and measure the concentration using nanodrop.
21. It is recommended that all RecN be recovered in the CssDNA beads fraction. This can be determined by analyzing a supernatant by SDS-PAGE.
22. Based on the current experimental setup, a highly efficient strand exchange reaction is observed when RecA is present at concentrations exceeding 1 μM.

Acknowledgments

We are indebted to Dr. Kenji Keyamura for plasmids, proteins, and other reagents. This work was supported by Grants-in-Aid for Scientific Research from the Ministry of Education, Culture, Sports, Science, and Technology (MEXT), Japan.

References

1. Amundsen SK, Smith GR (2003) Interchangeable parts of the Escherichia coli recombination machinery. Cell 112:741–744
2. Wyman C, Ristic D, Kanaar R (2004) Homologous recombination-mediated double-strand break repair. DNA Repair (Amst) 3:827–833
3. Cromie GA, Connelly JC, Leach DR (2001) Recombination at double-strand breaks and DNA ends: conserved mechanisms from phage to humans. Mol Cell 8:1163–1174
4. Kowalczykowski SC (2015) An overview of the molecular mechanisms of Recombinational DNA repair. Cold Spring Harb Perspect Biol 7:a016410
5. Singleton MR, Dillingham MS, Gaudier M et al (2004) Crystal structure of RecBCD enzyme reveals a machine for processing DNA breaks. Nature 432:187–193
6. Dillingham MS, Kowalczykowski SC (2008) RecBCD enzyme and the repair of double-stranded DNA breaks. Microbiol Mol Biol Rev 72:642–671
7. Cox MM (2007) Regulation of bacterial RecA protein function. Crit Rev Biochem Mol Biol 42:41–63
8. Bell JC, Kowalczykowski SC (2016) RecA: regulation and mechanism of a molecular search engine. Trends Biochem Sci 41:491–507
9. Lesterlin C, Ball G, Schermelleh L et al (2014) RecA bundles mediate homology pairing between distant sisters during DNA break repair. Nature 506:249–253
10. Ghodke H, Paudel BP, Lewis JS et al (2019) Spatial and temporal organization of RecA in the Escherichia coli DNA-damage response. elife 8:e42761
11. Shinagawa H, Iwasaki H (1996) Processing the Holliday junction in homologous recombination. Trends Biochem Sci 21:107–111
12. Sharples GJ, Ingleston SM, Lloyd RG (1999) Holliday junction processing in bacteria: insights from the evolutionary conservation of RuvABC, RecG, and RusA. J Bacteriol 181: 5543–5550
13. Donaldson JR, Courcelle CT, Courcelle J (2006) RuvABC is required to resolve Holliday junctions that accumulate following replication on damaged templates in Escherichia coli. J Biol Chem 281:28811–28821
14. Cortes-Ledesma F, de Piccoli G, Haber JE et al (2007) SMC proteins, new players in the maintenance of genomic stability. Cell Cycle 6:914–918
15. Peters JM, Tedeschi A, Schmitz J (2008) The cohesin complex and its roles in chromosome biology. Genes Dev 22:3089–3114
16. Graumann PL, Knust T (2009) Dynamics of the bacterial SMC complex and SMC-like proteins involved in DNA repair. Chromosom Res 17:265–275
17. Hirano T (2012) Condensins: universal organizers of chromosomes with diverse functions. Genes Dev 26:1659–1678
18. Nolivos S, Sherratt D (2014) The bacterial chromosome: architecture and action of bacterial SMC and SMC-like complexes. FEMS Microbiol Rev 38:380–392
19. Jeppsson K, Kanno T, Shirahige K et al (2014) The maintenance of chromosome structure: positioning and functioning of SMC complexes. Nat Rev Mol Cell Biol 15:601–614
20. Murayama Y (2018) DNA entry, exit and second DNA capture by cohesin: insights from biochemical experiments. Nucleus 9:492–502
21. Finch PW, Chambers P, Emmerson PT (1985) Identification of the Escherichia coli recN gene product as a major SOS protein. J Bacteriol 164:653–658
22. Reyes ED, Patidar PL, Uranga LA et al (2010) RecN is a cohesin-like protein that stimulates intermolecular DNA interactions in vitro. J Biol Chem 285:16521–16529
23. Pellegrino S, Radzimanowski J, de Sanctis D et al (2012) Structural and functional characterization of an SMC-like protein RecN: new insights into double-strand break repair. Structure 20:2076–2089
24. Nagashima K, Kubota Y, Shibata T et al (2006) Degradation of Escherichia coli RecN aggregates by ClpXP protease and its implications for DNA damage tolerance. J Biol Chem 281: 30941–30946
25. Odsbu I, Skarstad K (2014) DNA compaction in the early part of the SOS response is dependent on RecN and RecA. Microbiology 160: 872–882
26. Noda S, Akanuma G, Keyamura K et al (2023) RecN spatially and temporally controls RecA-mediated repair of DNA double-strand breaks. J Biol Chem 299:105466
27. Keyamura K, Sakaguchi C, Kubota Y et al (2013) RecA protein recruits structural maintenance of chromosomes (SMC)-like RecN protein to DNA double-strand breaks. J Biol Chem 288:29229–29237
28. Vickridge E, Planchenault C, Cockram C et al (2017) Management of E. coli sister chromatid

cohesion in response to genotoxic stress. Nat Commun 8:14618

29. McLean EK, Lenhart JS, Simmons LA (2021) RecA is required for the assembly of RecN into DNA repair complexes on the nucleoid. J Bacteriol 203:e0024021
30. Chimthanawala A, Parmar JJ, Kumar S et al (2022) SMC protein RecN drives RecA filament translocation for in vivo homology search. Proc Natl Acad Sci USA 119: e2209304119
31. Keyamura K, Hishida T (2019) Topological DNA-binding of structural maintenance of chromosomes-like RecN promotes DNA double-strand break repair in Escherichia coli. Commun Biol 2:413
32. Hishida T, Iwasaki H, Han YW et al (2003) Uncoupling of the ATPase activity from the branch migration activity of RuvAB protein complexes containing both wild-type and ATPase-defective RuvB proteins. Genes Cells 8:721–730
33. Haruta N, Kurokawa Y, Murayama Y et al (2006) The Swi5-Sfr1 complex stimulates Rhp51/Rad51- and Dmc1-mediated DNA strand exchange in vitro. Nat Struct Mol Biol 13:823–830

Chapter 15

Production and Purification of the Human SMC5/6 Core Complex from *S. cerevisiae*

Hemanta Adhikary and Damien D'Amours

Abstract

The human (h)SMC5/6 complex is one of the most enigmatic members of the SMC family of chromosome organizing complexes. It performs essential functions in the maintenance of genome stability and higher-order chromatin organization. Despite its functional significance, challenges in recombinant expression and purification of biologically active complex have limited structural and functional analysis of the human enzyme. Here, we present a robust yeast-based expression approach for high-level production and purification of recombinant hSMC5/6 core complex. We employed a subunit fusion approach coupled with affinity tag/linker removal steps to successfully express and purify near native hSMC5/6 complex from *Saccharomyces cerevisiae.* This bespoke purification approach takes advantage of triple-affinity tag purification steps and size exclusion chromatography to yield >95% pure protein with stoichiometric homogeneity. The workflow is scalable and enables efficient production of a functionally active recombinant enzyme, facilitating downstream structural and functional studies. The strategy developed to purify the hSMC5/6 core complex is broadly applicable to other difficult-to-express multi-subunit enzymes of similar size and complexity, demonstrating the utility of yeast-based microbial cell factories for the expression of human protein complexes.

Key words Human SMC5/6 complex, Recombinant protein expression, Protein purification, *Saccharomyces cerevisiae*, Affinity chromatography, Microbial cell factory

1 Introduction

The Structural Maintenance of Chromosomes 5/6 (SMC5/6) complex is a highly conserved eight-subunit enzyme that plays important roles in the organization and repair of eukaryotic chromosomes [1, 2]. Recent studies have shed light on important biological functions promoted by the SMC5/6 complex in viral DNA restriction [3, 4], cancer suppression [5, 6], and premature aging [7, 8]. Mutational inactivation of SMC5/6 complex subunits results in severe genome instability disorders in humans, including primordial dwarfism, LICS, and Atelís syndrome [9–11]. The SMC5/6 holoenzyme is capable of DNA loop extrusion in vitro

Yasuto Murayama and Yasutaka Kakui (eds.), *SMC Proteins: Methods and Protocols*, Methods in Molecular Biology, vol. 2991, https://doi.org/10.1007/978-1-0716-5005-9_15,

[12], but how this activity is harnessed by the SMC5/6 complex to promote genome stability and human health is not fully understood. At the molecular level, there is also a clear gap in knowledge regarding the mechanism of DNA engagement by the SMC5/6 complex and how individual subunits of the enzyme contribute to dynamic changes in DNA configuration. Resolving these and myriad other mechanistic questions will require in-depth structural, biochemical, and biophysical studies on the human enzyme. Importantly, performing these studies depends on the availability of effective methods to produce and purify large quantities of active hSMC5/6 complex.

The purification of the human SMC5/6 enzyme is a challenging task. Our laboratory and another group successfully achieved the purification of endogenous SMC complexes from yeast [13, 14]. However, the human SMC5/6 complex is particularly difficult to purify because its NSMCE4A (hNse4a) subunit contains several hydrophobic patches, making this protein highly prone to precipitation upon overexpression. To address this limitation, we developed a NSMCE4A-SMC5 co-translational folding approach to enhance the solubility of NSMCE4A and allow successful expression of active SMC5/6 complex in baker's yeast. With this approach, the SMC5 subunit of the complex is physically connected to NSMCE4A via a short flexible linker (described by Serrano et al. [13]) that can be conveniently removed by TEV cleavage following purification. We used *Saccharomyces cerevisiae* as an expression host because repeated attempts to produce the core subunits of the hSMC5/6 complex in *Escherichia coli* were unsuccessful (data not shown). *S. cerevisiae* is a popular eukaryotic host for the expression of foreign proteins [15, 16] and was accorded GRAS (generally recognized as safe) status by the United States Food and Drug Administration (FDA). Among various bioprocessing strategies used to manufacture therapeutic proteins (such as monoclonal antibody drugs, vaccines, and hormones), yeast-based platforms are among the fastest, most flexible, and economically viable for use in academic or industrial applications. Consistent with this, the *S. cerevisiae* protein expression strain used in our method grows quickly and allows effective scaling of culture volumes to increase protein purification yields, as required by downstream applications.

We describe below a workflow to produce milligram amounts of the recombinant hSMC5/6 core complex using S. *cerevisiae* as an expression factory/host and a multistep purification approach to isolate stoichiometric complex at high purity. The major steps of this workflow include (1) production of the recombinant hSMC5/6 core complex in yeast using a pilot-scale bioreactor (42 L), (2) an efficient cryo-lysis method to ensure optimal extraction of soluble proteins from yeast cells, and (3) a downstream purification strategy to maximize purity and homogeneity of the hSMC5/6 core complex (Fig. 1). The purification strategy utilizes three sequential

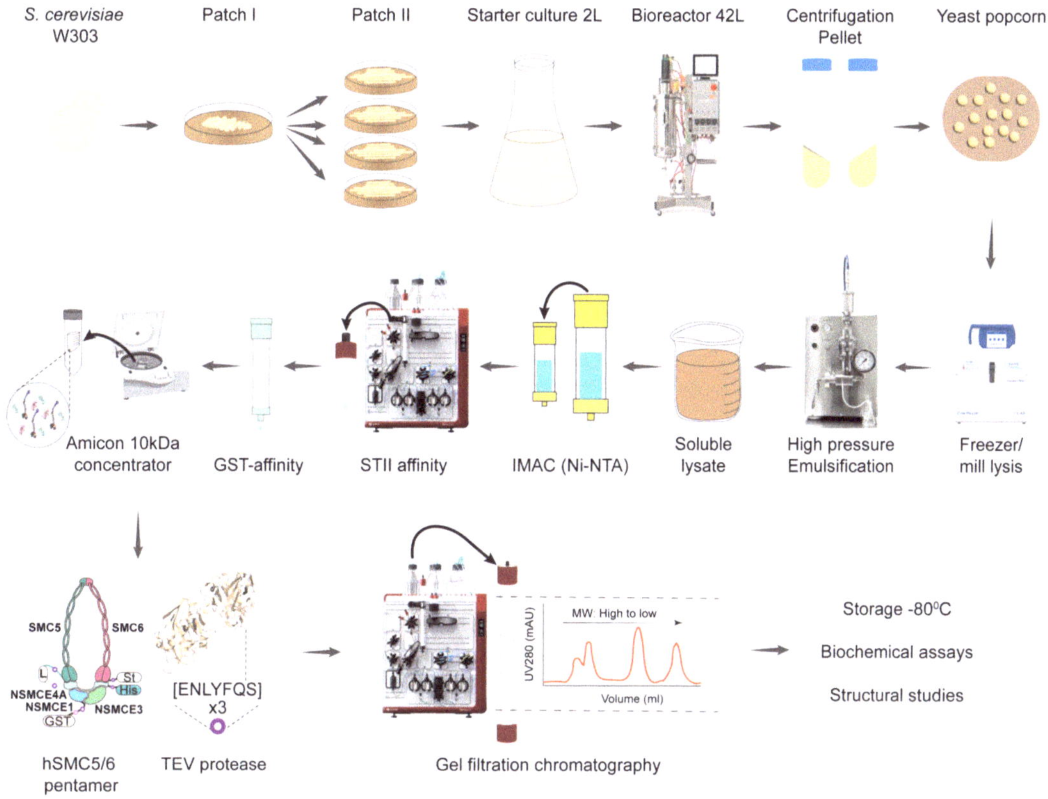

Fig. 1 Workflow for production of recombinant hSMC5/6 core complex in *S. cerevisiae* and downstream purification. A 42 L pilot-scale bioreactor (Techfors-S from Infors HT) is used to conduct the hSMC5/6 core complex production. The protocol ends with a size exclusion chromatography step to polish the complex for downstream applications. Representative structures were generated in Chimera X [21]

affinity purification steps coupled to a final stage of size exclusion chromatography (SEC; Fig. 1). The hSMC5/6 complex is first purified from a whole cell extract by standard nickel ion metal affinity chromatography (IMAC; Fig. 2b, c), followed by binding to a Strep-Tactin XT affinity column (via the 3xStrep-tag II on SMC6; Fig. 2d). The Strep-Tactin-eluted fractions are then loaded in batch mode on Glutathione Sepharose affinity matrix and purified via the GST tag fused to NSMCE1 (hNse1; Fig. 2e). Next, purification tags and the linker connecting NSMCE4A to SMC5 are removed via TEV protease-induced cleavage, followed by SEC (Fig. 3a, b). Notably, previous studies have demonstrated that the affinity tags used to purify the SMC5/6 complex do not interfere with the activity of the modified proteins [13, 17, 18]. Following TEV cleavage and SEC, the isolated recombinant proteins are >95% pure and yield a nearly homogeneous SMC5/6 pentameric core complex (Fig. 3a). Through this workflow, it is possible to produce an otherwise difficult-to-express multi-subunit protein complex in *S. cerevisiae* and obtain mg levels of soluble and

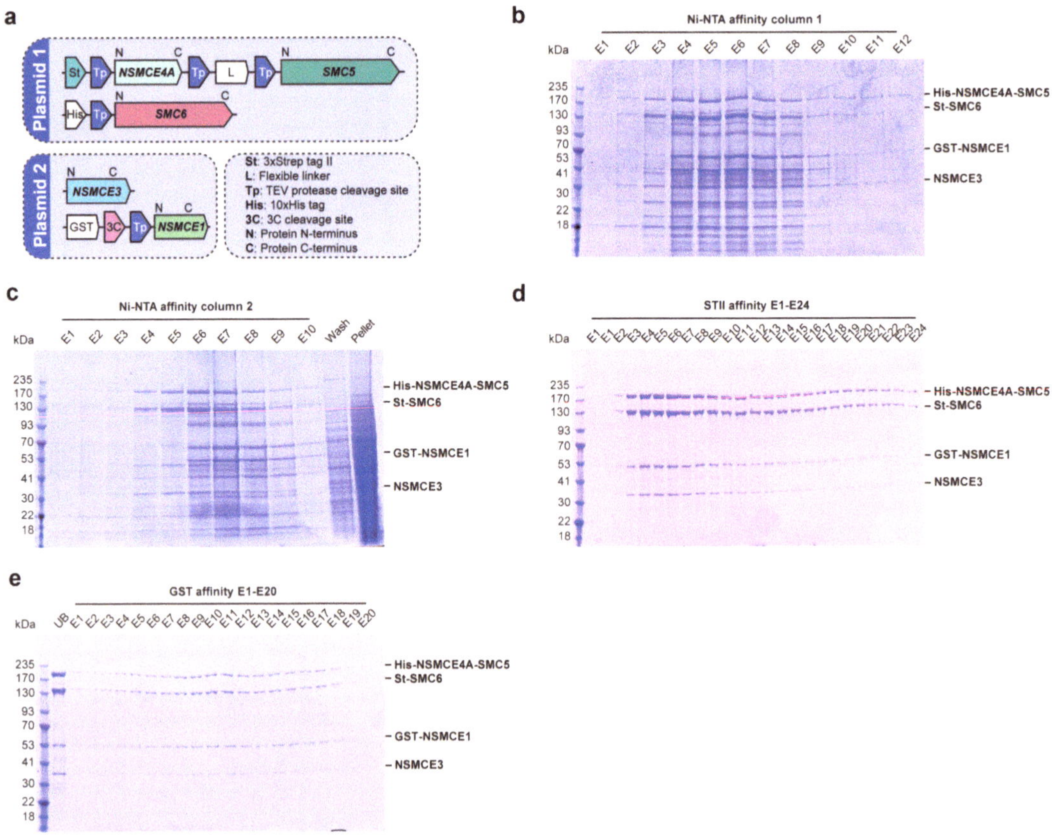

Fig. 2 Purification of the hSMC5/6 core complex. (**a**) Schematic representation of human open reading frames contained in the yeast overexpression plasmids as well as the tagging/fusion strategy used to express and purify soluble hSMC5/6 core complex. The yeast overexpression strain carries two plasmids: one for the overexpression of the NSMCE4A/SMC subunits and the other for the overexpression of NSMCE1 and NSMCE3. (**b**, **c**) IMAC purification steps. CBB-stained SDS-polyacrylamide gels showing hSMC5/6 complex fractions purified by Ni-NTA affinity chromatography. Elution fractions are indicated with horizontal bars. Specifically, E1–E12 represent 12 mL elution fractions from Ni-NTA column 1 (**b**), whereas E1–E10 represent 8 mL elution fractions from Ni-NTA column 2 (**c**). Wash and pellet lanes represent flow-through from washing step and insoluble fraction from lysed pellets, respectively. The loading volume of each sample was 40 μL. (**d**) Strep-Tactin XT affinity chromatography. Peak fractions are indicated with a horizontal bar (E1–E24; 1.5 mL each) and loading volume was 15 μL in each lane. (**e**) Glutathione Sepharose 4B (GS4B) affinity chromatography. Peak fractions are indicated with a horizontal bar (E1–E20; 2 mL each) and loading volume was 15 μL in each lane. Fraction UB represents 40 μL of GS4B unbound fraction, concentrated 20-fold.

stoichiometric hSMC5/6 core complex by the end of the purification procedure(Fig. 3c). Importantly, the recombinant SMC5/6 protein complex produced via this method has been used in several studies [13, 17], demonstrating its functionality. Furthermore, it is possible to downscale the purification setup, using smaller culture volumes, column sizes, and buffer volumes, among other parameters, without compromising the final purity of the hSMC5/6 complex. However, protein yield may vary based on the level of

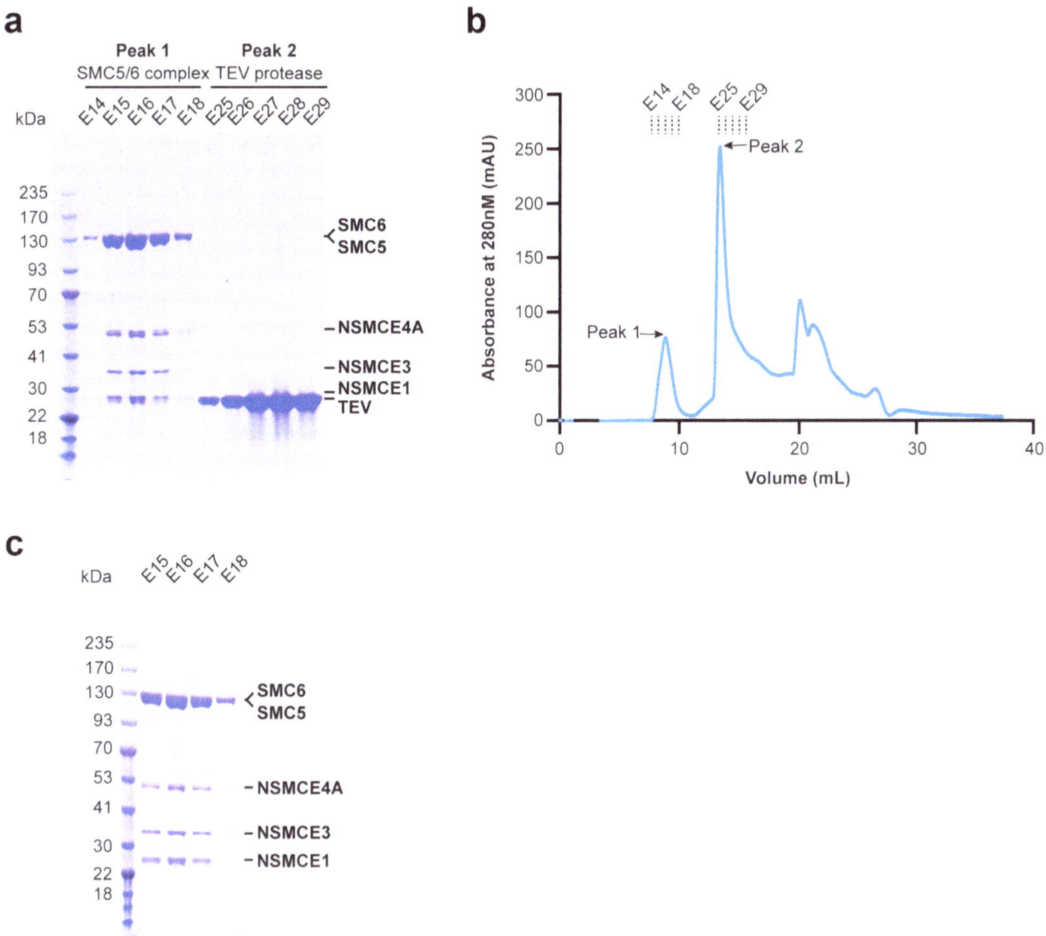

Fig. 3 Size exclusion chromatography (SEC) of the pentameric hSMC5/6 complex following TEV cleavage. (**a**) Peak fractions obtained from the Superdex 200 column were resolved by SDS-PAGE and stained with CBB. Fractions from peak 1 (E14–E18) represent the hSMC5/6 pentamer, whereas fractions from peak 2 (E25–E29) correspond to the excluded TEV protease. The volume of each SEC fraction was 2 mL, 20 μL of which was loaded in each lane of the polyacrylamide gel. (**b**) Elution profile of the hSMC5/6 pentamer from a Superdex 200 Increase 10/300 GL Tricon column. Protein elution was monitored by absorbance at 280 nm and is reported in milli-absorbance unit (AU). (**c**) CBB-stained polyacrylamide gel showing the final purified hSMC5/6 pentamer. 4x concentrated fractions from peak 1 (E15–E18; 0.5 mL each) are shown. The loading volume of each lane in the gel is 5 μL

overexpression with a scaled-down culture volume. The purification approach detailed in this chapter is equally effective with other recombinant proteins of similar or higher molecular weight (Fig. 4; [19]).

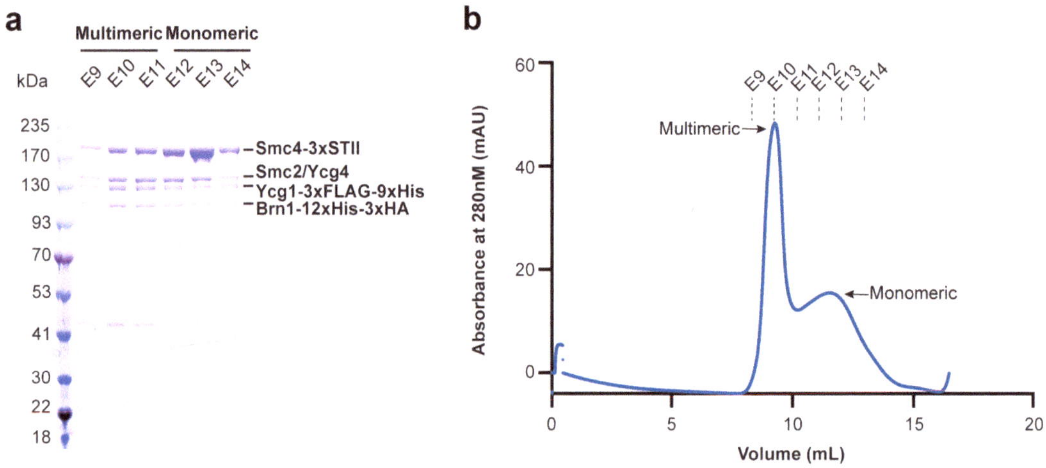

Fig. 4 Purification of *S. cerevisiae* condensin holoenzyme. The complex was purified using sequential steps of Ni-NTA, Strep-Tactin, and size exclusion chromatography, similar to the ones used for the purification of the hSMC5/6 core complex. (**a**) CBB stained SDS-polyacrylamide gel showing peak fractions from the final SEC purification step of condensin. Each fraction was 4x concentrated to 0.5 mL and the loading volume on the gel was 5 μL per lane. (**b**) Elution profile of condensin from the SEC column. Note the existence of two forms of purified condensin complex; one that is multimeric (E9–E11), while the other correspond in size to the monomeric enzyme (E12–E14). Protein elution was monitored by absorbance at 280 nm and is reported in milli-absorbance unit (AU)

2 Materials

2.1 Cell Culture and Upstream Process Development

1. *S. cerevisiae* strain D5819 [13] overexpressing the hSMC5/6 core complex (*see* **Note 1**).
2. AA-6 amino acid supplement powder mix: Weigh 10 g of each of the following 16 amino acids: L-alanine, L-arginine monohydrochloride, L-asparagine, L-aspartic acid, L-cysteine hydrochloride, L-glutamine, L-glutamic acid, L-glycine, L-isoleucine, L-methionine, L-phenylalanine, L-proline, L-serine, L-threonine, L-tyrosine, and L-valine. Mix the amino acids thoroughly (3–4 h) in a 1 L bottle to obtain a homogenous mixture. Do not add histidine, leucine, lysine, tryptophan, adenine, or uracil to the mixture.
3. 2.22x synthetic complete (SC) media: For 1 L, use 14.9 g Difco Bacto yeast nitrogen base (YNB) with ammonium sulfate but without amino acid (AA) and 4.44 g of AA-6 supplement powder mix without L-histidine, L-leucine, L-lysine, L-tryptophan, adenine, and uracil. Add ddH_2O to 1 L, dispense 225 mL in four 500 mL bottles, and sterilize.
4. 100x histidine stock solution: Dissolve 0.7 g L-histidine hydrochloride in 100 mL ddH_2O (33 mM). Sterilize by autoclaving.

5. 100x lysine solution: Dissolve 1.83 g L-lysine in 100 mL ddH_2O (100 mM). Sterilize by autoclaving.
6. 100x adenine solution: Dissolve 0.55 g adenine hemisulfate in 100 mL ddH_2O (30 mM). Sterilize by autoclaving.
7. SC media–Leu–Ura–Trp: For 500 mL of 1x SC media, mix 25 mL of 40% glucose, 5 mL each of 100x histidine, lysine, and adenine solutions, 225 mL of 2.22x SC stock solution, and 240 mL of ddH_2O. For solid 1x SC media, add 4% agar. Use the solid 1x SC agar media to propagate the yeast overexpression strain prior to large-scale culture.
8. Happy yeast media for starter culture: For 1 L media, combine 7.45 g yeast nitrogen base, 2.22 g AA-6 supplement powder mix, 12.5 mL of 80% lactic acid, 30 mL of 100% glycerol, 5 g glucose, 5 g NaOH, and 55 μl antifoam. Adjust volume of the media to 1 L with ddH_2O (*see* **Note 2**).
9. Standard laboratory equipment for small-scale microbial cultivation: Sterile 100 mL, 4 L, and 6 L Erlenmeyer flasks, temperature-controlled shaker incubator (30 °C).
10. Bioreactor for pilot-scale production: Techfors-S 42 L bioreactor, working volume 30 L with integrated eve® bioprocess platform software (Infors-HT).
11. Growth media for bioreactor (6x concentrate): Prepare 4 L of concentrated liquid media by combining 264 g yeast extract, 528 g peptone, 1.32 g adenine, 620.16 mL of 80% lactic acid, 720 mL of 100% glycerol, 280.08 g NaOH pellets, and 1.32 mL antifoam (*see* **Note 3**).
12. 20% (w/v) galactose: To prepare 1 L, dissolve 200 g of galactose in 700 ml of Milli-Q H_2O, then bring the final volume to 1 L, and sterilize with a 0.2 μm membrane filter (*see* **Note 4**).
13. Standard size conical tubes (15 mL and 50 mL; Falcon) and low-speed centrifugation bottles (1 L) for sample collection.
14. A spectrophotometer to measure the turbidity of yeast cultures at 600 nm.
15. Bucket centrifuge for cell pelleting.
16. Liquid nitrogen (N_2) to snap-freeze cell pellets.
17. −80 °C freezer for cell pellet storage.

2.2 Cell Lysis

Prepare all buffers without β-mercaptoethanol (βME) and protease inhibitors. Add these components to buffers just before use/when starting the experiment. Adjust the final pH of the stock buffers to pH 8 and filter-sterilize using a 0.2 μm membrane filter (*see* **Note 5**).

1. Potassium phosphate monobasic, 0.5 M: For 1 L, dissolve KH_2PO_4 68.045 g in 800 mL ultrapure water, make up the volume to 1 L, and filter-sterilize.
2. Potassium phosphate dibasic, 0.5 M: For 1 L, dissolve 114.11 g $K_2HPO_4.3H_2O$ in 800 mL ultrapure water, make up the volume to 1 L, and filter-sterilize.
3. E64 protease inhibitor solution, 200x: Dissolve 25 mg of E64 in 35 mL Milli-Q H_2O.
4. Pepstatin A protease inhibitor solution, 200x: Dissolve 25 mg of Pepstatin A in 18.25 mL 100% DMSO.
5. 4-(2-aminoethyl) benzene sulfonyl fluoride hydrochloride (AEBSF) protease inhibitor solution, 200x: Dissolve 1 g of AEBSF in 20.9 mL Milli-Q H_2O (*see* **Note 6**).
6. NS buffer (Ni-NTA stock): 50 mM potassium phosphate buffer pH 8, 50 mM Tris-HCl pH 8, 500 mM NaCl, 10% glycerol, 0.5% Triton X-100. For 1 L, measure 600 mL Milli-Q water in a 1 L beaker and add 28.36 mL 0.5 M KH_2PO4 and 21.64 mL 0.5 M K_2HPO_4, 6.057 g Tris base, 29.22 g NaCl, 100 mL 100% glycerol, 5 mL Triton X-100. Place on a magnetic stir plate to mix and adjust the pH to 8. Top up to a final volume of 1 L with Milli-Q water and store at 4 °C.
7. NBE buffer (Ni-NTA binding and equilibration): 1 L of NS buffer, supplemented with 20 mM imidazole, 2 mM βME, and protease inhibitor solutions (E64, Pepstatin A, and AEBSF) at 2x final concentration, adjust the final pH to 8 after adding imidazole.
8. Setup to make yeast popcorn for cryogrinding: Liquid N_2, Styrofoam box, 60 mL syringe, and 1 L polypropylene beaker.
9. Freezer/Mill CG-500: Programmable automated cryogenic grinder for milling and grinding yeast cells, mid-size polycarbonate center cylinder, mid-size stainless steel impactor (6881P), mid-size stainless steel end plugs, and vial opener (all parts are from Cole-Parmer).
10. Liquid N_2 dewars (230 L) with LP pressure building regulator.
11. High-pressure homogenizer (optional): EmulsiFlex-C3 (Avestin).
12. High-speed centrifuge.
13. Nalgene Oak Ridge high-speed PPCO centrifuge tubes (50 mL).

2.3 Protein Purification by Affinity Chromatography

2.3.1 IMAC

1. Ni-NTA Superflow affinity resin (Cytiva), reusable column.
2. NBE buffer: Same as 2.2 material 7.
3. NW buffer (Ni-NTA wash): 1 L NS buffer, supplemented with 60 mM imidazole, 2 mM βME, and 1X protease inhibitor solutions (E64, Pepstatin A, and AEBSF), pH 8.
4. SBE buffer (Strep-Tactin binding and equilibration): 50 mM Tris-HCl pH 8, 500 mM NaCl, 10% glycerol, and 0.5% Tween-20; adjust the pH 8 by adding 10 N NaOH dropwise.
5. NE buffer (Ni-NTA elution): SBE buffer supplemented with 500 mM imidazole, 2 mM βME, and 1X protease inhibitor solutions (E64, Pepstatin A, and AEBSF), adjust the pH of the buffer to 8 after adding immidazole.

2.3.2 Strep-Tag II Affinity

1. 5 mL Strep-Tactin XT Sepharose prepacked column (Cytiva).
2. SBE buffer (Strep-Tactin binding and equilibration): Same as Subheading 2.3.1, **item 4**.
3. SW buffer (Strep-Tactin wash): SBE buffer, supplemented with 0.5% Triton X-100, pH 8.
4. 10x PBS stock solution: For 1 L, dissolve 17.8 g Na_2HPO_4, 2.4 g KH_2HPO_4, 80 g NaCl, 2 g KCl in 700 mL Milli-Q H_2O and adjust the final volume to 1 L.
5. GBE buffer (GSTrap binding and equilibration): 150 mM NaCl, 10% glycerol in1x PBS.
6. STE buffer (Strep-Tactin elution): GBE buffer, supplemented with 50 mM biotin, 2 mM βME, 1X protease inhibitor solutions (E64, Pepstatin A, and AEBSF), pH 8 (*see* **Note 7**).
7. ÄKTA pure FPLC purification system.

2.3.3 GST Affinity

1. Glutathione Sepharose 4B resin.
2. Empty glass column for packing resin.
3. GBE buffer: Same as Subheading 2.3.2, **item 5**, supplement the buffer with 1 mM dithiothreitol (DTT) and 1x protease inhibitor solutions (E64, Pepstatin A, and AEBSF) (*see* **Note 8**).
4. GTE buffer (GSTrap elution): GBE buffer, supplemented with 20 mM L-glutathione reduced.

2.4 TEV Protease Cleavage

1. Purified TEV protease (*see* **Note 9**).
2. TEV protease cleavage buffer: 50 mM Tris-HCl pH 8.0, 0.5 mM EDTA, 1 mM DTT.
3. 50 mL centrifugal concentrator (10 kDa cutoff).

2.5 Size Exclusion Chromatography (SEC)

1. Superdex 200 Increase 10/300 GL prepacked Tricon glass column (Cytiva).
2. 0.5 mL centrifugal concentrator (10 kDa cutoff).
3. SEC buffer (Superdex equilibration and elution): GBE buffer, but substitute 1 mM DTT for 2 mM βME.
4. ÄKTA pure FPLC purification system (Cytiva).

3 Methods

This procedure was developed for pilot-scale production of the hSMC5/6 core complex in *S. cerevisiae* and its purification and quality analysis.

3.1 Bioreactor Culture

This part of the method section describes the steps of the fermentation experiment carried out in a 42 L pilot-scale Techfors-S bioreactor with an initial working volume of 24 L (pre-inoculation). We recommend reading the Techfors-S operating manual for more detailed instructions regarding the operation and safe and efficient handling of the bioreactor, which is out of the scope of this chapter. Safe operation of bioreactors requires proper training on the equipment. The operating manual can be downloaded from the following link:

https://infors-ht.com/getmedia/dddbd635-900b-440e-98f1-90d6cf99c185/BA-Techfors-S-2-1-en-GB

3.1.1 Revival of Yeast Strain from Long-Term Storage: 1–2 Weeks Prior to Production

1. Take out the yeast overexpression strain from long-term storage (i.e., −80 °C glycerol stock; *see* **Note 1**). Make a ~ 2 × 2 cm mini patch on the surface of solid selective medium (SC –Leu –Ura –Trp). Incubate yeast at 30 °C for 48–72 hours (h) (*see* **Note 10**).
2. Repatch the grown cells onto eight new plates (~10 × 10 cm patch on SC –Leu –Ura –Trp medium) and grow for another 72 h at 30 °C.

3.1.2 Starter Culture Preparation: Day 1

1. Four liters of starter culture are required to start a 24 L fermenter run. Liquid medium for this starter culture is prepared in two autoclaved/sterilized 6 L Erlenmeyer flasks containing 2 L each of happy yeast media. Store flasks containing the sterilized media at room temperature (RT) until use (*see* **Note 11**).
2. Prior to use, add histidine, lysine, and adenine (from 100x stock solutions) to the starter culture media. Also, add ampicillin to a final concentration of 0.1 mg/mL as a prophylactic measure to prevent media contamination. Keep 10 mL of the

reconstituted liquid media to use as a blank for turbidity/ OD_{600} measurements (*see* **Note 11**).

3. Aliquot 50 mL of the reconstituted media into a sterilized 100 mL flask. Collect yeast cells from the surface of 8 plates (i.e., large patches prepared at Subheading 3.1.1, **step 2**) using sterilized toothpicks and suspend cells in liquid media using a sterile 25 mL pipette to create a homogeneous inoculum.
4. Transfer 25 mL of the inoculum to 2 L of happy yeast media (containing all required supplements) in the sterilized 6 L flasks. Incubate overnight (approximately 21 ± 1 h) at 30 °C with shaking at 220 rpm.

3.1.3 In Situ Sterilization of Bioreactor and Culture Medium: Day 1

1. Prepare 4 L of concentrated (6X) growth media for the bioreactor (described in Subheading 2.1, **item 11**) to set up a 24 L culture. You can scale down the growth media based on requirements (*see* **Note 12**).
2. Connect the pH probe and dissolved oxygen (DO) sensor to the respective inlets near the bottom of the vessel.
3. Uncouple the impeller motor and place in the holder on the lifting device of the bioreactor.
4. Unscrew the vessel top plate and lift it to a convenient height using the lifting device to create enough room to pour the media in the vessel.
5. Add 4 L of concentrated (6X) growth media into the vessel, and make up the volume to 24 L with ddH_2O.
6. Lower down the vessel top plate again using the lifting device and tighten the screw to seal the top.
7. Connect the inlet air filter, exit gas cooler, and exit gas filter to the flanges on the vessel top plate with a clamp and a flat gasket (*see* **Note 13**).
8. Plug the impeller motor onto the drive hub of the vessel top plate.
9. Adjust the antifoam sensor in the vessel top plate to the desired mounting depth.
10. Connect the antifoam sensor cable by plugging the red banana connector into the lateral hole of the antifoam sensor and the black banana connector into the drill hole in the vessel top plate for ground connection.
11. Make sure all the inlet gas and exit gas lines are open. Open the main water supply inlet, heating device, and other two water supply inlets to the fermenter vessel (*see* **Note 14**).
12. Bring the valve to the STER position and fully open the rotameter valve.

13. Using the touch screen monitor, start the sterilization process. This should take 2–4 h.

3.1.4 Inoculation of Bioreactor, Yeast Culture, Induction, and Harvesting: Day 2

1. Set the bioreactor in operation mode. Make sure to open all the necessary water inlets, including the cooling water bath and inlet gas lines.
2. Program the bioreactor to operate at 30 °C and agitate medium at 200 rpm using the control panel. Manually set the rotameter based on the culture volume (*see* **Note 15**).
3. In the morning, measure the turbidity of the overnight starter culture (diluted five-fold) and calculate the volume of the culture required to inoculate the 24 L of media in the bioreactor. The starting turbidity/OD_{600} should be 0.2. Sample out 10 mL media from the bioreactor and use it as a blank for OD_{600} measurement.
4. Once the media in the bioreactor reaches the set temperature, add ampicillin (0.1 mg/mL; as prophylactic measure) and transfer the desired volume of inoculum into the vessel through the inoculation valve on the top of the vessel (*see* **Note 16**).
5. Immediately extract 10 mL of media from the bioreactor through the harvest port at the bottom end of the vessel. Place 1 mL of culture into a clean cuvette and measure the turbidity at 600 nm. Also, check for spills under the microscope, if any.
6. Run the bioreactor until the OD_{600} reaches 0.7–1.0. It usually takes 8–10 h to reach this value range. Measure OD_{600} every 2 h (*see* **Note 17**).
7. Once the turbidity/OD_{600} reaches between 0.7 and 1.0, set the temperature of the bioreactor at 18 °C.
8. Add galactose to a final concentration of 1.5% in the culture medium to start overexpression of the hSMC5/6 core complex (i.e., approximately 2 L of 20% galactose, depending on the volume of inoculum used in Subheading 3.1.4, **step 3**).
9. Continue incubation for 16 h.
10. Once the fermenter run is complete, set the temperature to 4 ° C so that the metabolic rate of yeast is slowed down until cells are fully harvested from the culture (*see* **Note 18**).
11. To transfer the culture out of the bioreactor, turn the steam trap counterclockwise and remove it from the harvest valve. Open the harvest valve clockwise to collect the culture in 1 L centrifuge bottles (in rounds of 6 bottles, 2 sets).
12. Pre-cool the centrifuge to 4 °C. Collect cells from the culture by centrifugation at 5020 g for 1 h. Once completed, decant the media supernatant from each centrifuge bottle.

13. Repeat steps 11 and 12 until all yeast from the 30 L culture are collected (*see* **Note 19**).
14. Resuspend yeast cell pellets by adding 13 mL of ice-cold water to each of the 30 bottles. Combine all the resuspended yeast cells into a single centrifuge bottle. Distribute the mixture equally into eight 50 mL Falcon tubes and centrifuge again at 3220 g for 45 min, decant the supernatant, and snap-freeze the pellets in liquid N_2.
15. Store the pellet at −80 °C until further use (*see* **Note 20**).
16. Clean the fermenter (*see* **Note 21**).

3.2 Cell Lysis

Perform all steps on ice or at 4 °C unless otherwise mentioned.

3.2.1 Freezer Milling: Day 3 (See Note 22)

1. Remove cell pellets from −80 °C storage and place on ice.
2. Add 25 mL of NBE buffer to each tube and let them thaw.
3. Get liquid N_2 in two separate Styrofoam boxes.
4. Pre-cool a 500 mL polypropylene beaker by immersing it in one of the liquid N_2 baths.
5. Pre-cool Freezer/Mill accessories (i.e., polycarbonate center cylinder, stainless steel impactor, and stainless-steel end plugs) in the other liquid N_2 bath. Make sure to close one end of the cylinder with the end plug and insert the metal impactor inside before soaking in the liquid N_2 bath.
6. Resuspend the cell pellet from step 2 by pipetting up and down and/or vortexing at regular intervals to get a homogeneous mixture. Minimize exposure of the cells to room temperature (RT).
7. Transfer the yeast suspension into a 60 mL syringe and slowly squirt the suspension dropwise in liquid N_2 (contained in a beaker that is also partly submerged in a liquid N_2 bath), forming small balls of yeast similar in appearance to popcorn. Repeat the step with the remaining yeast cell suspension.
8. Transfer the "popcorned" yeast to the Freezer/Mill cylinders. A 30 L culture will generate enough yeast popcorn for a total of 8 cylinders, which are filled up to half volume/capacity or less. Carefully hammer in the metal plug to close the open end of the cylinder (*see* **Note 23**).
9. Immediately submerge closed cylinders back into a liquid N_2 bath. At this point, the cylinders are ready to be loaded into the Freezer/Mill in pairs of two.
10. Connect the liquid N_2 Dewar to the Freezer/Mill and start the program. The machine will start drawing liquid N_2 automatically to pre-cool the apparatus. This will take approximately 15 min. Do not load the cylinders before pre-cooling.

11. Once pre-cooling of the Freezer/Mill is complete, load two cylinders in the apparatus, close the lid, and let the cylinders cool in liquid N_2 for an additional 2 min. After this step, the yeast popcorn is lysed/milled for 12 cycles each of 2 min at an impactor rate of 14 cps.
12. Repeat lysis protocol described above (**step 11**) for another cycle. Tighten the metal cap in between cycles. Each cycle takes 45 min.
13. Transfer the Freezer/Mill cylinders back into the liquid N_2 bath.
14. Open the cylinder carefully using the vial opener and transfer the yeast powder into an empty beaker standing in a liquid nitrogen bath (*see* **Note 24**). When completed, immediately transfer the beaker to a −80 °C freezer. Repeat the Freezer/Mill lysis steps until all yeast popcorn has been ground to a powder form.
15. Monitor the effectiveness of the cell lysis using a conventional light microscope. To do so, resuspend a pinch of yeast "powder" in water, distribute 5 μL on a microscope glass slide, add a coverslip, and observe lysed cells under the microscope. Usually 90% cell lysis is considered optimal.

3.2.2 Emulsification: Day 4 (Optional)

This step is optional. However, if you decide not to perform high-pressure emulsification, we highly recommend that one extra cycle of Freezer/Mill lysis be conducted (i.e., Subheading 3.2.1, **step 12**). After doing so, complete step 1 below and skip step 2, then move directly to Subheading 3.2.2, **step 3** (*see* **Note 25**).

1. Resuspend the yeast cell powder obtained after the Freezer/Mill step in a maximum of 500 mL of NBE buffer (*see* **Note 26**).
2. Apply the lysate to the entry port of an Avestin EmulsiFlex-C3 apparatus for high-pressure homogenization. Use an operating pressure of 25,000 psi.
3. Clear the lysate using high-speed centrifugation (75,000 g for 1 h).
4. Collect the supernatant and store on ice at 4 °C until further use.

3.3 Affinity Purification

The soluble lysate is purified over three affinity columns as follows:

3.3.1 IMAC : Day 4

Perform all steps on ice or at 4 °C unless otherwise stated.

1. Pack two Ni-NTA columns, the first with 80 mL of 50% slurry and the second containing 50 mL of 50% slurry. Rinse the resin

with ultrapure sterilized ice-cold water and equilibrate both columns with 3 column volume (CV) NBE buffer. Perform this step when you start high-speed centrifugation above (*see* Subheading 3.2.2, **step 3**).

2. Pass the soluble lysate through the first Ni-NTA affinity column by gravity flow at a rate not exceeding 2.5 mL/min. A slow flow rate will enhance binding of the hSMC5/6 complex on the Ni-NTA matrix.
3. The flow-through from the first column is applied to the second Ni-NTA column and processed as described above (*see* **Note 27**).
4. Wash each column with 10 CV of NW buffer at a flow rate of 5 mL/min.
5. Elute proteins from the columns with 5 CV of NE buffer, collecting 12 and 8 mL fraction volumes from column 1 and 2, respectively.
6. Load 40 μL of each fraction onto 4–12% gradient polyacrylamide gels, separate proteins by electrophoresis, and stain gels with Coomassie Brilliant Blue (CBB) for quality analysis (Fig. 2b, c).
7. Combine peak fractions from both the Ni-NTA columns and store at 4 °C overnight. Do not freeze.

3.3.2 Strep-Tactin XT Affinity Chromatography: Day 5

Perform all steps on ice or at 4 °C unless otherwise stated.

1. Attach a 5 mL Strep-Tactin XT column to the FPLC system.
2. Wash the column with 20 CV of ultrapure water and equilibrate the column with 10 CV of SE buffer at a flow rate of 2.5 mL/min. Perform steps 1 and 2 on day 4.
3. Hook the specific buffer inlet lines of the FPLC system to the respective buffers and samples according to the method design.
4. Equilibrate the column with 5 CV of SE buffer at a flow rate of 2.5 mL/min.
5. Load the combined Ni-NTA fractions onto the column at a flow rate of 1.5 mL/min. Monitor absorbance at 280 nm to track protein binding. Typically, A_{280} reaches 2000 mAU in our experiment.
6. Wash with 10 CV of SW buffer at a flow rate of 2.5 mL/min.
7. Elute bound protein using 12 CV of STE buffer (containing 50 mM biotin) at a flow rate of 1.5 mL/min.
8. Collect 1.5 mL fractions at 1.5 mL/min using a fraction collector and monitor absorbance of fractions at 280 nm.
9. Load 15 μL of each fraction into a 4–12% polyacrylamide gel, separate proteins by electrophoresis, and stain gels with CBB to

monitor the quality of purified proteins. The hSMC5/6 complex is usually 90% pure after this step (Fig. 2d).

10. Run Strep-Tactin XT column cleaning-in-place (CIP) protocol to remove the residual-bound protein and store the column in 20% ethanol until the next use (*see* **Note 28**).

3.3.3 GST Affinity Chromatography and TEV Cleavage: Day 6 (See Note 29)

1. Rinse 15 mL of Glutathione Sepharose 4B resin with ice-cold ultrapure water in a 50 mL Falcon tube followed by equilibration with 10 bed volumes of GE buffer. Pellet resin at each step by centrifugation at 800 g for 5 min at 4 °C.
2. Combine the peak Strep-Tactin XT eluted fractions and dilute six-fold with GE buffer (*see* **Note 30**).
3. Add the diluted fractions to the resin and incubate with gentle rotation at 4 °C for 2 h.
4. Transfer the resin into a 50 mL glass column and let the supernatant pass through the column completely under gravity flow. Adjust the flow rate to 2 mL/min. Collect the flow-through to analyze the percentage of unbound SMC5/6 complex (*see* **Notes 31** and **32**).
5. Wash the Glutathione resin-bound hSMC5/6 complex (attached via the GST-tagged NSMCE1 subunit) with 10 CV GW buffer under gravity at 5 mL/min.
6. Elute bound proteins in 2 mL fractions using 5 CV of GTE buffer containing 20 mM L-glutathione reduced at 1 mL/min.
7. Load 15 μL of each fraction into a 4–12% gradient polyacrylamide gel, separate proteins by electrophoresis, and stain purified fractions with CBB for quality analysis.
8. Remove the purification tags from the hSMC5/6 complex by mixing the peak fractions from the GStrap elution with TEV protease at a ratio of 100:1 (w/w). For example, we had success (i.e., cleavage approaching 100%) when incubating 1 mg of hSMC5/6 recombinant complex with 10 μg of TEV protease.
9. Digest overnight at 4 °C without rocking.
10. Load 15 μL of each fraction onto a 4–12% gradient polyacrylamide gel, separate proteins by electrophoresis, and stain cleaved proteins with CBB to monitor tag removal.
11. Combine fractions that have lost their affinity tags/linkers and concentrate to approximately 1–2 mL using a 50 mL centrifugal concentrator (10 kDa cutoff). Mix the sample gently in between each centrifugation to avoid protein precipitation.

3.4 Size Exclusion Chromatography (SEC): Day 7

1. Attach the Superdex 200 10/300G size exclusion chromatography column to the FPLC apparatus. Wash with 5 CV of water

and equilibrate with 5 CV of SEC buffer. Perform this step at the end of day 6.

2. Equilibrate the column with 1 CV of SEC buffer at a flow rate of 1 mL/min.
3. Load 1 mL of the concentrated fractions from the GST-affinity chromatography step through the injection loop at a flow rate of 0.5 mL/min.
4. Collect 1.5 mL fractions by running 2 CV SEC buffer. Keep the flow rate constant at 0.5 mL/min. Monitor absorbance at 280 nm.
5. Load 15 μL of each fraction into a 4–12% gradient polyacrylamide gel, separate proteins by electrophoresis, and stain gels with CBB to analyze the quality of the hSMC5/6 complex. The complex is >95% pure after this step (Fig. 3).
6. Run SEC column CIP protocol to remove residual proteins and store the column in 20% ethanol until the next use (*see* **Note 33**).
7. Concentrate the fractions between 200 and 500 μL using a 0.5 mL centrifugal concentrator (10 kDa cutoff).
8. Quantify protein concentration by measuring absorbance at 280 nm. The calculated molar extinction coefficient of hSMC5/6 core complex is 3.75 x10^5 L mol^{-1} cm^{-1}.
9. Make 20 μL and 50 μL aliquots in 1.5 mL tubes, snap-freeze in liquid N_2, and store tubes at −80 °C (*see* **Note 34**).

4 Notes

1. Yeast strain D5819 was used for hSMC5/6 core complex overexpression [13]. This strain contains a P_{GAL1}-expressed Gal4 transcription factor to increase protein overexpression and a deletion of the gene encoding Tdh3 because it is a common contaminant in this type of purification. The genotype of strain D5819 is *MATa tdh3::kanMX6 lys2::*P_{GAL1}*-GAL4::lys2 [2 μ LEU2d TRP1* P_{GAL10}*-hNSE1* P_{GAL1}*-hNSE3] [2 μ LEU2d URA3* P_{GAL10}*-hNSE4A-L-hSMC5* P_{GAL1}*-hSMC6]*.
2. Do not fill the flask with more than one-third volume of liquid media prior to sterilizing. This ratio helps achieve homogeneous and faster growth of the culture because of optimal aeration during incubation. For example, we use 2 L of liquid media in a 6 L flask.
3. The composition of the media is optimized for growth of *S. cerevisiae* in a bioreactor. At 1x concentration, this media contains 1.1% (w/v) yeast extract, 2.2% (w/v) peptone, 0.3 mM adenine, 3% (v/v) glycerol, and 2% (v/v) lactic acid.

When preparing the concentrated (6x) growth media, dissolve the yeast extract, peptone, and adenine first and then add glycerol and lactic acid. NaOH pellets are used to adjust the pH of the media to 7.0 at the penultimate step, and after the pellets are fully dissolved, antifoam is added last. Usually, ≥24 L of media results in the production of a yeast pellet sufficiently large for the purification of the hSMC5/6 core complex. We recommend utilizing the bioreactor at full capacity (24 L of 1x growth medium, excluding the volume of the starter culture [~4 L] and 20% galactose [~2 L] added later to the media) because its operation can be laborious and time-consuming.

4. Galactose does not dissolve easily at room temperature water, so we recommend dissolving the galactose powder in lukewarm water while preparing 20% galactose. Do not forget to filter-sterilize the solution using a 0.2 mm membrane filter. The final concentration of galactose used for induction is 1.5%.
5. Purification buffers can be stored for several days to weeks at 4 ° C. Prepare the buffer well in advance. Consider pre-cooling the buffer at 4 °C before adjusting the pH to avoid the effect of temperature on pH, as all the purification steps should be performed at 4 °C or on ice. Filtration through 0.2 mm filter extends the shelf life of buffers and removes contaminations and non-dissolved precipitants. Carefully inspect your buffers for any contamination before use and monitor the pH again if stored longer than 2 weeks.
6. We recommend preparing protease inhibitor stock solutions fresh to ensure optimal results. Never store protease inhibitor stock solutions at 4 °C. They should be stored at −20 °C for upto a maximum of 2 weeks, and we recommend avoiding freeze-thaw cycles. Prepare 200x stock solutions of protease inhibitors and use 1x as working concentration (unless specified otherwise). Using highly concentrated protease stock solutions will minimize dilution effects on purification buffers and reduce the concentration of DMSO during the purification procedure (i.e., Pepstatin A is dissolved in 100% DMSO). The final 1x working concentrations of protease inhibitors are 10 μM E64, 10 μM Pepstatin A, and 1 mM AEBSF.
7. Do not prepare the STE buffer in advance. Prepare on the day of purification. You need to adjust the pH again with NaOH after adding biotin. Filter-sterilize the buffer at the final stage of preparation.
8. Higher salt concentrations are inhibitory to GST binding on Glutathione Sepharose beads. Therefore, the final concentration of NaCl should be adjusted to no more than 250 mM in GBE stock buffer.

9. We purify TEV protease in our laboratory using an *E. coli* BL21 strain overexpressing TEV-MBP-6xHis and a protocol described by Tropea et al. [20]. Purified TEV protease stocks (1 mg/mL) are stored at −80 °C in 100 μL aliquots. Our in-house purified TEV protease is as active as the commercially available enzyme but considerably more economical.
10. Always revive yeast strains freshly before performing hSMC5/6 complex overexpression. Plates containing yeast patches should not be stored at 4 °C for more than 2 days before use.
11. Do not add media supplements during the autoclave sterilization step as they are heat labile. We prepare 100x stock solutions of most supplements and filter sterilize with 0.2 μm membranes. In addition to lactic acid and glycerol, the happy yeast media contain 0.5% glucose as a carbon source to minimize growth delay. It is recommended to inoculate 2 L of media with four 10 × 10 cm patches of yeast. Do not forget to add the required supplements while reconstituting the culture media immediately prior to inoculation.
12. The Techfors-S apparatus is a bioreactor with a total volume of 42 L, but the recommended maximum working volume for yeast cultures is 30 L. When preparing growth medium directly in the bioreactor, consider the contribution of the inoculum and galactose solution to determine the actual volume of growth media that needs to be prepared and sterilized. For example, we sterilize 24 L growth medium and add ~4 L inoculum and ~ 2 L of 20% galactose in one fermentation run.
13. Make sure no water is trapped inside the filter. A wet gas filter may get contaminated by microorganisms, which might spread through the exit gas line to the vessel, contaminating the culture.
14. Check multiple times whether all the water connections are open. Also, check the inlet and exit gas lines, including filters. A blocked inlet air or exit gas (or a clogged filter) may result in poor venting/gassing of the culture and overpressure in the vessel.
15. VVM is a measurement of volume (V) of air in liter per unit of the culture medium (V) in liter, per unit time in minutes (M). For 30 L culture, the recommended setting for the rotameter is 30 vvm.
16. Inoculate the cultures to an initial OD_{600} of 0.2. The vessel is under pressure during operation. Always depressurize the vessel before manipulating the top portion of the vessel.
17. When the culture turbidity/OD_{600} reaches the 0.7–1.0 range, the DO should drop below 30% and pH should be around

5.5–5.8. You can consider these values as a trigger/indicator for the time to induce protein expression with galactose.

18. The temperature of the bioreactor will not go below 12 °C.
19. Centrifugation time depends on the culture volume. To harvest cells from 1 L of culture, 1 h is usually sufficient with a Sorvall centrifuge (Beckman Coulter J6-MI high-capacity centrifuge) at a maximum speed of 4000 rpm (JS-4.2SM rotor). It takes approximately 5–7 h to harvest a total of 30 L of culture at the end of the fermentation run.
20. Do not store yeast pellets at −80 °C for more than 3 months.
21. The bioreactor is an expensive piece of equipment. At the end of the operation, make sure to sterilize again after filling the vessel with ddH_2O. At the end of this sterilization cycle, drain out all the water, rinse the vessel tank thoroughly, and dismantle all the accessories and hook into the designated slot.
22. The process involves handling large volumes of liquid N_2. Wear all necessary personal protective equipment for your safety.
23. Make sure to decant all the remaining liquid nitrogen from the cylinder before closing. This will avoid generating vapor pressure and blast while opening the metal cap later. Use minimal force while hammering the cap to close; excessive pressure will break the polycarbonate cylinder.
24. This step is technically critical because of the vapor generated while cryogrinding that itself creates pressure to blast open the cap. You may lose a lot of yeast sample if that happens. Load the cylinder to the opener quickly and start opening the cap slowly by putting minimum force on the handle tip. Wait 1 min after each trial.
25. High-pressure emulsification always helps in homogenization and faster lysis for larger culture volumes. In contrast, freezer milling of pellets from large culture volumes can be time-consuming as one cycle takes 45 min. However, when a cryogenic cell disruptor (i.e., Freezer/Mill) is available, we always recommend the freezer milling cell lysis method due to the maintenance of ultralow temperatures throughout the grinding process and faster cell lysis, preventing protein denaturation, degradation, and/or aggregation. We typically perform either a combination of two cycles of freezer milling with two cycles of emulsification or three cycles of freezer milling without emulsification. We found that proteins purified from emulsified preparations tends to be more "sticky" compared to proteins obtained from three cycles of freezer milling.
26. The volume of lysis buffer may be adjusted; however, we recommend using as low as possible a volume to ensure that the size of the final lysate is reasonable for downstream

processing steps. This will save time when binding lysate proteins to the Ni-NTA matrix. We start to dissolve the yeast powder with 200 mL of Lysis buffer and, when necessary, add additional lysis buffer to a maximum volume of 500 mL to generate a homogeneous suspension.

27. Because of the large volume of soluble lysate, the first Ni-NTA column will leave a good amount of hSMC5/6 complex unbound. To achieve a maximum yield of hSMC5/6 complex purification, it is important to pass the unbound fraction from the first Ni-NTA column on a second column of the same type (Fig. 2c).
28. Strep-Tactin XT column cleaning-in-place (CIP) protocol: Wash the column with 10 CV of 0.5 M NaOH followed by 20 CV of Milli-Q H_2O.
29. hSMC5/6 complex (specifically GST-NSMCE1) does not bind efficiently to the Glutathione Sepharose resin, thus leading to a 30–40% loss of the complex at this step. This corresponds to a very significant loss, but this step remains essential because it ensures the proper stoichiometry of the purified complex (Fig. 2e).
30. Salt concentrations above 250 mM are inhibitory to NSMCE1 binding to the Glutathione Sepharose resin. It is essential to dilute the Strep-Tactin XT fractions before loading to the glutathione column to maximize binding.
31. Carefully examine the supernatant to ensure all beads have transferred to the column. To make sure, rinse the tube with 20 mL of GBE buffer.
32. At this point, the flow-through from the Glutathione Sepharose column represents the unbound fraction. This partially purified hSMC5/6 complex fraction may be saved and concentrated to 2 mL using a 50 mL centrifugal concentrator (50 kDa cutoff). This fraction is active in many assays and can be used for downstream applications.
33. SEC column CIP protocol: Wash the column with 2 CV of 0.5 M NaOH followed by 5 CV of Milli-Q H_2O.
34. The protein can be snap-frozen in liquid nitrogen and stored at −80 °C for 1–3 months. We strongly recommend avoiding freeze-thaw cycles. Aliquot the protein in volumes/size consistent with downstream assay requirements.

Acknowledgments

Research in D.D.'s laboratory is supported by a grant from CIHR (FDN-167265) and by the Canada Research Chair in Chromatin Dynamics & Genome Architecture (CRC-2023-00405).

References

1. Roy S, Adhikary H, D'Amours D (2024) The SMC5/6 complex: folding chromosomes back into shape when genomes take a break. Nucleic Acids Res 52(5):2112–2129
2. Peng XP, Zhao X (2023) The multi-functional Smc5/6 complex in genome protection and disease. Nat Struct Mol Biol 30(6):724–734
3. Irwan ID, Bogerd HP, Cullen BR (2022) Epigenetic silencing by the SMC5/6 complex mediates HIV-1 latency. Nat Microbiol 7(12): 2101–2113
4. Decorsière A, Mueller H, van Breugel PC et al (2016) Hepatitis B virus X protein identifies the Smc5/6 complex as a host restriction factor. Nature 531(7594):386–389
5. Crowner A, Smith K, DeSmet M (2024) Regulation of R-loops in DNA tumor viruses. Pathogens (Basel) 13(10):863
6. Roy S, Zaker A, Mer A et al (2023) Large-scale phenogenomic analysis of human cancers uncovers frequent alterations affecting SMC5/6 complex components in breast cancer. NAR cancer 5(3):zcad047
7. Moradi-Fard S, Mojumdar A, Chan M et al (2021) Smc5/6 in the rDNA modulates lifespan independently of Fob1. Aging Cell 20(6): e13373
8. Moradi-Fard S, Sarthi J, Tittel-Elmer M et al (2016) Smc5/6 is a telomere-associated complex that regulates Sir4 binding and TPE. PLoS Genet 12(8):e1006268
9. van der Crabben SN, Hennus MP, McGregor GA et al (2016) Destabilized SMC5/6 complex leads to chromosome breakage syndrome with severe lung disease. J Clin Invest 126(8): 2881–2892
10. Grange LJ, Reynolds JJ, Ullah F et al (2022) Pathogenic variants in SLF2 and SMC5 cause segmented chromosomes and mosaic variegated hyperploidy. Nat Commun 13(1): 6664
11. Payne F, Colnaghi R, Rocha N et al (2014) Hypomorphism in human NSMCE2 linked to primordial dwarfism and insulin resistance. J Clin Invest 124(9):4028–4038
12. Pradhan B, Kanno T, Umeda Igarashi M et al (2023) The Smc5/6 complex is a DNA loop-extruding motor. Nature 616(7958):843–848
13. Serrano D, Cordero G, Kawamura R et al (2020) The Smc5/6 core complex is a structure-specific DNA binding and compacting machine. Mol Cell 80(6):1025–1038.e5
14. Gutierrez-Escribano P, Hormeño S, Madariaga-Marcos J et al (2020) Purified Smc5/6 complex exhibits DNA substrate recognition and compaction. Mol Cell 80(6): 1039–1054.e6
15. Wang G, Huang M, Nielsen J (2017) Exploring the potential of Saccharomyces cerevisiae for biopharmaceutical protein production. Curr Opin in Biotechnol 48:77–84
16. Buckholz RG, Gleeson MA (1991) Yeast systems for the commercial production of heterologous proteins. Biotechnology 9(11): 1067–1072
17. Roy S, Adhikary H, Isler S et al (2024) The Smc5/6 complex counteracts R-loop formation at highly transcribed genes in cooperation with RNase H2. elife 13:e96626
18. Roy MA, Siddiqui N, D'Amours D (2011) Dynamic and selective DNA-binding activity of Smc5, a core component of the Smc5-Smc6 complex. Cell Cycle 10(4):690–700
19. St-Pierre J, Douziech M, Bazile F et al (2009) Polo kinase regulates mitotic chromosome condensation by hyperactivation of condensin DNA supercoiling activity. Mol Cell 34(4): 416–426
20. Tropea JE, Cherry S, Waugh DS (2009) Expression and purification of soluble His(6)-tagged TEV protease. Methods Mol Biol 498: 297–307
21. Pettersen EF, Goddard TD, Huang CC et al (2021) UCSF ChimeraX: structure visualization for researchers, educators, and developers. Protein Sci 30(1):70–82

Part IV

Assays of Chromosomal Structures Mediated by SMC Proteins

Chapter 16

ChIA-PET Analysis for Investigating SMC-Mediated 3D Genomic Contacts

Kyoung-Dong Kim and Ken-ichi Noma

Abstract

We present a detailed ChIA-PET benchwork protocol optimized for investigating cohesin- and condensin-mediated chromatin contacts. This method combines chromatin immunoprecipitation (ChIP), proximity ligation, EcoP15I digestion, and sequencing library construction to achieve high specificity in detecting SMC-dependent genomic contacts. Key quality control measures, including ChIP DNA validation, linker ligation strategies, and PCR cycle optimization, are outlined to enhance data reliability. By providing a genome-wide view of SMC-driven chromatin contacts, this method offers valuable insights into the fundamental principles of genome folding and maintenance.

Key words ChIA-PET, SMC, Condensin, Cohesin, 3D genome organization

1 Introduction

Chromatin structure plays a fundamental role in genome organization and gene regulation. Among the key mediators of higher-order chromatin architecture are Structural Maintenance of Chromosomes (SMC) complexes, such as cohesin and condensin, which form chromatin contacts linked to gene regulation, DNA replication, and genome stability [1, 2]. Understanding how these chromatin organizers mediate genomic contacts is crucial for elucidating gene regulatory networks and chromatin dynamics.

Chromatin Interaction Analysis with Paired-End Tag Sequencing (ChIA-PET) is a powerful technique for genome-wide mapping of protein-mediated chromatin contacts. Unlike traditional chromosome conformation capture (3C)-based techniques, ChIA-PET provides high-resolution, protein-specific chromatin contact profiling, allowing precise identification of genomic contacts associated with regulatory elements (Fig. 1) [3–5]. Cohesin and condensin, two key SMC complexes, contribute to chromatin organization via loop extrusion and diffusion capture mechanisms

Yasuto Murayama and Yasutaka Kakui (eds.), *SMC Proteins: Methods and Protocols*, Methods in Molecular Biology, vol. 2991, https://doi.org/10.1007/978-1-0716-5005-9_16,

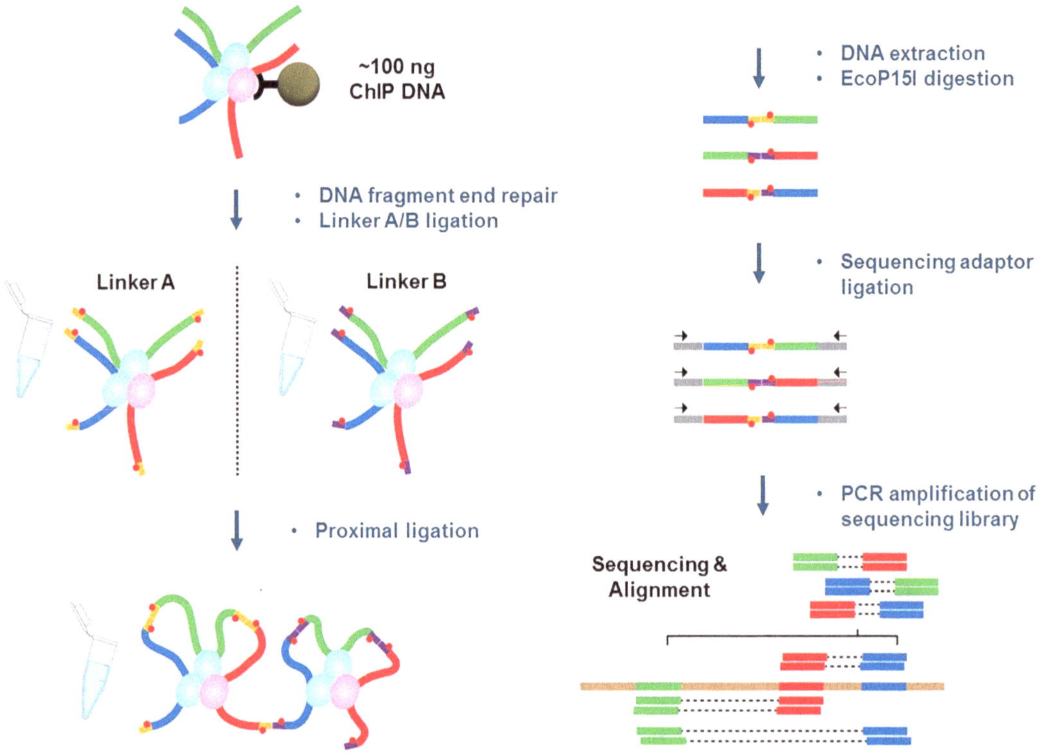

Fig. 1 Workflow of the ChIA-PET procedure for mapping SMC-mediated chromatin contacts. A schematic overview illustrates the Chromatin Interaction Analysis with Paired-End Tag (ChIA-PET) workflow for detecting cohesin- and condensin-mediated chromatin interactions. The process begins with chromatin immunoprecipitation (ChIP) to capture protein-mediated chromatin complexes. DNA ends are processed and ligated with Linkers A/B to distinguish self-ligation and proximal ligation events. Proximity ligation connects interacting chromatin fragments while preserving structural information. After DNA extraction and EcoP15I digestion, sequencing adaptors are ligated, followed by PCR amplification of the library. Finally, high-throughput sequencing and computational analysis reconstruct chromatin contact networks, providing insights into SMC-dependent genome organization

and play a critical role in anchoring chromatin loops near regulatory elements such as enhancers and promoters, thereby facilitating coordinated transcriptional regulation [6–8].

Recent studies have demonstrated that cohesin plays a key role in organizing topologically associating domains (TADs), while condensin is primarily responsible for large-scale chromatin compaction and mitotic chromosome organization [9–11]. Using ChIA-PET, protein-mediated chromatin contacts can be directly mapped, offering insights into how SMC complexes contribute to genome architecture. Its proximity ligation strategy ensures that only protein-bound DNA fragments are ligated and sequenced, improving the specificity and accuracy of detecting chromatin contacts [3, 4]. Given the growing focus on 3D genome organization, applying ChIA-PET to study cohesin- and condensin-mediated contacts provides valuable insights into chromatin structure and its functional organization [9].

This article presents a comprehensive ChIA-PET protocol optimized for studying cohesin- and condensin-mediated chromatin contacts. We outline key steps, including linker ligation, EcoP15I digestion, and sequencing library construction, which are essential for capturing SMC-dependent chromatin contacts. With advancements in sequencing technologies and bioinformatics analysis, ChIA-PET remains a critical tool for unraveling the 3D genome architecture regulated by SMC complexes.

2 Materials

2.1 Reagents

1. Starting Material: 100–200 ng ChIP DNA on Dynabeads Protein G (Thermo Fisher) and suspended in TE buffer (*see* **Note 1**).
2. End-blunting reaction mix (1 mL): 880 μL nuclease-free water, 100 μL 10X T4 DNA polymerase buffer, 10 μL 10 mM dNTP mix, and 10 μL T4 DNA polymerase (9.7 U/μL).
3. A-tailing reaction mix (500 μL): 444.25 μL nuclease-free water, 2 μL 10 mM dATP, 50 μL 10X NEBuffer2, and 3.75 μL Klenow exo- (NEB).
4. Linker A/B ligation reaction mix (500 μL): 386 μL nuclease-free water, 10 μL 100 ng/μL biotinylated half-linker (A or B; premixed with nuclease-free water), 100 μL 5X T4 DNA ligase buffer with PEG, and 4 μL T4 DNA ligase (30 U/μL; Fermentas).
5. 5′-end phosphorylation reaction mix (1 mL): 880 μL nuclease-free water, 100 μL 10X T4 DNA ligase buffer (no PEG, NEB), and 20 μL T4 DNA polynucleotide kinase (10 U/μL; NEB).
6. DNA circularization reaction mix (500 μL): 433.3 μL nuclease free water, 50 μL 10X T4 DNA ligase buffer (NEB), and 16.7 μL T4 DNA ligase (30 U/μL; Fermentas, EL0013).
7. Elution buffer: TE with 0.1% SDS.
8. EcoP15I digestion mix (60 μL): 34 μL DNA in Elution buffer, 6 μL 10X NEBuffer3, 6 μL 10 mM ATP, 0.6 μL 10 μg/μL BSA, 0.6 μL 10 mM Sinefungin, 2 μL EcoP15I (NEB), and 10.8 μL nuclease-free water.
9. ChIA-PET PCR reaction mix (10 μL): 0.6 μL nick-repaired DNA on beads, 0.75 μL 2 μM Paired-End (PE) PCR primers (1.0/2.0), 3.65 μL nuclease-free water, and 5 μL Phusion High-Fidelity PCR Mater Mix (NEB).
10. Proteinase K.
11. PCI (Phenol:Chloroform:Isoamyl alcohol; 25:24:1).

12. 3 M sodium acetate (pH 5.2).
13. Glycoblue.
14. Isopropanol.
15. 75% ethanol.
16. Sequencing adaptor ligation reagents: 50 μL resuspended Streptavidin M280 beads, Illumina adaptors, and primers.
17. Quick Ligation Kit (NEB): 2X Quick Ligation buffer and Quick Ligase.
18. Wash buffer (50 mL): 10 mM Tris-HCl pH 7.5 (500 μL 1 M), 1 mM EDTA (100 μL 0.5 M), 2 M NaCl (5 mL 5 M), and 44 mL DW.
19. 2X Binding buffer (BB, 15 mL scale): 10 mM Tris-HCl pH 7.5 (150 μL 1 M), 1 mM EDTA (30 μL 0.5 M), 2 M NaCl (6 mL 5 M), and 8.82 mL DW.
20. Tween wash buffer (TWB, 10 mL scale): 5 mL 2X BB, 5 μL 0.05% Tween 20, and 5 mL DW.

2.2 Equipment

1. Magnet stand for 1.5 mL tubes.
2. Rotator for 1.5 mL tubes.
3. Air incubator (37 °C, 16 °C).
4. Low-binding tubes.
5. Microfuge for 1.5 mL tubes both with refrigeration or in a cold room (4 °C).
6. Thermocycler.
7. −80 °C freezer.
8. Vortex mixer.
9. Qubit assay system.
10. DNA gel electrophoresis system.

3 Methods

Figure 1 shows the schematic representation of the ChIA-PET procedure.

3.1 End Preparation of ChIP DNA

3.1.1 End-Blunting

1. Pellet Dynabeads-bound ChIP DNA using a magnetic stand. The ChIP DNA was prepared according to a previously published protocol [9].
2. Remove the supernatant completely and place the tube on ice.
3. Resuspend the Dynabeads-bound ChIP DNA in the End-blunting reaction mix.
4. Incubate at 37 °C with rotation for 20 min.

3.1.2 A-Tailing

1. Pellet the Dynabeads-bound ChIP DNA using a magnet stand.
2. Discard the supernatant and wash the Dynabeads-bound ChIP DNA three times with 1 mL of ice-cold wash buffer.
3. Resuspend the Dynabeads-bound ChIP DNA in the A-tailing reaction mix.
4. Incubate at 37 °C for 1 h with rotation.

3.2 Linker A/B Ligation

1. Pellet the Dynabeads-bound ChIP DNA using a magnet stand.
2. Discard the supernatant and wash the Dynabeads-bound ChIP DNA three times with 1 mL of ice-cold wash buffer.
3. Split the sample into two low-binding tubes.
4. Resuspend the Dynabeads-bound ChIP DNA in the Linker A/B ligation reaction mix (*see* **Note 2**).
5. Incubate overnight (~16 h) at 16 °C with gentle rotation at low speed (*see* **Note 3**).

3.3 Proximal Ligation

3.3.1 5′-End Phosphorylation

1. Combine the two tubes of linker-ligated DNA in a single 1.5 mL low-binding tube.
2. Remove excess linkers by washing three times with 1 mL of ice-cold wash buffer using the magnetic stand.
3. Resuspend the Dynabeads-bound ChIP DNA in the 5′-end phosphorylation reaction mix.
4. Incubate at 37 °C with rotation for 30 min.

3.3.2 DNA Circularization

1. Wash the Dynabeads-bound ChIP DNA twice with 1 mL of ice-cold wash buffer using the magnetic stand.
2. Prepare the DNA circularization reaction mix on ice.
3. Add the DNA circularization reaction mix to beads, and mix thoroughly.
4. Incubate overnight (~16 h) at 16 °C with gentle rotation at low speed.

3.4 DNA Extraction

3.4.1 Elution of Chromatin Complex

1. Pellet the beads and discard the supernatant.
2. Resuspend the beads in 400 μL of Elution buffer.
3. Add 20 μL of Proteinase K and mix thoroughly.
4. Incubate overnight (~16 h) at 37 °C with gentle rotation at low speed.
5. The following day, add an additional 5 μL Proteinase K and incubate for 2 more hours.

3.4.2 DNA Purification

1. Perform phenol-chloroform extraction using 450 μL of PCI.
2. Add 1/10 volume of 3 M sodium acetate (pH 5.2) to a final concentration of 0.3 M, 1 μL of GlycoBlue, and 1 volume of isopropanol. Mix thoroughly by pipetting or gentle inversion.
3. Incubate at −80 °C for at least 1 h.
4. Thaw for ~10 min, then centrifuge at 13,000 rpm for 30 min at 4 °C.
5. Carefully decant the supernatant and wash the DNA pellet twice with 1 mL of 75% ethanol.
6. Dissolve the DNA pellet in 34 μL of Elution buffer, ensuring thorough resuspension.
7. Quantify DNA yield (*see* **Note 4**).

3.5 EcoP15I Digestion (See Note 5)

Figure 2a shows the expected reaction during EcoP15I digestion.

1. Prepare the EcoP15I digestion mix on ice.
2. Add the EcoP15I digestion mix to the DNA sample and incubate at 37 °C overnight in a Thermocycler (*see* **Note 6**).
3. The following day, supplement the reaction with an additional 2 μL EcoP15I, 6 μL 10 mM ATP, and 0.6 μL 10 mM

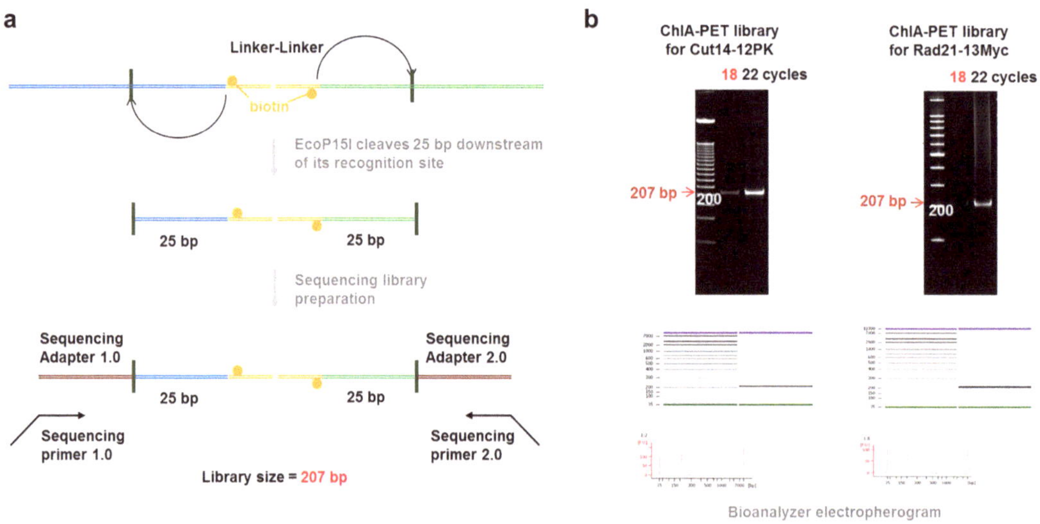

Fig. 2 EcoP15I digestion and validation of ChIA-PET library fragments. This figure illustrates the EcoP15I digestion step in the ChIA-PET procedure and the validation of the expected 207 bp fragment size. (**a**) A schematic representation of EcoP15I digestion, which cleaves ~25 bp downstream of its recognition site, generating DNA fragments flanked by sequencing adapters. These fragments are further processed for sequencing library construction. (**b**) Gel electrophoresis images display distinct 207 bp bands from sequencing libraries, confirming successful digestion and library preparation. Bioanalyzer electropherograms further validate the fragment size distribution, showing a predominant peak at 207 bp, ensuring high-quality sequencing data

Sinefungin (*see* **Note 7**) and incubate at 37 °C for 2 additional hours.

3.6 Sequencing Adapter Ligation

3.6.1 DNA Immobilization on Streptavidin Beads

1. Vortex Streptavidin M280 beads thoroughly.
2. Transfer 50 μL of resuspended Streptavidin M280 beads to a 1.5 mL low-binding tube.
3. Wash the beads sequentially with 500 μL of 1X TWB, followed by 500 μL of 1X BB using the magnetic stand.
4. Resuspend the washed beads in 58 μL of 2X BB buffer.
5. Add 58 μL of digestion mix, vortex briefly, and incubate at room temperature with rotation for 30 min to facilitate the binding of biotinylated Linker A/B to the Streptavidin M280 beads.
6. Wash the beads three times with 1X TWB to remove unbound components using the magnetic stand.

3.6.2 Ligation of Illumina Adapters

1. Prepare the ligation buffer, excluding DTT and ATP.
2. Add 97 μL of 1X Quick Ligation buffer.
3. Add 1 μL of 25 μM Illumina adaptor 1, 1 μL of 25 μM Illumina adaptor 2, and 1 μL of Quick ligase.
4. Incubate at 20 °C for 30 min in a Thermocycler (*see* **Note 6**).

3.7 PCR Amplification of ChIA-PET Library

1. Prepare the ChIA-PET PCR reaction mix.
2. Perform PCR:
   - 98 °C for 30 s
   - 16–22 cycles of:

     98 °C for 10 s (denaturation)

     65 °C for 30 s (annealing)

     72 °C for 30 s (extension).
   - 72 °C for 7 min (final extension).
3. Verify the PCR amplicon size to confirm proper fragment enrichment (*see* **Note 8**).

   Figure 2b shows a typical ChIA-PET library.

4 Notes

1. Since ChIA-PET analysis begins with ChIP DNA, assessing both its quality and quantity is essential. ChIP DNA quality can be evaluated using qPCR with specific target primers, while its quantity should be approximately 100 ng to ensure sufficient material for downstream processing. To obtain over

100 ng of ChIP DNA, 1 liter of cells expressing the condensin subunit Cut14 and the cohesin subunit Rad21 were cultured for our published study [9].

2. In ChIA-PET experiments, two distinct linkers, referred to as Linkers A and B, are utilized to differentiate between true chromatin contacts and potential artifacts arising from random ligations. Each linker contains a unique nucleotide barcode, enabling the identification of ligation products based on their linker composition. Paired-end tags (PETs) with identical linkers (AA or BB) are considered to originate from valid chromatin contacts, while PETs with mixed linkers (AB or BA) are indicative of random ligations between different chromatin complexes (Fig. 1). This strategy enhances the accuracy of detecting genuine chromatin contacts by effectively distinguishing them from background noise [3, 4].
3. To ensure uniform mixing without bead precipitation, place the tubes on a rotator compatible with 1.5 mL tubes and incubate them in a 16 °C incubator.
4. Before DNA extraction, it is not possible to measure the sample DNA amount since it remains bound to the beads. However, at this stage, the quantity of ChIA-PET intermediate DNA can be estimated since it has been extracted from the beads. DNA concentration was measured using the Qubit dsDNA HS Assay Kit (Thermo Fisher, Q32851). Based on our experience, approximately 10 ng of DNA is required for successful downstream processing [9].
5. EcoP15I cleaves DNA at a fixed distance (27 bp) downstream of its recognition site (Fig. 2a). This property allows EcoP15I digestion to generate short paired-end tags (PETs) optimized for sequencing, ensuring that the fragments contain enough sequence information to be uniquely mapped back to the genome [12, 13].
6. To ensure precise temperature control, transfer the sample to a PCR tube and incubate it using a thermocycler.
7. Sinefungin acts as a SAM analog that stabilizes EcoP15I cleavage activity while inhibiting its methylation activity, ensuring efficient and specific DNA cleavage in the ChIA-PET procedure [12].
8. In the ChIA-PET procedure, DNA fragments typically exhibit a single dominant size due to the precise ligation, restriction enzyme cleavage, and sequencing adapter addition steps in the protocol. This standardized fragment size ensures consistent sequencing and accurate chromatin contact mapping, making it a critical quality checkpoint in the ChIA-PET experiment. Additionally, minimizing PCR cycles helps reduce redundant sequencing reads and prevents overamplification biases. Based

on our experience, limiting library amplification to 16 cycles is recommended for optimal results (Fig. 2b) [9].

Acknowledgments

This work was supported by the National Research Foundation of Korea (NRF) grant funded by Korean government (MSIT) (No. 2022R1A2C100442312) to KK. This work was also supported by the National Institutes of Health/National Institute of General Medical Sciences R01GM124195, JSPS KAKENHI grant JP20K23376, Grant for Basic Science Research Projects from the Sumitomo Foundation, the Uehara Memorial Foundation, Takeda Science Foundation, the Mitsubishi Foundation, the NOVARTIS Foundation (Japan) for the Promotion of Science, and the Photo-Excitonix Project in Hokkaido University and the Grant for Joint Research Program of the Institute for Genetic Medicine, Hokkaido University (to KN).

References

1. Nasmyth K, Haering CH (2009) Cohesin: its roles and mechanisms. Annu Rev Genet 43: 525–558
2. Hirano T (2012) Condensins: universal organizers of chromosomes with diverse functions. Genes Dev 26(15):1659–1678
3. Fullwood MJ, Liu MH, Pan YF et al (2009) An oestrogen-receptor-alpha-bound human chromatin interactome. Nature 462(7269):58–64
4. Li G, Fullwood MJ, Xu H et al (2010) ChIA-PET tool for comprehensive chromatin interaction analysis with paired-end tag sequencing. Genome Biol 11(2):R22
5. Tang Z, Lu OJ, Li X et al (2015) CTCF-mediated human 3D genome architecture reveals chromatin topology for transcription. Cell 163(7):1611–1627
6. Rao SSP, Huang SC, Hilaire BG et al (2017) Cohesin loss eliminates all loop domains. Cell 171(2):305–320 e24
7. Davidson IF, Peters JM (2021) Genome folding through loop extrusion by SMC complexes. Nat Rev Mol Cell Biol 22(7):445–464
8. Higashi TL, Uhlmann F (2022) SMC complexes: lifting the lid on loop extrusion. Curr Opin Cell Biol 74:13–22
9. Kim KD, Tanizawa H, Iwasaki O et al (2016) Transcription factors mediate condensin recruitment and global chromosomal organization in fission yeast. Nat Genet 48(10): 1242–1252
10. Noma KI (2017) The yeast genomes in three dimensions: mechanisms and functions. Annu Rev Genet 51:23–44
11. Tanizawa H, Kim KD, Iwasaki O et al (2017) Architectural alterations of the fission yeast genome during the cell cycle. Nat Struct Mol Biol 24(11):965–976
12. Moncke-Buchner E, Rothenberg M, Reich S et al (2009) Functional characterization and modulation of the DNA cleavage efficiency of type III restriction endonuclease EcoP15I in its interaction with two sites in the DNA target. J Mol Biol 387(5):1309–1319
13. Raghavendra NK, Bheemanaik S, Rao DN (2012) Mechanistic insights into type III restriction enzymes. Front Biosci (Landmark Ed) 17(3):1094–1107

Chapter 17

High-Resolution Chromosome Conformation Capture Method to Study Archaeal SMC Complexes

Kodai Yamaura and Naomichi Takemata

Abstract

Structural Maintenance of Chromosomes (SMC) ATPases are key regulators of 3D genome architecture in diverse organisms. Although SMC homologs are widely found in the domain Archaea, their roles in archaeal genome organization remains largely unknown. Recently, we and others have successfully applied the Chromosome Conformation Capture (3C) sequencing methods Hi-C and 3C-seq to various archaeal model organisms. These studies have revealed that archaeal SMC proteins exhibit both common and unique functional properties compared to those in eukaryotes and bacteria. Here, we present a high-resolution 3C-seq protocol that is optimized for mapping genome organization in the hyperthermophilic model archaeon *Thermococcus kodakarensis*.

Key words Hi-C, 3C-seq, Archaea, SMC, *Thermococcus kodakarensis*

1 Introduction

SMC ATPases assemble with Kleisin and Kleisin-interacting proteins (Hawks in eukaryotes and Kites in prokaryotes) to form ring-shaped complexes that play critical roles for the spatial organization of genomes [1, 2]. Growing evidence suggests that these SMC complexes act as DNA motors that use the energy of ATP hydrolysis to extrude DNA loops [3, 4]. The eukaryotic SMC complex cohesin, best known for establishing mitotic sister chromatid cohesion, is hypothesized to use the loop extrusion mechanism to organize interphase chromosomes into arrays of Topologically Associating Domains (TADs) [5–8]. This model also posits that TAD boundaries are determined by "roadblock" proteins, such as CTCF and RNA polymerases, that stall the translocation of cohesin [9–12]. Furthermore, loop extrusion provides a compelling explanation for how the eukaryotic SMC complex condensin and the bacterial SMC complexes Smc-ScpAB and MukBEF promote the

Yasuto Murayama and Yasutaka Kakui (eds.), *SMC Proteins: Methods and Protocols*, Methods in Molecular Biology, vol. 2991, https://doi.org/10.1007/978-1-0716-5005-9_17,

compaction and individualization of sister chromosomes without wrongly mediating inter-sister interactions [1, 13]. However, several studies suggest that cohesin and condensin may employ alternative, loop extrusion-independent mechanisms for TAD formation and mitotic chromosome assembly, respectively [14–16].

SMC proteins are widespread in both eukaryotes and prokaryotes. Over the past three decades, our understanding of the SMC family has been greatly advanced by studying representative model organisms of the domains *Eukarya* and *Bacteria* [1, 3, 4]. What is often overlooked, however, is that SMC proteins are also prevalent in the third domain of life—*Archaea*. According to the latest International Code of Nomenclature of Prokaryotes (ICNP), the domain *Archaea* consists of four kingdoms: *Methanobacteriati* (or the phylum *Euryarchaeota*), *Nanobdellati* (or the DPANN superphylum), *Thermoproteati* (or the TACK superphylum), and *Promethearchaeati* (or the Asgard archaea) [17, 18]. Most archaea in these kingdoms possess homologs of bacterial Smc-ScpAB subunits: the Smc ATPase, the Kleisin protein ScpA, and the Kite protein ScpB [19]. Interestingly, certain archaeal lineages in *Methanobacteriati* have Smc and ScpA but lack ScpB, despite its essential role in the bacterial Smc-ScpAB complex [19]. Even more surprisingly, canonical SMC proteins are absent in many members of *Thermoproteati* and some of *Methanobacteriati* [19, 20]. Instead, a number of these SMC-lacking archaea possess noncanonical SMC proteins such as coalescin, Sph, and Arcadin-4, the former two of which belong to the recently proposed SMC clade Archaea-Specific SMC-Related Proteins (ASRPs) [20–23]. Notably, *Aeropyrum pernix*, a hyperthermophilic archaeon from *Thermoproteati*, lacks both canonical SMC and ASRP, yet its genome is highly organized in 3D [24]. Given the unique phylogenetic distribution of Smc and SMC-related proteins in archaea, investigating their roles could yield novel insights into the functions of the SMC family.

SMC and SMC-related proteins in archaea remain poorly characterized, partly because the harsh growth conditions of most model archaea hinder the application of many in vivo experimental methods. Recently, we and others have successfully applied genome-wide Chromosome Conformation Capture (3C) techniques, namely, Hi-C and 3C-seq, to model archaea from various extreme habitats [20, 24–28]. 3C detects in vivo DNA–DNA contacts via proximity ligation of crosslinked chromosomal fragments generated by restriction digestion. Hi-C, which includes a biotin pulldown step to enrich for ligation junctions, and 3C-seq, which lacks this enrichment step, quantify the DNA contact frequency on a genome-wide scale by paired-end sequencing [25, 29].

The first archaeal Hi-C work has revealed that the SMC-like protein coalescin, unlike any canonical SMC proteins, facilitates the formation of transcriptionally active and less active compartments, known as A/B compartments [20]. Another study has shown that Smc is required for the formation of certain domain boundaries in the halophilic archaeon *Haloferax volcanii* [26]. More recently, we have uncovered that Smc-ScpAB in the hyperthermophilic anaerobic archaeon *Thermococcus kodakarensis* sculpts TAD-like domains by collaborating with the presumed roadblock protein TrmBL2 [30]. Thus, 3C-based technologies are powerful tools not only for studying eukaryotic and bacterial SMC complexes but also for uncovering the largely unexplored functions of the archaeal SMC proteins.

Here, we present a high-resolution 3C-seq protocol for *T. kodakarensis*, one of the best-studied archaea due to its natural competence and well-established genetic systems. Whereas typical Hi-C/3C-seq experiments use a single restriction enzyme for chromosome digestion, this method employs the two four-base cutters AluI (AG|CT) and HaeIII (GG|CC), which generate blunt ends that can be directly ligated. We have found that the combination of these enzymes allows for high-resolution detection of genomic contacts in diverse archaea including *T. kodakarensis* ([30] and our unpublished results). One limitation of this approach is that blunt ends are incompatible with the biotin incorporation strategy implemented in Hi-C to capture ligation junctions. However, this drawback can be mitigated by performing proximity ligation exclusively on the insoluble fraction, which corresponds to the "in situ" ligation procedure adopted by eukaryotic Hi-C protocols. In prokaryotic Hi-C, this refinement has been shown to markedly increase the number of valid reads for contact map generation [26]. Additionally, omitting the biotin labeling and enrichment steps reduces the time required to optimize the protocol for another species. Another key feature of our protocol is the use of the long-arm crosslinker disuccinimidyl glutarate (DSG) in addition to formaldehyde. This dual crosslinking strategy enhances the capture of chromosomal contacts, even for small DNA fragments [31]. The method presented here will help to elucidate how archaeal SMCs contribute to 3D genome organization in vivo.

2 Materials

2.1 Strains

Our 3C-seq protocol is optimized for the *T. kodakarensis* wild-type strain KOD1 and its uracil-auxotrophic derivative KU216 (Δ*pyrF*) [32]. We have also successfully applied this method to a variety of KU216 derivative mutants generated via pop-in/pop-out recombination using the *pyrF* marker (*see* **Note 1**).

2.2 Growth Medium

1. 4x ASW (Artificial Sea Water)-m1: 80 g/L NaCl, 12 g/L $MgCl_2 \cdot 6H_2O$, 24 g/L $MgSO_4 \cdot 7H_2O$, 4 g/L $(NH_4)_2SO_4$, 0.8 g/L $NaHCO_3$, 1.2 g/L $CaCl_2 \cdot 2H_2O$, 2 g/L KCl, 1.68 g/L KH_2PO_4, 0.2 g/L NaBr, 0.08 g/L $SrCl_2 \cdot 6H_2O$, 0.04 g/L $Fe(NH_4)$ citrate, 0.0166 g/L KI, 0.00618 g/L H_3BO_3, 0.0119 g/L $NiCl_2 \cdot 6H_2O$. This stock solution is used to make ASW-YT-m1-S^0 medium. Do not autoclave (*see* **Note 2**).
2. 0.2% (w/v) resazurin sodium salt. Do not autoclave.
3. 5% (w/v) Na_2S. Filter sterilize and store in an anaerobic chamber.
4. ASW-YT-m1-S^0 medium: 0.8x ASW-m1, 0.5% (w/v) yeast extract, 0.5% (w/v) tryptone, 0.2% (w/v) sulfur powder, trace amount of 5% (w/v) Na_2S. *See* Subheading 3.1 for how to mix the components. Store at room temperature if needed, but the medium can be kept only for a couple of days.
5. A stock culture of *T. kodakarensis* cells (*see* **Note 3**).

2.3 Cell Crosslinking

1. Fixative Solution 1: 4.3 mL of 37% formaldehyde solution (containing methanol as a stabilizer), 6.4 mL of 4x ASW-m1, and 21.3 mL of ultrapure water. Mix in each of eight 50-mL tubes immediately before use.
2. 2.5 M glycine. Sterilize by autoclaving.
3. Disuccinimidyl glutarate (DSG): 300 mM stock solution dissolved in DMSO. Store the stock at −20 °C.
4. Fixative Solution 2: 60 μL of 300 mM DSG, 1200 μL of 4x ASW-m1, and 4740 μL of ultrapure water. Mix immediately before use.
5. 1x phosphate-buffered saline (PBS). pH is not adjusted. Sterilize by autoclaving. Chill before use.

2.4 DNA Preparation

1. Wash Buffer: 10 mM Tris-HCl (pH 8.0), 50 mM NaCl, 10 mM $MgCl_2$. Sterilize by autoclaving. Split into two aliquots and store them at 4 °C and room temperature, respectively.
2. 10x NEBuffer 2 (New England Biolabs). Store at −20 °C. Dilute immediately before use.
3. 10% (w/v) SDS. Filter-sterilize.
4. 10% (v/v) Triton X-100. Filter-sterilize and keep the solution in the dark (*see* **Note 4**).
5. Digestion Buffer: 43.8 μL of 10x NEBuffer 2, 100 μL of 10% (v/v) Triton X-100, and 243.7 μL of ultrapure water. Mix immediately before use.
6. 10 U/μL AluI. Store at −20 °C.
7. 10 U/μL HaeIII. Store at −20 °C.

8. 10x T4 DNA Ligase Reaction Buffer: 500 mM Tris-HCl (pH 7.5), 100 mM $MgCl_2$, 10 mM ATP, and 100 mM DTT. Store single-use aliquots at −20 °C and avoid repeated freeze-thaw cycles (*see* **Note 5**). Dilute immediately before use.
9. 400 U/μL T4 DNA Ligase. Store at −20 °C.
10. 1x TE buffer: 10 mM Tris-HCl (pH 8.0) and 1 mM EDTA. Sterilize by autoclaving.
11. 0.5 M EDTA (pH 8.0).
12. Decrosslinking Buffer: 1x TE buffer, 10% (w/v) SDS, and 0.5 M EDTA at a volume ratio of 20:2:1. Mix immediately before use.
13. 800 U/mL proteinase K. Store at −20 °C.
14. 3 M sodium acetate (pH 5.2).
15. 100% ethanol.
16. 70% ethanol. Store at 4 °C.
17. 20 mg/mL glycogen.
18. Phenol:chloroform:isoamyl alcohol (49.5:49.5:1). Store at 4 °C.
19. 10 mM Tris-HCl, pH 8.0.
20. RNase A. Store at −20 °C.
21. Agarose.
22. 1x TAE buffer: 4.85 g/L Tris(hydroxymethyl)aminomethane, 0.372 g/L EDTA· 2Na, and 1.14 mL/L acetic acid.

2.5 Library Construction

1. NEBNext Ultra II DNA Library Prep Kit for Illumina (New England Biolabs). The kit contains NEBNext Ultra II End Prep Enzyme Mix, NEBNext Ultra II End Prep Reaction Buffer, NEBNext Ultra II Ligation Master Mix, NEBNext Ligation Enhancer, and NEBNext Ultra II Q5 Master Mix.
2. Qubit 1x dsDNA HS Assay Kit (Thermo Fisher Scientific).
3. SPRIselect (Beckman Coulter).
4. 80% ethanol. Prepare on the day of use.
5. 0.1x TE, pH 8.0.
6. NEBNext 96 Unique Dual Index Primer Pairs (New England Biolabs).
7. D1000 ScreenTape (Agilent).
8. D1000 Reagents (Agilent).

2.6 Equipment

1. Medium container. Use a pressure-resistant laboratory glass bottle or a glass vial that can be closed with an aluminum cap and a rubber stopper.

2. Anaerobic chamber.
3. An incubator that can heat to 85 °C for cultivation of *T. kodakarensis*.
4. Thermomixer.
5. microTUBE AFA Fiber Pre-Slit Snap-Cap 6 × 16 mm (Covaris).
6. M220 Focused-ultrasonicator (Covaris).
7. Qubit 4 Fluorometer (Thermo Fisher Scientific).
8. Thermocycler.
9. DNA low retention tube.
10. Magnetic stand.
11. 4150 TapeStation System (Agilent).

3 Methods

3.1 Cell Cultivation

1. Prepare ASW-YT-m1-S^0 medium and pour it to a pressure-resistant laboratory glass bottle or a glass vial that can be closed with an aluminum cap and a rubber stopper (*see* **Note 6**, Fig. 1a). We typically use a bottle with 200 mL to 1 L capacity for large-scale culture and a vial with 20–100 mL capacity for small-scale culture. When a glass vial is used, gently place a rubber stopper on the mouth (make sure not to completely close it) and then wrap the mouth and the stopper in foil (Fig. 1b,c). Do not add the sulfur powder at this stage.
2. Add 0.2% (w/v) resazurin sodium salt to the medium at a volume ratio of 1:4000 as a redox indicator.

a b c

Fig. 1 Containers used for cultivation of *T. kodakarensis*. (**a**) Shown are different sizes of pressure-resistant laboratory glass bottles and glass vials with aluminum caps and rubber stoppers. (**b**) When autoclaving a glass vial, gently place a rubber stopper on the mouth of the vial without pushing it in. (**c**) Then wrap the mouth and the rubber stopper in foil

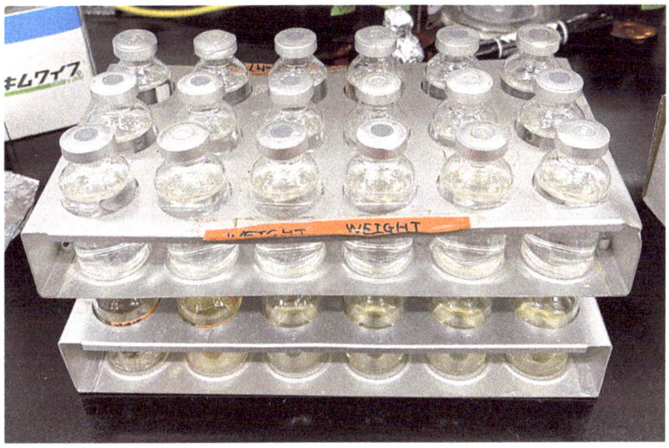

Fig. 2 Setup to prevent media from boiling over vials

3. Autoclave for 20 min at 121 °C.
4. Wrap the required amount of sulfur powder in a weighing paper during the autoclaving.
5. Remove the vial/bottle from the autoclave when the temperature drops to 80–90 °C and the pressure returns to atmospheric pressure. Close the vial/bottle quickly. For the vial, squeeze the rubber stopper into the mouth without removing the foil. Be careful not to get burned.
6. Place the vial/bottle and the wrapped sulfur powder in an anaerobic chamber. As the medium could boil over during the gas replacement process, make sure that the vial/bottle is completely closed. In case of using a vial, it is recommended to place it in a rack and put a weight on top of them (Fig. 2).
7. Before the medium cools down, add 5% (w/v) Na_2S to the medium at a volume ratio of 1:4000. The color of resazurin will become clear in a while. If not, add more Na_2S.
8. Add the sulfur powder. It will not dissolve but that is okay.
9. Inoculate the medium with a stock culture at 1% (v/v).
10. After closing up the container with the cap or rubber stopper, remove it from the anaerobic chamber. In case of using a glass vial, cap it with an aluminum cap outside.
11. Grow the cells overnight at 85 °C without shaking.
12. Dilute the overnight culture to an OD_{660} of 0.015 in fresh ASW-YT-m1-S^0 medium. At least 80 mL of the diluted culture are required. Grow the cells until mid-to-late log phase (OD_{660}: 0.2–0.25) at 85 °C without shaking (*see* **Notes 7** and **8**).

3.2 Fixation

1. Remove the culture from the incubator. Immediately dispense 8 mL of it into each of eight 50-mL tubes containing 32 mL of Fixative Solution 1 (*see* **Note 9**).
2. Incubate for 30 min at 25 °C with agitation at 75 rpm on a shaker.
3. Quench the crosslinking reaction by adding 10 mL of 2.5 M glycine to each tube and incubating for 10 min at room temperature.
4. Centrifuge for 15 min at 8000 *g*, 25 °C.
5. Remove the supernatants with ~2 mL per tube left for resuspension.
6. Resuspend the cells by vortex, then combine the suspensions, and dispense it into four 5 mL tubes.
7. Centrifuge for 5 min at 10,000 *g*, 25 °C.
8. Remove the supernatants and resuspend every two cell pellets together in 3 mL of Fixative Solution 2.
9. Incubate for 40 min at 25 °C with agitation at 75 rpm on a shaker.
10. Quench the crosslinking reaction by adding 750 μL of 2.5 M glycine to each tube and incubating for 5 min at room temperature.
11. Centrifuge for 5 min at 10,000 *g*, 4 °C.
12. Remove the supernatants and resuspend all the cell pellets together in 1 mL of chilled 1x PBS.
13. Centrifuge for 5 min at 10,000 *g*, 4 °C.
14. Remove the supernatant and resuspend the cell pellet in 1 mL of chilled 1x PBS.
15. Centrifuge for 5 min at 10,000 *g*, 4 °C.
16. Store the pellet at −80 °C until use (*see* **Note 10**).

3.3 DNA Preparation

1. Resuspend a frozen pellet in 300 μL of chilled Wash Buffer. Dilute the cell suspension to an OD_{660} of 3 by adding chilled Wash Buffer.
2. Centrifuge 1.2 mL of the cell suspension for 10 min at 10,000 *g*, 4 °C.
3. Remove the supernatant and resuspend the pellet in 50 μL of 1x NEBuffer 2.
4. Transfer 50 μL of the cell suspension to a 1.5-mL tube.
5. Mix the transferred suspension with 5.55 μL of 10% (w/v) SDS.
6. Incubate for 15 min at 65 °C with agitation at 600 rpm on a thermomixer to lyse the cells.

7. Immediately cool down the lysate on ice for 90 sec (*see* **Note 11**). After that, store the lysate at room temperature until use.
8. Mix well 37.5 μL of the lysate with 232.5 μL of Digestion Buffer for restriction reaction (*see* **Note 12**). Mix 12.5 μL of the remainder of the lysate with 82.5 μL Digestion Buffer for purification of undigested control DNA.
9. Add 15 μL each of 10 U/μL HaeIII and 10 U/μL AluI to the sample for restriction reaction. Add 5 μL of ultrapure water to the undigested control.
10. Incubate the samples for 3 h at 37 °C with agitation at 600 rpm on a thermomixer. Mix by inverting the tube every 30 min.
11. Add 10% (w/v) SDS to the restriction reaction and the control reaction at a volume ratio of 1:9. Incubate for 10 min at room temperature.
12. Collect the insoluble material by centrifuging for 10 min at 16,000 *g*, 25 °C.
13. Remove the supernatants and resuspend each pellet in 500 μL of room temperature Wash Buffer.
14. Centrifuge for 10 min at 16,000 *g*, 25 °C.
15. Resuspend the pellets from the restriction reaction and the control reaction in 1185 μL and 395 μL of 1.01 × T4 DNA Ligase Reaction Buffer, respectively.
16. Mix 790 μL of the digested sample with 10 μL of 400 U/μ L T4 DNA Ligase. Add 5 μL of ultrapure water to the rest of the digested sample (395 μL) for purification of un-ligated control DNA. Also add 5 μL of ultrapure water to the undigested control.
17. Incubate the samples for 3 h at 16 °C with agitation at 600 rpm on a thermomixer. Mix by inverting the tube every 30 min.
18. Collect the insoluble material by centrifuging for 10 min at 16,000 *g*, 25 °C.
19. Resuspend each pellet in 115 μL of Decrosslinking Buffer.
20. Add 3 μL of 800 U/mL proteinase K to each sample.
21. De-crosslink by incubating the samples overnight at 37 °C with agitation at 600 rpm on a thermomixer (*see* **Note 13**).
22. Further add 3 μL of 800 U/mL proteinase K to each sample.
23. Incubate for 4 h at 37 °C with agitation at 600 rpm on a thermomixer.
24. Transfer 30 μL of the ligation sample to another tube as ligated control DNA and mix it with 90 μL of Decrosslinking Buffer.

25. Dilute 90 μL of the ligation sample by adding 30 μL of Decrosslinking Buffer. Transfer the dilution to a microTUBE AFA Fiber Pre-Slit Snap-Cap 6 × 16 mm.
26. Sonicate the transferred sample (*see* **Note 14**).
27. Adjust the volumes of the sonicated sample and the three controls (undigested control DNA, un-ligated control DNA, and ligated control DNA) to 200 μL with 1x TE buffer. Extract DNA from the samples with phenol:chloroform:isoamyl alcohol.
28. For the sonicated sample only, repeat **step 27** once for a total of two extraction cycles.
29. Add 3 M NaOAc to each sample at a volume ratio of 1:10. Then add 2.5 × volume of 100% ethanol to each sample.
30. Add 2 μL of 20 mg/mL glycogen to each sample as a DNA carrier.
31. Mix well and incubate for 1 h at −20 °C.
32. Centrifuge for 30 min at 20,400 *g*, 4 °C.
33. Remove the supernatants and wash the pellets with chilled 70% ethanol.
34. Centrifuge for 5 min at 20,400 *g*, 4 °C.
35. Remove the supernatants.
36. Spin down briefly and remove residual ethanol.
37. Air-dry the pellets for 5 min at 37 °C.
38. Dissolve each pellet in 30 μL of 10 mM Tris-HCl containing 0.05 mg/mL of RNase A.
39. Incubate for 30 min at room temperature.
40. Perform DNA gel electrophoresis in a 1.5% agarose/1x TAE gel to check DNA size. Run 3 μL of the ligated and sonicated DNA and 10 μL each of the other samples. Typical results are shown in Fig. 3.

3.4 Library Construction

The protocol essentially follows the instruction manual for NEBNext Ultra II DNA Library Prep Kit for Illumina.

1. Quantify the concentration of the sheared DNA (*see* **Note 15**). Use 220 ng DNA for library construction.
2. Adjust the volume of the DNA solution to 50 μL by adding 10 mM Tris-HCl.
3. Mix with 7 μL of NEBNext Ultra II End Prep Reaction Buffer and 3 μL of NEBNext Ultra II End Prep Enzyme Mix in a PCR tube.

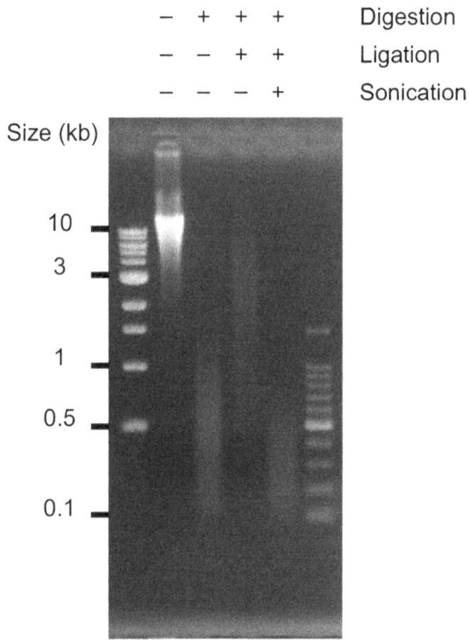

Fig. 3 Quality check of 3C DNA by gel electrophoresis. 1 kb DNA ladder (New England Biolabs) and 100 bp DNA ladder (Takara Clontech) were used as markers

4. Place the tube in a thermocycler with the heated lid set to ≥75 °C. Run the following program: 20 °C for 30 min, 65 °C for 30 min, and hold at 4 °C.
5. Mix the reaction with 2.5 μL of NEBNext Adaptor for Illumina, 1 μL of NEBNext Ligation Enhancer, and 30 μL of NEBNext Ultra II Ligation Master Mix in a 1.5-mL DNA low retention tube (*see* **Note 16**).
6. Incubate for 15 min at 20 °C on a thermomixer.
7. Add 3 μL of USER Enzyme to the reaction. Incubate for 15 min at 37 °C on a thermomixer.
8. Vortex SPRIselect well and add 25 μL of the bead suspension to the reaction. Mix them by vortex.
9. Incubate for 5 min at room temperature. Briefly spin down the tube and place it on a magnetic stand. After 5 min, transfer the supernatant to a new 1.5-mL DNA low retention tube.
10. Vortex SPRIselect well and add 10 μL of the bead suspension to the supernatant. Mix them by vortex.
11. Incubate for 5 min at room temperature. Briefly spin down the tube and place it on a magnetic stand. After 5 min, remove the supernatant.
12. Add 200 μL of 80% ethanol to the tube on a magnetic stand. Incubate for 30 sec and then remove the supernatant.
13. Repeat **step 12** once for a total of two washes.

14. Briefly spin down the tube and place it on a magnetic stand. Remove residual ethanol with a pipette tip.
15. Air-dry the beads for up to 5 min on a magnetic stand with the lid open (*see* **Note 17**).
16. Resuspend the beads in 15 μL of 0.1x TE by vortex.
17. Incubate for 2 min at room temperature. Briefly spin down the tube and place it on a magnetic stand. After 5 min, transfer 15 μL of the supernatant to a PCR tube.
18. Mix with 10 μL of a primer mix from NEBNext 96 Unique Dual Index Primer Pairs and 25 μL of NEBNext Ultra II Q5 Master Mix (*see* **Note 18**).
19. Perform PCR amplification with the following conditions:
    - Initial denaturation: 98 °C for 30 s.
    - 7 cycles of 98 °C for 10 s.
      65 °C for 75 s.
    - Final extension: 65 °C for 5 min.
    - Hold at 4 °C.
20. Transfer the PCR reaction to a 1.5-mL DNA low retention tube.
21. Vortex SPRIselect well and add 45 μL of the bead suspension to the reaction. Mix them by vortex.
22. Incubate for 5 min at room temperature. Briefly spin down the tube and place it on a magnetic stand. After 5 min, remove the supernatant.
23. Add 200 μL of 80% ethanol to the tube on a magnetic stand. Incubate for 30 sec and then remove the supernatant.
24. Repeat **step 23** once for a total of two washes.
25. Briefly spin down the tube and place it on a magnetic stand. Remove residual ethanol with a pipette tip.
26. Air-dry the beads for up to 5 min on a magnetic stand with the lid open (*see* **Note 17**).
27. Resuspend the beads in 25 μL of 0.1x TE by vortex.
28. Incubate for 2 min at room temperature. Briefly spin down the tube and place it on a magnetic stand. After 5 min, transfer the supernatant to a new 1.5-mL DNA low retention tube.
29. Check the size and concentration of the library DNA (*see* **Note 19**). The size distribution of a typical 3C-seq library is shown in Fig. 4.
30. Perform paired-end sequencing on the Illumina platform.

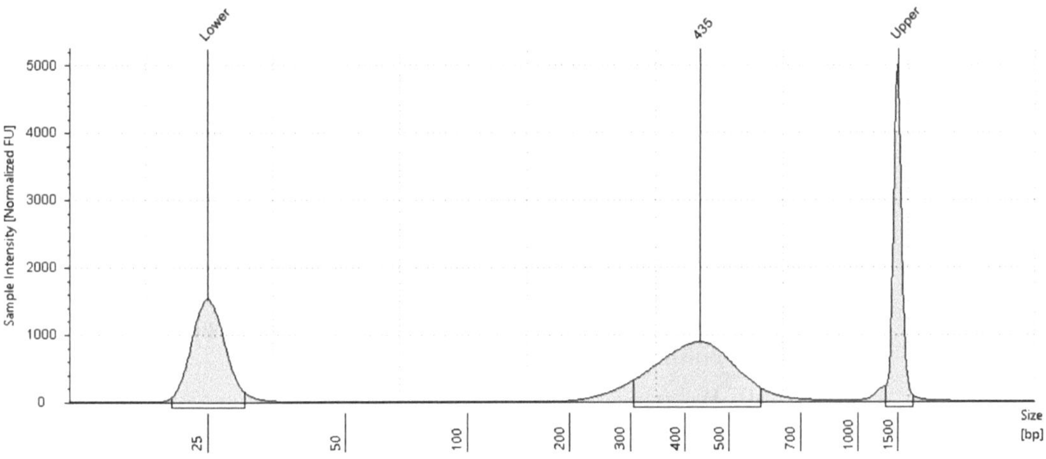

Fig. 4 TapeStation electrophoresis report from a typical 3C-seq library

4 Notes

1. For genetic analysis, some readers may instead wish to use an agmatine-auxotrophic strain carrying a mutation in the arginine decarboxylase gene *pdaD*. However, Δ*pdaD* cells may not be suitable for the 3C-seq protocol presented here, because we have failed to collect insoluble protein–DNA complexes from crosslinked Δ*pdaD* cells after restriction digestion. This observation may be relevant to the fact that PdaA catalyzes the first step in the biogenesis pathway of polyamines, positively charged aliphatic compounds that can affect DNA structure [33].
2. A precipitate will form during storage. It will dissolve when diluted to make ASW-YT-m1-S^0 medium.
3. In most cases, liquid cultures of *T. kodakarensis* can be stored at room temperature for at least 1 year. A frozen cell stock can be made by mixing a culture and 80% (v/v) glycerol containing 0.8x ASW-m1 (autoclaved and stored in an anaerobic chamber) at a volume ratio of 7:1 and flash freeze the mixture in liquid N_2.
4. Triton X-100 is light sensitive. It is recommended to prepare a fresh stock solution every few months.
5. T4 DNA Ligase Reaction Buffer contains ATP and DTT, both of which degrade after multiple freeze-thaw cycles.
6. For efficient removal of oxygen, it is recommended to use a 0.8x or larger volume of the medium relative to that of the container.

7. In case a glass vial with a rubber stopper is used, the culture can be withdrawn by inserting a syringe into the rubber. This can be performed outside of the anaerobic chamber. Since the pressure inside the vial is elevated due to the hydrogen sulfide generated by the cells, make sure to vent the vial before withdrawing a culture (e.g., by inserting a syringe without a plunger). The venting should also be performed in a fume hood.
8. Cell density can be measured with OD_{600} instead of OD_{660}.
9. Remove as much sulfur powder as possible when pipetting the culture.
10. We typically use the pellet in a week after cell fixation. The pellet could be stored longer.
11. Longer incubation could lead to precipitate formation.
12. Be careful to avoid bubble formation.
13. During decrosslinking, DNA may be degraded by endogenous nucleases in *T. kodakarensis*, which are active even in the presence of SDS. If your undigested control DNA is highly degraded, consider lowering the temperature to no less than 30 °C. You can proceed to library construction if you observe smearing of the undigested control but also a significant amount of large DNA (> 10 kb) in it.
14. We sonicate DNA for 200 sec at 7 °C using an M220 Focused-ultrasonicator (Covaris) with the peak power set to 50, the duty factor set to 20, and the cycles/burst set to 200.
15. We measure the DNA concentration using a Qubit 4 Fluorometer and Qubit 1x dsDNA HS Assay Kit.
16. In case the amount of input DNA does not exceed 100 ng, dilute the adaptor according to the manufacturer's instructions.
17. Be careful not to overdry the beads. When the beads turn lighter brown and start to crack, they must have been overdried.
18. We typically analyze eight to nine 3C-seq libraries per run on an Illumina NovaSeq 6000 System using SP Reagent Kit v1.5 (100 cycles), aiming to obtain approximately 100 million read pairs per library. Whereas NovaSeq 6000 can generate a substantial number of reads suitable for high-resolution analysis, it also has an increased rate of index hopping. To mitigate this, we use dual index primer pairs from NEBNext 96 Unique Dual Index Primer Pairs. Different types of index primer kits can be selected based on the sequencer in use.
19. For the assessment of DNA size, we use a D1000 ScreenTape and D1000 reagents on an Agilent 4150 TapeStation System.

If the library is contaminated with remaining primers or adaptors, perform an additional clean-up step by mixing the sample with vortexed SPRIselect at a volume ratio of 10:9 and perform **steps 22–29**.

Acknowledgments

K.Y. was supported by JST, the establishment of university fellowships toward the creation of science technology innovation (JPMJFS2123) and SPRING (JPMJSP2110). N.T. was supported by JST PRESTO (JPMJPR20K7), JST FOREST (JPMJFR224V), JSPS Grants-in-Aid for Young Scientists Start-up (JP21K20636), JSPS Grants-in-Aid for Transformative Research Areas (JP23H04281), Takeda Science Foundation, and the Uehara Memorial Foundation.

References

1. Yatskevich S, Rhodes J, Nasmyth K (2019) Organization of chromosomal DNA by SMC complexes. Annu Rev Genet 53:445–482. https://doi.org/10.1146/annurev-genet-112618-043633
2. Burmann F, Lowe J (2023) Structural biology of SMC complexes across the tree of life. Curr Opin Struct Biol 80:102598. https://doi.org/10.1016/j.sbi.2023.102598
3. Davidson IF, Peters JM (2021) Genome folding through loop extrusion by SMC complexes. Nat Rev Mol Cell Biol 22(7): 445–464. https://doi.org/10.1038/s41580-021-00349-7
4. Kim E, Barth R, Dekker C (2023) Looping the genome with SMC complexes. Annu Rev Biochem 92:15–41. https://doi.org/10.1146/annurev-biochem-032620-110506
5. Rao SSP, Huang SC, Glenn St Hilaire B et al (2017) Cohesin loss eliminates all loop domains. Cell 171(2):305–320 e324. https://doi.org/10.1016/j.cell.2017.09.026
6. Davidson IF, Bauer B, Goetz D et al (2019) DNA loop extrusion by human cohesin. Science 366(6471):1338–1345. https://doi.org/10.1126/science.aaz3418
7. Kim Y, Shi Z, Zhang H et al (2019) Human cohesin compacts DNA by loop extrusion. Science 366(6471):1345–1349. https://doi.org/10.1126/science.aaz4475
8. Fudenberg G, Imakaev M, Lu C et al (2016) Formation of chromosomal domains by loop extrusion. Cell Rep 15(9):2038–2049. https://doi.org/10.1016/j.celrep.2016.04.085
9. Nora EP, Goloborodko A, Valton AL et al (2017) Targeted degradation of CTCF decouples local insulation of chromosome domains from genomic compartmentalization. Cell 169(5):930–944 e922. https://doi.org/10.1016/j.cell.2017.05.004
10. Banigan EJ, Tang W, van den Berg AA et al (2023) Transcription shapes 3D chromatin organization by interacting with loop extrusion. Proc Natl Acad Sci USA 120(11): e2210480120. https://doi.org/10.1073/pnas.2210480120
11. Jeppsson K, Sakata T, Nakato R et al (2022) Cohesin-dependent chromosome loop extrusion is limited by transcription and stalled replication forks. Sci Adv 8(23):eabn7063. https://doi.org/10.1126/sciadv.abn7063
12. Valton AL, Venev SV, Mair B et al (2022) A cohesin traffic pattern genetically linked to gene regulation. Nat Struct Mol Biol 29(12): 1239–1251. https://doi.org/10.1038/s41594-022-00890-9
13. Goloborodko A, Imakaev MV, Marko JF et al (2016) Compaction and segregation of sister chromatids via active loop extrusion. elife 5. https://doi.org/10.7554/eLife.14864
14. Kinoshita K, Tsubota Y, Tane S et al (2022) A loop extrusion-independent mechanism contributes to condensin I-mediated chromosome shaping. J Cell Biol 221(3). https://doi.org/10.1083/jcb.202109016

15. Guerin TM, Barrington C, Pobegalov G et al (2024) An extrinsic motor directs chromatin loop formation by cohesin. EMBO J. https://doi.org/10.1038/s44318-024-00202-5
16. Gerguri T, Fu X, Kakui Y et al (2021) Comparison of loop extrusion and diffusion capture as mitotic chromosome formation pathways in fission yeast. Nucleic Acids Res 49(3): 1294–1312. https://doi.org/10.1093/nar/gkaa1270
17. Goker M, Oren A (2024) Valid publication of names of two domains and seven kingdoms of prokaryotes. Int J Syst Evol Microbiol 74(1). https://doi.org/10.1099/ijsem.0.006242
18. Imachi H, Nobu MK, Kato S et al (2024) Promethearchaeum syntrophicum gen. Nov., sp. nov., an anaerobic, obligately syntrophic archaeon, the first isolate of the lineage 'Asgard' archaea, and proposal of the new archaeal phylum Promethearchaeota phyl. nov. and kingdom Promethearchaeati regn. nov. Int J Syst Evol Microbiol 74(7). https://doi.org/10.1099/ijsem.0.006435
19. Jeon JH, Lee HS, Shin HC et al (2020) Evidence for binary Smc complexes lacking kite subunits in archaea. IUCrJ 7(Pt 2):193–206. https://doi.org/10.1107/S2052252519016634
20. Takemata N, Samson RY, Bell SD (2019) Physical and functional compartmentalization of archaeal chromosomes. Cell 179(1):165–179 e118. https://doi.org/10.1016/j.cell.2019.08.036
21. Herrmann U, Soppa J (2002) Cell cycle-dependent expression of an essential SMC-like protein and dynamic chromosome localization in the archaeon Halobacterium salinarum. Mol Microbiol 46(2):395–409. https://doi.org/10.1046/j.1365-2958.2002.03181.x
22. Ettema TJ, Lindas AC, Bernander R (2011) An actin-based cytoskeleton in archaea. Mol Microbiol 80(4):1052–1061. https://doi.org/10.1111/j.1365-2958.2011.07635.x
23. Yoshinaga M, Nakayama T, Inagaki Y (2022) A novel structural maintenance of chromosomes (SMC)-related protein family specific to archaea. Front Microbiol 13:913088. https://doi.org/10.3389/fmicb.2022.913088
24. Badel C, Bell SD (2023) Chromosome architecture in an archaeal species naturally lacking structural maintenance of chromosomes proteins. Nat Microbiol. https://doi.org/10.1038/s41564-023-01540-6
25. Takemata N, Bell SD (2021) Multi-scale architecture of archaeal chromosomes. Mol Cell 81 (3):473–487 e476. https://doi.org/10.1016/j.molcel.2020.12.001
26. Cockram C, Thierry A, Gorlas A et al (2021) Euryarchaeal genomes are folded into SMC-dependent loops and domains, but lack transcription-mediated compartmentalization. Mol Cell 81 (3):459–472 e410. https://doi.org/10.1016/j.molcel.2020.12.013
27. Badel C, Samson RY, Bell SD (2022) Chromosome organization affects genome evolution in Sulfolobus archaea. Nat Microbiol 7(6): 820–830. https://doi.org/10.1038/s41564-022-01127-7
28. Pilatowski-Herzing E, Samson RY, Takemata N, Badel C et al (2024) Capturing chromosome conformation in Crenarchaea. Mol Microbiol. https://doi.org/10.1111/mmi.15245
29. Lieberman-Aiden E, van Berkum NL, Williams L et al (2009) Comprehensive mapping of long-range interactions reveals folding principles of the human genome. Science 326(5950):289–293. https://doi.org/10.1126/science.1181369
30. Yamaura K, Takemata N, Kariya M et al (2025) Chromosomal domain formation by archaeal SMC, a roadblock protein, and DNA structure. Nat Commun 16(1):1312. https://doi.org/10.1038/s41467-025-56197-y
31. Akgol Oksuz B, Yang L et al (2021) Systematic evaluation of chromosome conformation capture assays. Nat Methods 18(9):1046–1055. https://doi.org/10.1038/s41592-021-01248-7
32. Sato T, Fukui T, Atomi H et al (2005) Improved and versatile transformation system allowing multiple genetic manipulations of the hyperthermophilic archaeon Thermococcus kodakaraensis. Appl Environ Microbiol 71(7): 3889–3899. https://doi.org/10.1128/AEM.71.7.3889-3899.2005
33. Takemata N (2024) How do thermophiles organize their genomes? Microbes Environ 39(5). https://doi.org/10.1264/jsme2.ME23087

Correction to: Assessing SMC Complex Function in Replication Fork Progression with DNA Fiber Assays

Neus P. Lorite, Snehlata, Neus Colomina, Jordi Torres-Rosell, Sara Rodríguez-Acebes, and Juan Méndez

Correction to:
Chapter 10 in: Yasuto Murayama and Yasutaka Kakui (eds.), *SMC Proteins: Methods and Protocols*, Methods in Molecular Biology, vol. 2991,
https://doi.org/10.1007/978-1-0716-5005-9_10

The original version of the chapter - **Assessing SMC Complex Function in Replication Fork Progression with DNA Fiber Assays** was inadvertently published without including two contributing authors **Sara Rodríguez-Acebes** and **Juan Méndez**. The correct author details have been updated in the chapter

The updated version of this chapter can be found at
https://doi.org/10.1007/978-1-0716-5005-9_10

Yasuto Murayama and Yasutaka Kakui (eds.), *SMC Proteins: Methods and Protocols*, Methods in Molecular Biology, vol. 2991,
https://doi.org/10.1007/978-1-0716-5005-9_18,

INDEX

Yasuto Murayama and Yasutaka Kakui (eds.), *SMC Proteins: Methods and Protocols*, Methods in Molecular Biology, vol. 2991, https://doi.org/10.1007/978-1-0716-5005-9,

N

O

P

R

S

T

V

W

MIX
Papier aus verantwortungsvollen Quellen
Paper from responsible sources
FSC® C105338

If you have any concerns about our products,
you can contact us on
ProductSafety@springernature.com

In case Publisher is established outside the EU,
the EU authorized representative is:
Springer Nature Customer Service Center GmbH
Europaplatz 3, 69115 Heidelberg, Germany

Printed by Libri Plureos GmbH
in Hamburg, Germany